Cancer Drug Discovery and Development

Series Editor
Beverly A. Teicher
National Cancer Institute, Rockville, MD, USA

For further volumes:
http://www.springer.com/series/7625

Frank E. Koehn
Editor

Natural Products and Cancer Drug Discovery

Humana Press

Editor
Frank E. Koehn
Natural Products, Oncology Worldwide Medicinal Chemistry
Pfizer Inc
Groton, CT, USA

ISBN 978-1-4614-4653-8 ISBN 978-1-4614-4654-5 (eBook)
DOI 10.1007/978-1-4614-4654-5
Springer New York Heidelberg Dordrecht London

Library of Congress Control Number: 2012950045

Printed on acid-free paper

Humana Press is a brand of Springer
Springer is part of Springer Science+Business Media (www.springer.com)

Preface

Since the dawn of human medicine, compounds derived from animals, plants, and microbes have been used to treat disease. The treatment of cancer in particular has been profoundly impacted. No other disease area, with the exception of anti-infectives, has seen a greater proportion of its small molecule drugs sourced directly from natural products. The list of natural product cancer therapeutics is impressive—*Vinca* alkaloids, anthracycline antitumor antibiotics, camptothecins, epothilones, podophyllotoxins, rapamycin mTOR inhibitors, taxanes, to name a few. Indeed, entire drug families, chemical and mechanistic classes of anticancer agents, have their origins in the secondary metabolites produced by microbes, plants, and now recently even marine organisms. Many of these agents were discovered on the basis of their striking antitumor activity in phenotypic assays, or in vivo efficacy screens. These successes drove an emphasis on natural products discovery beginning in the 1950s extending into the 1980s, which saw virtually every major pharma company utilize natural products as a major platform for drug lead generation.

But then came a change. Beginning in the 1980s the pharmaceutical industry gradually turned away from efficacy-driven drug discovery and instead adopted a process-based target-driven strategy, relying on high-throughput biochemical screening (HTS) of large collections of synthetic compounds or natural products extracts against defined targets, followed by rapid hit-to-lead and lead optimization phases to ultimately provide clinical candidates. This move was driven by a number of factors, including the desire for earlier understanding of drug mechanism, and a steady decline in the number of truly novel natural products being isolated in traditional pharma programs. Overall, natural products-based drug discovery is challenged to meet the rapid cycle times that are characteristic of the HTS approach, and has not thrived in the HTS arena. As a result, major Pharma has de-emphasized natural products discovery as a core platform for discovering new cancer drugs. We are now in a landscape where the programs that drive natural products discovery are not concentrated within large pharmaceutical companies, but instead are distributed amongst various pharma, biotech, and academic groups. Remarkably, even with the greatly reduced pharma presence, we continue to see each year the approval of one or more natural product-derived anticancer drugs—a testimony to the privileged nature of these molecules. Ironically, new discoveries in the fundamental sciences relating to natural products discovery coupled with dramatic advancements in enabling technologies have placed natural product researchers into a position to redefine the impact on cancer drug discovery in a major way. These include (1) advances in synthetic chemistry that make it possible to produce complex natural product molecules at a scale that is sufficient for clinical advancement, (2) new appreciation of targets that are uniquely druggable by complex natural product structures, (3) advances in genomics, molecular biology, and cultivation of natural product producing organisms which provide a means of accessing new chemotypes, optimizing natural product lead structures, and enabling large-scale production of drug candidates, (4) new methods of delivering natural product cytotoxins to tumors, and (5) advances in chemical biology which enable natural products to be effective tools for identification of new drug targets. The challenge will be to leverage these advances in the new landscape of cancer drug discovery.

The purpose of this book is to provide some perspective on key areas where natural products will impact cancer drug discovery, both today and in the future. The volume is comprised of chapters written by leading researchers in drug discovery, natural products chemistry, cancer biology, and chemical biology. The volume begins with a chapter by Ganesan, who concisely describes the unique structural and physicochemical features of natural products which impart biological activity and chemical space complementarity to synthetic drugs. Four chapters then follow which relate to specific cancer targets and therapies. Hooper, Loganzo, May, and Gerber discuss in detail the identification and development of natural product drugs which interfere with tumor growth by targeting tumor vasculature. They describe current preclinical and clinical natural product drug candidates and provide an analysis of the challenges and opportunities which lie ahead in this class. Longley then presents a highly illustrative case study on the vascular targeting agent, discodermolide, describing the myriad challenges which must be met in order to successfully advance an anticancer marine natural product from initial discovery to the clinic. Salvador and Luesch then review the discovery of natural product drugs in a newly emerging class of anticancer agents—histone deacetylase inhibitors. They enumerate the challenges in the development of these agents and provide insights in the continuing role of natural products in the discovery of HDAC inhibitors and other epigenetic modulators. In the final therapeutically related chapter, Koehn describes the attributes of natural product cytotoxins which make them uniquely suited as payloads for anticancer antibody drug conjugates.

The second section of the book details advances in the science underlying natural products drug discovery. Mang and Hausteadt provide an in-depth review into sources, structural features, and structure–activity relationships of a selection of natural product chemical scaffolds frequently used in cancer therapy. Their detailed case studies show that the total synthesis of highly complex natural products is now an approach that can deliver clinical candidates. For microbial natural products, genetic engineering of the producing organism is becoming an increasingly important means of enabling the discovery of novel anticancer agents. Unsin, Rajski, and Shen highlight in their chapter recent genetic engineering advances, such as metabolic engineering and combinatorial biosynthesis, that have been successfully applied to the development of microbial natural product anticancer agents. This is followed by a chapter by Leone and Roberts who examine the growing capabilities of plant combinatorial biosynthesis and cell culture in the production of plant metabolites, and highlight the cases of paclitaxel and cyclopamine—developments which have fueled new optimism for future plant anticancer agents. In the concluding chapter, Beutler examines a special attribute of natural products, namely the identification of new anticancer targets.

Most researchers will agree that the fields of cancer drug discovery and natural products are in the midst of significant change. While we might agree on these points, there is less clarity around precisely what the path toward the future will look like. The goal of this volume is to assess the current landscape and to help point the way to the future of natural products in cancer drug discovery. The journey should prove exciting.

Groton, CT, USA

Frank E. Koehn

Contents

Contributors

John A. Beutler Molecular Targets Laboratory, Center for Cancer Research, National Cancer Institute, Frederick, MD, USA

A. Ganesan School of Pharmacy, University of East Anglia, Norwich, UK

Hans-Peter Gerber Bioconjugate Discovery and Development, Oncology Research, Pfizer Worldwide Research and Development, Pearl River, NY, USA

Lars Ole Haustedt AnalytiCon Discovery GmbH, Potsdam, Germany

Andrea T. Hooper Bioconjugate Discovery and Development, Oncology Research, Pfizer Worldwide Research and Development, Pearl River, NY, USA

Frank E. Koehn Natural Products, Oncology Worldwide Medicinal Chemistry, Pfizer World-Wide Research and Development, Groton, CT, USA

Lisa M. Leone Department of Chemical Engineering, University of Massachusetts, Amherst, MA, USA

Frank Loganzo Bioconjugate Discovery and Development, Oncology Research, Pfizer Worldwide Research and Development, Pearl River, NY, USA

Ross E. Longley Lake Erie College of Osteopathic Medicine-Bradenton, Bradenton, FL, USA

Hendrik Luesch Department of Medicinal Chemistry, University of Florida, Gainesville, FL, USA

Christian P. Mang AnalytiCon Discovery GmbH, Potsdam, Germany

Chad May Bioconjugate Discovery and Development, Oncology Research, Pfizer Worldwide Research and Development, Pearl River, NY, USA

Scott R. Rajski School of Pharmacy, Division of Pharmaceutical Sciences, University of Wisconsin, Madison, WI, USA

Susan C. Roberts Department of Chemical Engineering, University of Massachusetts, Amherst, MA, USA

Lilibeth A. Salvador Department of Medicinal Chemistry, University of Florida, Gainesville, FL, USA

Ben Shen Departments of Chemistry and Molecular Therapeutics, Natural Products Library Initiative at The Scripps Research Institute, The Scripps Research Institute, Jupiter, FL, USA

Claudia Eva-Maria Unsin School of Pharmacy, Division of Pharmaceutical Sciences, University of Wisconsin, Madison, WI, USA

Part I
Natural Product Drugs and Cancer Therapeutics

Chapter 1
The Impact of Natural Products Upon Cancer Chemotherapy

A. Ganesan

Abstract Natural products are a historically valuable source of drugs. Their unique features such as evolutionary selection to possess biological activity and complementarity of chemical space to synthetic drugs are discussed. This is followed by an analysis of the physicochemical properties of all anticancer drugs that are natural products or derived from a natural product lead. Log P and the number of hydrogen bond donors are found to be the two most important parameters for bioavailability. Most natural product anticancer drugs maintain a Clog $P \leq 4$ and H-bond donors ≤ 5.

1.1 Introduction

Historically, crude natural product extracts were the major source of medications for mankind. In the nineteenth century, it became possible to purify active ingredients from such extracts. For example, alkaloids like morphine, quinine, atropine, and cocaine date from this period and are widely used to this day. Compared to an extract, a purified natural product has obvious advantages in terms of standardization of the dose and the avoidance of side effects due to other components present in the mixture. From this early beginning of investigating traditional medicines, the discovery of biologically active natural products progressed in the twentieth century to large-scale efforts involving mechanism-based screening. In this method, a library of crude extracts is tested in a biological assay and active hits are fractionated to identify individual compounds whose structure is then elucidated. Among others, the process has produced many of our clinical antibiotics, the cholesterol lowering statins, the immunosuppressant FK506, and the antiparasitic avermectins (Brown 2007; Cragg and

A. Ganesan (✉)
School of Pharmacy, University of East Anglia,
Norwich Research Park, Norwich NR4 7TJ, UK
e-mail: a.ganesan@uea.ac.uk

F.E. Koehn (ed.), *Natural Products and Cancer Drug Discovery*, Cancer Drug Discovery and Development, DOI 10.1007/978-1-4614-4654-5_1,

Newman 2007). Nevertheless, natural products are no longer our sole provider of medicines. Our increasing insight into disease targets at the molecular level has produced a large number of synthetic small molecule drugs made by medicinal chemists as well as large molecular weight biologics made by cell culture. Competition from these quarters and the intrinsic challenges in natural product screening lead to the reasonable question of whether it will continue to play an important role in the future (Rishton 2008; Li and Vederas 2009; Kingston 2011). At least in the pharmaceutical industry, the answer seems to be an unfortunate "No" as most organizations have either downsized or removed natural product screening from their lead discovery efforts. We have described (Ortholand and Ganesan 2004; Ganesan 2004, 2006) the unique features of natural products and three that are relevant to drug discovery are summarized below.

1.1.1 Predisposition to Biological Activity

Natural products are often made by multistep biosynthetic pathways that consume significant resources of the producing organism. It is likely that the retention of these pathways provides some positive benefit to the producing organism, either within itself or in its interactions with other species in its natural environment (Stone and Williams 1992; Maplestone et al. 1992; Haslam 1994). Furthermore, biologically relevant protein space is relatively small with a finite number of protein folds. Thus, a natural product that is designed to interact with a particular protein may coincidentally bind with high affinity to another protein of interest as a therapeutic target (Breinbauer et al. 2002; Eberhardt et al. 2011).

1.1.2 Complementary Chemical Space to Synthetic Compounds

The chemical reactions of natural product biosynthesis pathways are substantially different to those of the medicinal chemist. Consequently, it is not surprising that the two populate different areas of chemical space. Natural products tend to be rich in oxygen atoms while medicinal chemistry emphasizes nitrogen. Many of nature's building blocks are non-aromatic and the resulting secondary metabolites have a high proportion of sp^3 hybridized carbon—a metric recently recognized to be of value in drug design (Lovering et al. 2009). Often, natural products exist in rigid conformations with intramolecular hydrogen bonding—another useful concept in medicinal chemistry (Kuhn et al. 2010). Natural products are rich in stereochemical features such as chiral centers and bridging, fused and spirocyclic rings. Natural products achieve high structural diversity by branching out from a small set of biosynthetic intermediates to a large number of distinct scaffolds whereas synthetic compounds have relatively few scaffolds but more known examples within each scaffold. In terms of drug discovery, both natural products and synthetics are extensively populated by privileged scaffolds (Grabowski and Schneider 2007; Rosen et al. 2009; Welsch et al. 2010; Drewry and Macarron 2010; Khanna and Ranganathan 2011; López-Vallejo et al. 2012).

1.1.3 Continuous Evolution of Natural Product Chemistry

The process of natural product discovery is constantly evolving. Key are the recent advances in the rapid dereplication of hits (Wolfender et al. 2011), structure elucidation with smaller amounts of material (Molinski 2010), the activation of silent genes (Lewinsohn and Gijzen 2009; Yin and Keller 2011), and metabolic pathway engineering (Keasling 2010). With these improvements, some of the original factors that led to a declining interest in natural product screening no longer hold true. It is time for industry to reevaluate the role of natural products and the positive benefits they bring (Schmitt et al. 2011; Camp et al. 2012; Koehn 2012). The historical record certainly provides ample evidence of natural products used as tool compounds and drugs that are often first in class against a particular target (Pucheault 2008; Carlson 2010; La Clair 2010; Gerwick and Moore 2012).

1.2 Natural Products and Cancer Chemotherapy

The comprehensive surveys by Cragg and Newman highlight the important role played by natural products in drug development (Cragg et al. 1997; Newman et al. 2003; Newman and Cragg 2007). These reviews were the basis of our analysis of natural products isolated in 1970–2006 (Ganesan 2008). In this period there were 24 unique natural product scaffolds that led to an approved drug and these were clearly subdivided into two distinct families. One was largely compliant of physicochemical metrics of druglikeness such as Lipinski's Rule of Five while the other largely violated these rules. Even so, these compounds were compliant in terms of log P and number of H-bond donors, highlighting the importance of these two parameters. In this chapter, we will perform a similar analysis on *all* natural products irrespective of year of discovery that have produced a drug in the specific therapeutic area of cancer chemotherapy. Before doing so, it is helpful to appreciate the idiosyncratic features of oncology drugs derived from natural products.

1.2.1 Congruence Between the Needs of Producing Organism and Humans

A secondary metabolite with growth inhibitory or cytotoxic properties has evolutionary advantages for the producing organism, whether it is competition for resources or self-defense against predators. This cytotoxic property is the same that we wish to exploit to inhibit tumor cell proliferation and the mechanism of action is probably the same that the producing organism is targeting. The situation can be different in other therapeutic areas where this matching may not be present e.g., bacteria can produce a natural product that happens to target a human virus although this would be foreign to the natural environment of the producing organism.

1.2.2 *Ease of Discovery*

Nowadays, new natural products are usually discovered by bioassay guided fractionation of crude extracts. Thanks to the high-throughput screening efforts of the National Cancer Institute, robust in vitro assays for growth inhibition and cytotoxicity of cancer cell lines are available and are the most widely used assays by natural product chemists. Thus, as long as a compound is cell permeable one can reliably know if it has potential anticancer activity by looking at the data in the natural product isolation paper. Although the mechanism of action is usually unknown, at least the phenotypic activity can be observed. This situation is unique to cancer and to some extent antimicrobials. In other therapeutic areas, highly active natural products are undoubtedly missed as the appropriate screens are not routinely in operation.

1.2.3 *Unmet Health Care Need*

Cancer is a complex disease with survival rates ranging from nearly 0 to 100% depending on the type. As a result, oncology is a major therapeutic focus of the pharmaceutical industry and there is a constant need for new medicines from natural products, synthetics, and biologics. However, the severity of the disease allows drugs that may not be ideal to be tolerated. The majority of cancer chemotherapeutics are administered intravenously rather than the preferred oral route in other therapeutic areas. Furthermore, the toxicity of anticancer drugs is inseparable from their mechanism of action of cell killing. Clinically, these drugs are often employed close to the concentration at which dose limiting toxicity manifests in order to have the best chance of tumor regression.

1.3 The Chemical Space of Natural Product Anticancer Drugs

The above factors suggest that anticancer natural products are relatively easy to discover but they may be inferior to natural products for other indications in terms of oral bioavailability and side effects. However, for this very reason an analysis of anticancer drugs is illuminating as it will highlight the physicochemical properties that are most important and retained even in this therapeutic area. The starting point for the analysis is Newman and Cragg's comprehensive list of all anticancer drugs that received approval from the 1940s to 2010 (Newman and Cragg 2012). Their survey covers approvals in all countries globally and includes drugs that are now discontinued. Two filters were applied to the natural product anticancer drugs as follows:

1. Drugs that are synthesized from a natural product starting material but do not share a mechanism with the latter are excluded. For anticancer drugs, this criterion mainly excludes steroid drugs made from inexpensive plant steroids.

2. Drugs inspired by endogenous human metabolites are excluded. For anticancer drugs, this criterion mainly excludes drugs based on peptide hormones.

To some extent, the exclusions cancel out as the steroids tend to be lipophilic compounds whereas the peptide hormones are hydrophilic. After applying the filters, a total of 42 drugs remain and these are listed in Table 1.1 in order of ascending MW. The year of approval is indicated followed by "Type" where NP = natural product, ND = natural product derivative, and NS = synthetic based on a natural product lead. The majority of approved drugs are unmodified natural products while some are semisynthetic derivatives. Two are fully synthetic, the first being amrubicin an anthracycline related to daunorubicin and doxorubicin. Amrubicin is a synthetic second-generation anthracycline developed by Sumitomo and produced by total synthesis (Yamamoto et al. 2009). The second, eribulin (Havalen) is an excellent argument for the value of total synthesis (Huyck et al. 2011). The drug originates from Yoshito Kishi's extensive studies on the marine natural product halichondrin B and the biological testing of intermediates along the total synthesis route (Jackson et al. 2009). Surprisingly, it was discovered that activity was retained in simpler compounds that were only half of the halichondrin molecule—a result that could not be predicted by investigating the natural product alone. The scarcity of the natural product and the structural simplification made a total synthesis approach viable for drug discovery.

The "Target" column shows that anticancer natural product drugs have rather limited mechanisms of action and typically work on the microtubule or DNA rather than protein receptors or enzymes. Two recent exceptions are temsirolimus (Rini et al. 2007) a semisynthetic rapamycin analog that targets the kinase mTOR (mammalian target of rapamycin) and romidepsin (FK228) (VanderMolen et al. 2011) a depsipeptide natural product that inhibits HDACs (histone deacetylases). The column "Route" indicates the method of administration. As expected, these drugs are predominantly intravenous but it should be noted that there are six examples of oral drugs with MW > 450.

The physicochemical properties for the drugs in Table 1.1 were generated from their PubChem entries (http://pubchem.ncbi.nlm.nih.gov/). Shaded cells indicate values exceeding the accepted limits for druglike space: MW 500, XlogP 5, H-bond donors 5, H-bond acceptors 10, rotatable bonds 10, total polar surface area 140 Å^2, and heavy atoms 30. The results are striking: despite violations in some parameters, natural product anticancer drugs are highly compliant in terms of log P. The importance of log P and its correlation with other properties that are critical in drug development are now well recognized (Leeson and Springthorpe 2007; Leeson and Empfield 2010; Hann 2011). Nature has known this all along and shows us that it is possible to break some rules as long as the important ones are adhered to. After log P, the next most important physicochemical parameter is the number of H-bond donors. Once again, natural products are highly compliant in this parameter which may be related to recognition by efflux transporters and provide a mechanism for drug resistance (Raub 2006; Wager et al. 2010). The number of rotatable bonds is the third parameter that natural product anticancer drugs largely obey. This helps minimize the molecular flexibility and the potential for promiscuous off-target binding.

Table 1.1 Natural product anticancer drugs from the 1940s to 2010

Drug	Approval	Type	Target	Route	MW	XlogP3	HBD	HBA	Rot	PSA	HA
Sarkomycin	1954	NP	Unknown	iv	140	0.2	1	3	1	54	10
Arglabin	1999	NP	Farnesyltransferase	Oral	246	1.7	0	3	0	39	18
Streptozotocin	1977	NP	DNA	iv	265	−1.4	5	8	2	152	18
Pentostatin	1992	NP	Adenosine deaminase	iv	268	−2.1	4	5	2	112	19
Elliptinium acetate	1983	ND	Topoisomerase II	iv	277	4.6	2	1	0	40	21
Masoprocol	1992	NP	5-Lipoxygenase	Oral	302	4.3	4	4	5	81	22
Mitomycin C	1956	NP	DNA	iv, topical	334	−0.4	3	8	4	147	24
Topotecan	1996	ND	Topoisomerase I	Oral, iv	422	0.5	2	7	3	103	31
Belotecan	2004	ND	Topoisomerase I	Oral, iv	470	2	3	6	5	92	33
Amrubicin	2002	NS	Topoisomerase II	iv	484	0.9	5	10	3	177	35
Idarubicin	1990	ND	Topoisomerase II	Oral	498	1.9	5	10	3	177	36
Ixabepilone	2007	ND	Microtubule	Oral	507	3.6	3	6	2	140	35
Daunomycin	1967	NP	DNA	iv	528	1.8	5	11	4	186	38
Romidepsin	2010	NP	Histone deacetylase	iv	541	2.2	4	8	2	193	36
Doxorubicin	1966	NP	Topoisomerase II	Oral	544	1.3	6	12	5	206	39
Epirubicin	1984	ND	Topoisomerase II	iv	544	1.3	6	12	5	206	39
Irinotecan	1994	ND	Topoisomerase I	iv	587	3	1	8	5	113	43
Etoposide	1980	ND	Topoisomerase II	Oral, iv	589	0.6	3	13	5	161	42
Carzinophilin	1954	NP	DNA	iv	624	1.3	4	12	12	193	45
Pirarubicin	1998	ND	Topoisomerase I	iv	628	2.7	5	13	7	204	45
Teniposide	1967	ND	Microtubule	iv	657	1.2	3	14	6	189	46

Neocarzinostatin	1976	NP	DNA	iv	662	2.3	4	13	8	175	48
Valrubicin	1999	ND	Topoisomerase II	iv	724	4	5	16	11	215	51
Eribulin	2010	NS	Microtubule	iv	730	1.1	2	12	4	146	52
Vindesine	1979	ND	Microtubule	iv	754	2.7	5	10	7	165	55
Trabectedin	2007	NP	DNA	iv	762	3.4	4	15	4	194	54
Vinorelbine	1989	ND	Microtubule	Oral, iv	779	3.6	2	11	10	134	57
Docetaxel	1995	ND	Microtubule	iv	808	1.6	5	14	13	224	58
Vinblastine	1965	NP	Microtubule	iv	811	3.7	3	12	10	154	59
Aclarubicin	1981	NP	DNA	iv	812	3.8	4	16	10	217	58
Vinflunine	2010	ND	Microtubule	iv	817	4.4	2	13	10	134	59
Vincristine	1963	NP	Microtubule	iv	825	2.8	3	12	10	171	60
Cabazitaxel	2010	ND	Microtubule	iv	836	2.7	3	14	15	202	60
Paclitaxel	1993	NP	Microtubule	iv	854	2.5	4	14	14	221	62
Solamargine	1989	NP	Unknown	Topical	868	1.1	9	16	7	239	61
Temsirolimus	2007	ND	mTOR	iv	1,030	5.6	4	16	11	242	73
Mithramycin	1961	NP	DNA	iv	1,085	0.6	11	24	15	358	76
Chromomycin A3	1961	NP	DNA	iv	1,183	2.3	8	26	20	359	83
Actinomycin D	1964	NP	DNA	iv	1,255	3.8	5	18	8	356	90
Ozogamicin	2000	ND	DNA	iv	1,368	2	8	27	24	410	84
Peplomycin	1981	NP	DNA	iv	1,474	–6.7	21	26	38	696	102
Bleomycin	1966	NP	DNA	iv	1,513	–1.9	21	29	36	770	101
Average					700.199	1.82381	4.928571	12.57143	8.714286	208.2619	49.47619

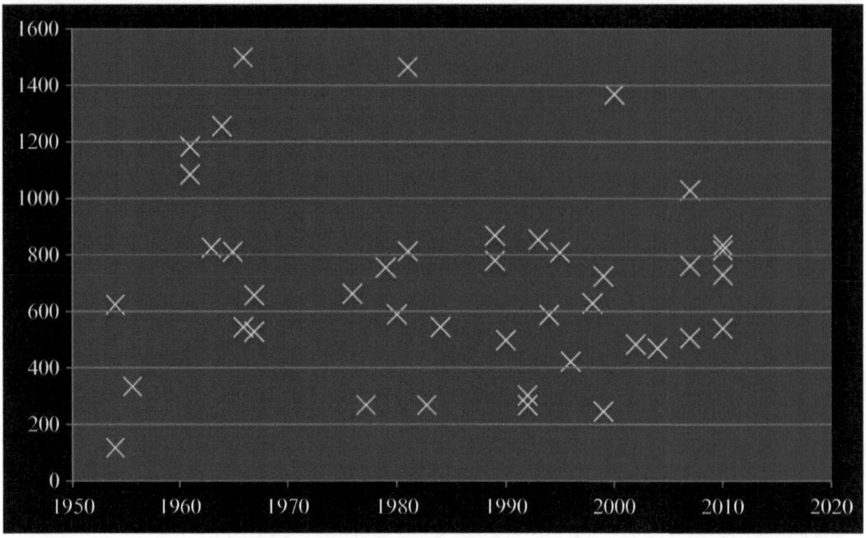

Fig. 1.1 MW of anticancer natural product drugs plotted against year of approval

Interestingly, the MW of the natural products is not correlated with the year of approval (Fig. 1.1) whereas synthetic drugs show a detectable rise in this parameter attributed to industry going after harder targets with time (Leeson and Davis 2004).

1.4 Natural Product Chemotypes

Table 1.1 lists all the anticancer drugs derived from natural products. The number of leads is smaller as multiple drugs have arisen from a common natural product. Performing this reduction yields a total of 25 unique natural product leads that have produced an approved anticancer drug (Table 1.2). Analysis of the leads shows that they are a little smaller than the derived drugs but equally highly compliant in terms of log *P* and H-bond donors.

The drugs classified as "ND and NS" in Table 1.1 are an interesting subset. These are the examples where the natural product itself was modified to a semisynthetic derivative or replaced by a synthetic compound. We can look at how physicochemical properties evolved between the natural product lead and the marketed drug (Table 1.3). The main improvement lies in the log *P* column where on average there was a reduction of 0.5 log units. Once again, this emphasizes the importance of log *P* in lead optimization. Less satisfactorily, the lower log *P* was typically engineered by the introduction of polar functional groups that increase the number of H-bond donors and acceptors. Since the natural product lead is typically low in H-bond donors to begin with, this enables the introduction of additional donors without exceeding the Lipinski limit.

Table 1.2 Successful natural product anticancer leads from the 1940s to2010

Lead	Derived drugs	MW	XlogP3	HBD	HBA	Rot	PSA	HA
Sarkomycin	Sarkomycin	140	0.2	1	3	1	54	10
Arglabin	Arglabin	246	1.7	0	3	0	39	18
Ellipticine	Ellipticine acetate	246	4.8	1	1	0	29	19
Streptozotocin	Streptozotocin	265	−1.4	5	8	2	152	18
Pentostatin	Pentostatin	268	−2.1	4	5	2	112	19
Masoprocol	Masoprocol	302	4.3	4	4	5	81	22
Mitomycin C	Mitomycin C	334	−0.4	3	8	4	147	24
Camptothecin	Belotecan, irinotecan, topotecan	348	1	1	5	1	80	26
Podophyllotoxin	Etoposide, teniposide	414	2	1	8	4	93	30
Epothilone B	Ixabepilone	508	4.2	2	7	2	138	35
Daunomycin	Aclarubicin, amrubicin, daunomycin, doxorubicin, epirubicin, idarubicin, pirarubicin, valrubicin	528	1.8	5	11	4	186	38
Romidepsin	Romidepsin	541	2.2	4	8	2	193	36
Carzinophilin	Carzinophilin	624	1.3	4	12	12	193	45
Neocarzinostatin	Neocarzinostatin	662	2.3	4	13	8	175	48
Trabectedin	Trabectedin	762	3.4	4	15	4	194	54
Vincristine	Vinblastine, vincristine, vindesine, vinflunine, vinorelbine	825	2.8	3	12	10	171	60
Paclitaxel	Cabazitaxel, docetaxel, paclitaxel	854	2.5	4	14	14	221	62
Solamargine	Solamargine	868	1.1	9	16	7	239	61
Rapamycin	Temsirolimus	914	6	3	13	6	195	65
Mithramycin	Mithramycin	1,085	0.6	11	24	15	358	76
Halichondrin B	Eribulin	1,111	3.6	3	19	4	216	79
Chromomycin A3	Chromomycin A3	1,183	2.3	8	26	20	359	83
Actinomycin D	Actinomycin D	1,255	3.8	5	18	8	356	90
Calicheamicin	Ozogamicin	1,368	2	8	27	24	410	84
Bleomycin	Bleomycin, peplomycin	1,513	−1.9	21	29	36	770	101
Average		686.56	1.924	4.72	12.36	7.8	206.44	48.12

1.5 A Middle Space for Anticancer Drug Discovery

We can derive Lipinski-type "Rules" that fit the majority of the natural product anticancer drugs in Table 1.1. For convenience, these are not taken to fit 90% of all drugs but a slightly smaller proportion for some parameters and give rise to the following guidelines:

The 4/5/10 rule for cancer chemotherapeutics

- Clog $P \leq 4$
- H-bond donors ≤5
- Rotatable bonds ≤10

Table 1.3 Changes in physicochemical properties from natural product lead to anticancer drug

Drug	Lead	ΔMW	ΔXlogP3	ΔHBD	ΔHBA	Δrot	ΔPSA	ΔHA
Elliptinium acetate	Ellipticine	31	−0.2	1	0	0	11	2
Topotecan	Camptothecin	74	−0.5	1	2	2	23	5
Belotecan	Camptothecin	122	1	2	1	4	12	7
Amrubicin	Daunorubicin	−44	−0.9	5	−1	−1	−9	−3
Idarubicin	Daunorubicin	−30	0.1	5	−1	−1	−9	−2
Ixabepilone	Epothilone B	−1	−0.6	1	−1	0	2	0
Epirubicin	Daunorubicin	16	−0.5	1	1	1	20	1
Irinotecan	Camptothecin	239	2	0	4	4	23	17
Etoposide	Podophyllotoxin	175	−1.4	2	5	1	68	12
Pirarubicin	Doxorubicin	84	1.4	−1	1	2	−2	−6
Teniposide	Podophyllotoxin	243	−0.8	2	6	2	96	16
Valrubicin	Doxorubicin	180	2.7	−1	4	6	9	12
Eribulin	Halichondrin B	−381	−2.5	−1	−7	4	−70	−27
Vindesine	Vinblastine	−57	−1	2	−2	−3	−9	−4
Vinorelbine	Vinblastine	−32	−0.1	−1	−1	0	−20	−1
Docetaxel	Paclitaxel	−46	−2.4	1	0	−1	3	−4
Vinflunine	Vinblastine	6	0.7	−1	1	0	−20	0
Cabazitaxel	Paclitaxel	−18	−1.3	−1	0	1	−19	−2
Temsirolimus	Rapamycin	116	−0.4	1	3	5	47	8
Peplomycin	Bleomycin	−444	−4.8	0	−3	2	−74	1
Average		11.65	−0.475	0.9	0.6	1.4	4.1	1.6

These Rules are identical to the Lipinski and Veber guidelines for small molecule oral bioavailability except for a more stringent restriction of $\log P \leq 4$. Once again, the key importance of $\log P$ as a measure of druglikeness is confirmed. There are two approaches to maintaining low $\log P$ relative to molecular size. One is to stick to small molecules with MW < 500 and this is the strategy classically employed by medicinal chemists. However, as drug discovery targets become more challenging and the low hanging fruit have already been picked, it becomes harder to find drugs within this MW range. An alternative approach is to go for medium-sized molecules and this is the strategy successfully used by nature. As long as the above three critical parameters are adhered to, cancer drug space becomes significantly larger than Lipinski space and the limits for MW and other properties are relaxed.

Secondary rules for cancer chemotherapeutics

- MW ≤ 1,000
- H-bond acceptors ≤15
- Polar surface area ≤200 Å^2
- Non-hydrogen atoms ≤60

Natural products thus provide valuable design considerations for medicinal chemists who need to make synthetic compounds at the edge of Lipinski space for difficult drug targets. A low $\log P$ is absolutely essential whereas MW up to 1,000 can be tolerated—provided care is taken to minimize H-bond donors and rotatable bonds. These last two requirements help to enforce rigid conformations with internal

H-bonds and reduce efflux and in compensation allow for molecules with a large polar surface area and more H-bond acceptors. Finally, it is likely that many natural products are recognized by transporters and uptake is by active transport (Dobson et al. 2009; Sugano et al. 2010).

This chapter clearly demonstrates the historical effectiveness of natural products in mankind's fight against cancer. There are many more exciting natural products that remain to be discovered and developed into medicines. The only way that we will not harness this untapped potential is if we stop screening natural products as a source for drug discovery.

References

Breinbauer R, Vetter IR, Waldmann H (2002) From protein domains to drug candidates-natural products as guiding principles in the design and synthesis of compound libraries. Angew Chem Int Ed 41:2879–2890

Brown D (2007) Overview of sources of new drugs. In: Kennewell P (ed) Comprehensive medicinal chemistry II, vol 1. Elsevier, Amsterdam

Camp D, Davis RA, Campitelli M, Ebdon J, Quinn RJ (2012) Drug-like properties: guiding principles for the design of natural product libraries. J Nat Prod 75:72–812

Carlson EE (2010) Natural products as chemical probes. ACS Chem Biol 5:639–653

Cragg GM, Newman DJ (2007) Natural product sources of drugs: plants, microbes, marine organisms, and animals. In: Kennewell P (ed) Comprehensive medicinal chemistry II, vol 1. Elsevier, Amsterdam

Cragg GM, Newman DJ, Snader KM (1997) Natural products in drug discovery and development. J Nat Prod 60:52–60

Dobson PD, Patel Y, Kell DB (2009) 'Metabolite-likeness' as a criterion in the design and selection of pharmaceutical drug libraries. Drug Discov Today 14:31–40

Drewry D, Macarron R (2010) Enhancements of screening collections to address areas of unmet medical need: an industry perspective. Curr Opin Chem Biol 14:289–298

Eberhardt L, Kumar K, Waldmann H (2011) Exploring and exploiting biologically relevant chemical space. Curr Drug Targets 12:1531–1546

Ganesan A (2004) Natural products as a hunting ground for combinatorial chemistry. Curr Opin Biotechnol 15:584–590

Ganesan A (2006) Natural products and combinatorial chemistry—an uneasy past but a glorious future. In: Boldi AM (ed) Combinatorial synthesis of natural product-based libraries. CRC Press, Boca Raton

Ganesan A (2008) The impact of natural products upon modern drug discovery. Curr Opin Chem Biol 12:306–317

Gerwick WH, Moore BS (2012) Lessons from the past and charting the future of marine natural products drug discovery and chemical biology. Chem Biol 19:85–98

Grabowski K, Schneider G (2007) Properties and architecture of natural products revisited. Curr Chem Biol 1:115–127

Hann MM (2011) Molecular obesity, potency and other addictions in drug discovery. Med Chem Commun 2:349–355

Haslam E (1994) Secondary metabolism—evolution and function: products or processes? Chemoecology 5:89–95

Huyck TK, Gradishar W, Manuguid F, Kirkpatrick P (2011) Eribulin mesylate. Nat Rev Drug Discov 10:173–174

http://pubchem.ncbi.nlm.nih.gov/

Jackson KL, Henderson JA, Phillips AJ (2009) The halichondrins and E7389. Chem Rev 109:3044–3079
Keasling JD (2010) Manufacturing molecules through metabolic engineering. Science 330:1355–1358
Khanna V, Ranganathan S (2011) Structural diversity of biologically interesting datasets: a scaffold analysis approach. J Cheminform 3:30
Kingston DGI (2011) Modern natural products drug discovery and its relevance to biodiversity conservation. J Nat Prod 74:496–511
Koehn FE (2012) Biosynthetic medicinal chemistry of natural product drugs. Med Chem Commun 3:854–865
Kuhn B, Mohr P, Stahl M (2010) Intramolecular hydrogen bonding in medicinal chemistry. J Med Chem 53:2601–2611
La Clair JJ (2010) Natural product mode of action (MOA) studies: a link between natural and synthetic worlds. Nat Prod Rep 27:969–995
Leeson PD, Davis AM (2004) Time-related differences in the physical property profiles of oral drugs. J Med Chem 47:6338–6348
Leeson PD, Empfield JR (2010) Reducing the risk of drug attrition associated with physicochemical properties. In: Macor JE (ed) Annual reports in medicinal chemistry, vol 45. Academic, San Diego
Leeson PD, Springthorpe B (2007) The influence of drug-like concepts on decision-making in medicinal chemistry. Nat Rev Drug Discov 6:881–890
Lewinsohn E, Gijzen M (2009) Phytochemical diversity: the sounds of silent metabolism. Plant Sci 176:161–169
Li JW, Vederas JC (2009) Drug discovery and natural products: end of an era or an endless frontier? Science 325:161–165
López-Vallejo F, Giulianotti MA, Houghten RA, Medina-Franco JL (2012) Expanding the medicinally relevant chemical space with compound libraries. Drug Discov Today 17:718–726
Lovering F, Bikker J, Humblet C (2009) Escape from flatland: increasing saturation as an approach to improving clinical success. J Med Chem 52:6752–6756
Maplestone RA, Stone MJ, Williams DH (1992) The evolutionary role of secondary metabolites—a review. Gene 115:151–157
Molinski TF (2010) NMR of natural products at the 'nanomole-scale'. Nat Prod Rep 27:321–329
Newman DJ, Cragg GM (2007) Natural products as sources of new drugs over the last 25 years. J Nat Prod 70:461–477
Newman DJ, Cragg GM (2012) Natural products as sources of new drugs over the 30 years from 1981 to 2010. J Nat Prod 75:311–335
Newman DJ, Cragg GM, Snader KM (2003) Natural products as sources of new drugs over the period 1981–2002. J Nat Prod 66:1022–1037
Ortholand JY, Ganesan A (2004) Natural products and combinatorial chemistry—back to the future. Curr Opin Chem Biol 8:271–280
Pucheault M (2008) Natural products: chemical instruments to apprehend biological symphony. Org Biomol Chem 6:424–432
Raub TJ (2006) P-glycoprotein recognition of substrates and circumvention through rational drug design. Mol Pharm 3:3–25
Rini B, Kar S, Kirkpatrick P (2007) Temsirolimus. Nat Rev Drug Discov 6:599–600
Rishton GM (2008) Natural products as a robust source of new drugs and drug leads: past successes and present day issues. Am J Cardiol 101:43D–49D
Rosen J, Gottfries J, Muresan S, Backlund A, Oprea TI (2009) Novel chemical space exploration via natural products. J Med Chem 52:1953–1962
Schmitt EK, Moore CM, Krastel P, Petersen F (2011) Natural products as catalysts for innovation: a pharmaceutical industry perspective. Curr Opin Chem Biol 15:497–504
Stone MJ, Williams DH (1992) On the evolution of functional secondary metabolites (natural products). Mol Microbiol 6:29–34

Sugano K, Kansy M, Artursson P, Avdeef A, Bendels S, Di L, Ecker GF, Faller B, Fischer H, Gerebtzoff G, Lennernaes H, Senner F (2010) Coexistence of passive and carrier-mediated processes in drug transport. Nat Rev Drug Discov 9:597–614

VanderMolen KM, McCulloch W, Pearce CJ, Oberlies NH (2011) Romidepsin (Istodax, NSC 630176, FR901228, FK228, depsipeptide): a natural product recently approved for cutaneous T-cell lymphoma. J Antibiot 64:525–531

Wager TT, Chandrasekaran RY, Hou X, Troutman MD, Verhoest PR, Villalobos A, Will Y (2010) Defining desirable central nervous system drug space through the alignment of molecular properties, in vitro ADME, and safety attributes. ACS Chem Neurosci 1:420–434

Welsch ME, Snyder SA, Stockwell BR (2010) Privileged scaffolds for library design and drug discovery. Curr Opin Chem Biol 14:1–15

Wolfender JL, Marti G, Ferreira Queiroz E (2011) Advances in techniques for profiling crude extracts and for the rapid identification of natural products: dereplication, quality control and metabolomics. Curr Org Chem 14:1808–1832

Yamamoto M, Takakura A, Masuda N (2009) Next-generation anthracycline for the management of small cell lung cancer: focus on amrubicin. Drug Des Devel Ther 2:189–192

Yin W, Keller NP (2011) Transcriptional regulatory elements in fungal secondary metabolism. J Microbiol 49:329–339

Chapter 2
Identification and Development of Vascular Disrupting Agents: Natural Products That Interfere with Tumor Growth

Andrea T. Hooper, Frank Loganzo, Chad May, and Hans-Peter Gerber

Abstract Despite rapid progress made in understanding the biological mechanism regulating developmental and tumor vasculature, the clinical success of targeted therapeutics that interfere with key molecular and cellular pathways of angiogenesis has been limited by a lack of improved patient survival when these agents were administered as single agents. These findings indicate a redundancy of mechanisms regulating blood vessel formation, leading to tumor refractoriness to anti-angiogenic treatment. In contrast, vascular disruptive agents (VDAs) induced rapid regression of established tumors in preclinical studies. For several classes of VDAs, including the natural products flavonoids, colchicines and vincas, it was demonstrated that the effects on the vascular compartment diverge from that on the tumor parenchyma and are both anti-angiogenic (vascular stasis/prevention) and vascular disrupting (vascular regression)—resulting in catastrophic insults to the nascent vasculature. Vascular endothelial cells (ECs) are exquisitely dependent on the tubulin cytoskeleton for essential functions, including invasion, proliferation, migration, tube formation, and cell signaling—hallmarks of tumor angiogenesis. However, the clinical development of VDAs was frequently hampered by their narrow therapeutic index, limiting the use of these compounds due to side effects in the normal vasculature and/or other normal tissues. In this chapter we review the progress made to identify VDAs in

A.T. Hooper • F. Loganzo • C. May • H. Gerber (✉)
Bioconjugate Discovery and Development, Oncology Research, Pfizer Worldwide Research and Development, 401 North Middletown Road, Pearl River, NY 10965, USA
e-mail: hanspeter.gerber@pfizer.com

F.E. Koehn (ed.), *Natural Products and Cancer Drug Discovery*, Cancer Drug Discovery and Development, DOI 10.1007/978-1-4614-4654-5_2,

oncology, with focus on preclinical and clinical studies. The identification of novel natural products and therapeutic compounds with improved potency and safety characteristics continue to keep great promise to overcome the limitations of first generation VDAs.

2.1 Introduction

2.1.1 *Identification of Natural Products Targeting Tumor Vasculature*

The prerequisite for the success of anti-angiogenic paradigm is the identification of pathophysiological differences between ECs of normal vasculature and tumor vessels, which can be harnessed therapeutically. A large variety of molecular and cellular differences between normal and tumor vasculature have been described, including the remodeling of tumor vessels, their reliance on a tubulin cytoskeletal network for functional integrity, the lack of pericytes, and increased vascular permeability in tumor vs. normal vasculature (Pluda 1997; Darland and D'Amore 1999; Jain 2003, 2005; Neri and Bicknell 2005; Carmeliet and Jain 2011a, b). Tumor associated ECs proliferate faster compared to ECs within the vasculature of normal tissues, which display an average doubling time of several months (Denekamp 1982; Denekamp and Hobson 1982). Natural products targeting tubulin formation display various degrees of selectivity for tumor vasculature. Among the most promising tubulin interfering agents are derivatives of combretastatins, colchicines, and dolastatin-10 (Tozer et al. 2005).

In addition to the increased proliferation of ECs in the tumor vasculature, a variety of changes between normal, resting vasculature and tumor vasculature, were reported. Tumor vasculature is abnormal in a variety of structural and functional aspects. Blood vessels within tumors are heterogeneous, tortuous, with irregular branching patterns and vascular lumens. There is increased interstitial fluid pressure within tumor due to the leakiness of tumor vessels and escaping fluid (Jain 2005). As a consequence, blood flow is heterogeneous and blood cells, hematopoietic cells and nutrients are distributed unevenly, causing a milieu of stress, including hypoxia, hypoglycemia, and low pH. In summary, the magnitude and the variety of the molecular and cellular vascular changes in tumor vs. normal tissues provide a wealth of opportunities for therapeutic interference, including the identification of compounds targeting metabolic and cell signaling pathways or structural components such as tubulin. Ideally, novel therapeutic compounds selectively interfere with mechanisms critical for the development of pathologic tumor vasculature, but not with the maintenance of established, quiescent vasculature in normal tissues. Harnessing the differences between contact inhibited, resting ECs and tumor associated, rapidly proliferating ECs, yields great promise to identify a new generation of natural products interfering selectively with tumor vasculature.

2.1.2 Pharmacological Advantages of Compounds Targeting Tumor Vasculature

The recent FDA approvals of targeted therapeutics, or vascular targeting agents (VTAs) interfering with vascular endothelial growth factor (VEGF) induced angiogenesis for the treatment of solid tumors has validated the anti-angiogenic approach as a viable therapeutic strategy for the treatment of solid tumors (Ferrara et al. 2007). Given these clinical successes, the focus of future clinical and preclinical oncology research is likely to include vascular targeting strategies. One of the key advantages of VTAs, including VDAs, is that they do not depend on penetration of the many cellular and membrane layers within a tumor mass to exert their pharmacological effects. Vascular ECs are in direct contact with the circulatory system and therapeutic agents, thus the frequently described limited tumor perfusion of high-molecular-weight compounds into the tumor mass does not limit the exposure to VTAs to their respective targets. Such favorable exposure/efficacy relationship of VTAs may ultimately provide an important advantage for drug development by potentially limiting undesired on- and/or off-target toxicities. In addition, the vascular targeting approach should be applicable to most solid tumor types, based on their common dependence on angiogenesis for tumor progression (Folkman 1985).

2.2 Flavonoids

2.2.1 Identification and Mechanisms of Action

Flavonoids, which belong to a large family of polyphenolic compounds, are compromised of more than 5,000 compounds that can be classified into ten chemical groups. These compounds are found in diverse botanical sources, including a variety of fruits and vegetables, as well as green tea and red wine. They have been reported to exhibit a variety of pharmacological properties including inhibition of tumor cell proliferation and anti-angiogenic activities (Wahl et al. 2011).

Early in the developmental program of the National Cancer Institute (NCI), more than 200 flavonoids were evaluated for their antitumor activities in vivo. At the time none of these compounds demonstrated activity. However, the plant-derived flavonoid quercetin was eventually rescreened in 1971 using different in vivo models and demonstrated moderate antitumor activity (Folkman 1985) (Fig. 2.1). The results from these observations led the NCI to screen a series of flavones from Lyonnaise Industrielle Pharmaceutique (Venditti et al. 1984). Flavone acetic acid (FAA) ester (LM985) emerged as a lead compound from this screen after demonstrating modest antitumor activity (Plowman et al. 1986) (Fig. 2.1). LM985 was selected for a Phase I clinical trial in the UK, however it did not progress due to the negative results of the Phase I data. It was reported during this time that LM985 appeared to act as a prodrug for FAA, the compound which was thought to be responsible for the antitumor activity in mice (Double et al. 1986; Bibby et al. 1987) (Fig. 2.1).

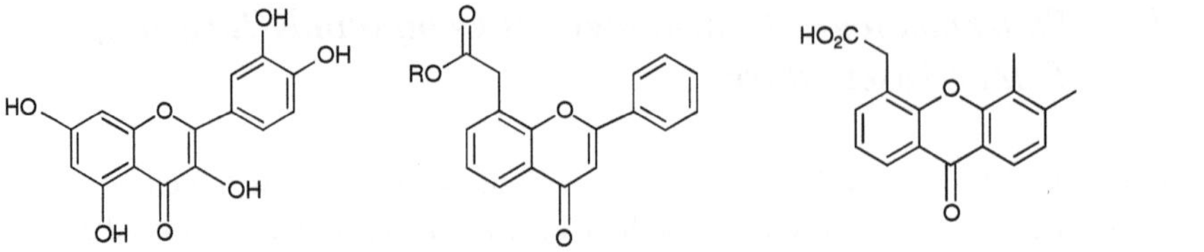

Fig. 2.1 Flavonoids

Evidence from follow-up studies suggested FAA inhibited tumor growth by reducing blood flow to the tumor as a result of vascular disruption (Bibby et al. 1989; Zwi et al. 1989) without effecting systemic blood flow (Bibby and Double 1993). FAA induced the release of cytokines from inflammatory cells, including tumor necrosis factor (TNF), and these factors were proposed to play an important role in the antitumor activity of FAA (Chabot et al. 1993). Further data suggested that FAA-mediated nitric oxide (NO) production within the tumor also contributed to the observed cytotoxic activity (Thomsen et al. 1990, 1991, 1992; Harris and Thorgeirsson 1997).

2.2.2 Clinical Development of Flavonoids as Vascular Disrupting Agents

A number of unsuccessful Phase I and II clinical trials were carried out with FAA (Havlin et al. 1991; Pratt et al. 1991; Siegenthaler et al. 1992). The negative results led to a search for more active drugs with structures related to FAA. This effort led to the identification of a tricyclic analog xanthenone-4-acetic acid (XAA), and eventual synthesis of the highly potent XAA derivative 5,6-dimethylxanthenone-4-acetic acid (DMXAA, ASA404), which displayed a significant increase in activity as compared to FAA (Rewcastle et al. 1989) (Fig. 2.1).

While the precise mechanism of action of ASA404 is largely unknown, the results of preclinical studies suggest ASA404-mediated vascular disruption may occur by direct and indirect anti-vascular activities. Experimental data from murine studies have reported a number of responses following ASA404 administration, namely, increased tumor-vascular permeability (Zhao et al. 2005; Chung et al. 2008), increased tumor endothelial cell apoptosis (Ching et al. 2002, 2004), and decreased tumor blood flow (Zwi et al. 1994; Lash et al. 1998). Similar to what was observed with FAA, evidence suggests vascular disruption may be occurring by increasing intra-tumoral concentrations of TNF and NO following administration (Joseph et al. 1999; Ching et al. 2002). TNF appears to be a major contributor to the induction of vascular collapse. This was demonstrated by a significant reduction in anti-vascular activity in TNF or TNF-receptor (TNFR) knockout mice following treatment with ASA404 (Ching et al. 1999; Zhao et al. 2002). Likewise, induction of NO after administration of ASA404 is considered an important factor in mediating increased vascular permeability and inhibition of blood flow (Thomsen et al. 1990). In response to tumor hypoxia resulting from the disruption of tumor blood flow ASA404 has also been reported to increase VEGF production (Baguley and Siemann 2010). TNF has been shown to enhance VEGF-mediated endothelial cell permeability (Clauss et al. 2001), therefore the activity of ASA404 may involve the combined effects of all these factors. The effects of ASA404 culminate in the breakdown of vasculature and hemorrhagic tumor necrosis. ASA404 has also shown to have either additive or synergistic antitumor effects when combined with several cytotoxic chemotherapeutic agents (Siim et al. 2003). The preclinical demonstration of ASA404 activity in combination with taxanes and platinum-based anticancer drugs provided a rationale for combination clinical trials. Three Phase II trials were designed for patients with non-small cell lung cancer NSCLC, ovarian cancer, and prostate cancer.

In the Phase II trial of carboplatin and paclitaxel with or without ASA404 conducted in patients with NSCLC, ASA404 plus chemotherapy appeared to improve efficacy over chemotherapy alone in terms of overall response rate (ORR), median time to progression (TTP), and median overall survival (OS) (McKeage et al. 2008). To further verify those results a single-arm phase II trial of patients with advanced NSCLC was performed to evaluate ASA404 at a higher dose, again in combination with carboplatin and paclitaxel. Again efficacy appeared to be improved with ASA404 in ORR, TTP, and OS. However, these increases did not reach statistical significance (McKeage et al. 2009). Both Phase II trials designed for patients with ovarian cancer and prostate cancer demonstrated increased response rates when treatment was combined with ASA404, but did not prolong survival (Pili et al. 2010).

The Phase II results led to the randomized, double blind, placebo-controlled Phase III trials in advanced NSCLC, ATTRACT-1 (Antivascular Targeted Therapy: Researching ASA404 in Cancer Treatment) as first-line therapy, and ATTRACT-2 as second-line therapy. ATTRACT-1 included patients with stage IIIB/IV NSCLC who had not previously received systemic therapy for metastatic disease. These patients were treated with ASA404 plus carboplatin and paclitaxel vs. placebo plus carboplatin and paclitaxel (Lara et al. 2011).

In 2010 independent data monitoring committees reviewed the interim results from the ATTRACT-1 and ATTRACT-2 Phase III trials. The committees recommended halting both trials when interim data analyses demonstrated ASA404 when used in combination with chemotherapy would not likely meet their primary endpoints of significantly extending overall survival (Lara et al. 2011) and the pharmaceutical company leading this effort announced that it would discontinue further development of the tumor-vascular disrupting agent ASA404.

2.3 Colchicines

2.3.1 Colchicines as Vascular Disrupting Agents

Of the three major classes of microtubule-binding agents, those that bind to the colchicine binding site are probably the least identified as traditional anticancer therapies; however, the colchicines may have the most promise as VDAs—resulting in severe interruption of tumor blood flow and necrosis to the tumor cell compartment. As a class, colchicinoid microtubule-destabilizing agents are structurally related to the well-studied tubulin binding agent colchicine and bind to the colchicine binding site of tubulin, located at the interface between α and β subunits of the tubulin dimer, thereby inhibiting microtubule polymerization. So well is the colchicine site characterized, that previously tubulin was referred to as a "colchicine binding protein." Members of this class have been coined as mitotic or spindle poisons due to their rapid and deleterious effects on the cytoskeleton. A significant advantage of colchicinoid VDAs, is dual targeting of both the stromal and parenchymal populations within a given tumor microenvironment, due to the concomitant EC and tumor cell sensitivity to this class. The downstream effects of colchicinoids occur just a few minutes after binding, starting with disruption in blood flow and intravascular thrombosis, followed by direct assault on the ECs (Dark et al. 1997). Events after drug administration include morphological changes and disruption of interphase microtubules, cytoskeleton collapse, mitotic catastrophe—EC death, subsequent vascular injury, platelet activation, thrombosis, vascular blockage, plasma protein leakage into interstitial space, vasoconstriction, neutrophil recruitment, reduction in blood flow, avascularization, and hemorrhagic necrosis (Parkins et al. 2000; Galbraith et al. 2001; Tozer et al. 2001; Kanthou and Tozer 2002; Prise et al. 2002; Brooks et al. 2003).

The ancient natural product colchicine, for which the class is named, is one of the oldest known microtubule inhibitors and was noted for its spindle poisoning properties by Pernice, a Sicilian pathologist in 1889 (Stafford et al. 2005) (Fig. 2.2). Colchicine is a highly soluble tricyclic alkaloid isolated from *Colchicum autumnale*, also known as autumn crocus or meadow saffron. Colchicine has long been demonstrated to have anti-vascular effects (Ludford 1948; Baguley et al. 1991), and has recently been demonstrated to be more precisely anti-angiogenic in an endothelial sprouting assay; however, significant inhibitory effect blocking the initiation of angiogenesis was only seen at higher concentrations of colchicine, whereas lower, clinically more relevant dose levels were ineffective (Stafford et al. 2005). Therefore these data and others indicate that, although colchicine induces vascular damage, it is only at doses that are limited by toxicity—in fact, there is no clear-cut distinction between nontoxic, toxic and lethal doses (Finkelstein et al. 2010). As a result, colchicine is FDA approved exclusively for its anti-inflammatory activities in the prevention and treatment of gout and familial Mediterranean fever. Notwithstanding, several groups have shown that at least some of the anti-inflammatory properties of colchicine

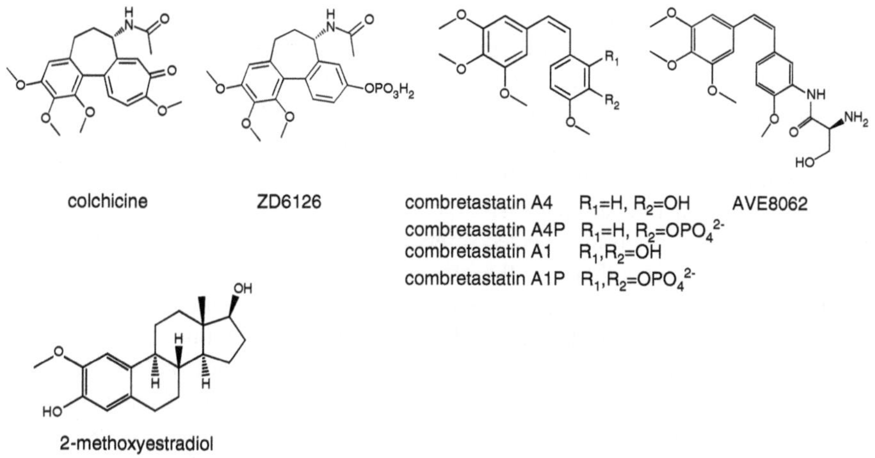

Fig. 2.2 Colchicines and 2-methoxyestradiol

are due to the regulation of adhesion molecules, such as selectins and VCAM, on the EC surface, preventing neutrophil rolling, attachment, and extravasation (Cronstein et al. 1995; Asahina et al. 2001). Interestingly, a prodrug of an N-acetylcolchinol, related to colchicine, ZD6126 (also known as ANG453) (Fig. 2.2), has been in clinical trials as a treatment for cancer as a VDA resulting in vascular occlusion and tumor necrosis. ZD6126 had shown promising results in vitro and in vivo in tumor-bearing mice although due to dosing concerns and limited dose related cardio-cytotoxicity in human patients further development was halted (Micheletti et al. 2003; LoRusso et al. 2008). Notably, cardiotoxicity related to tubulin inhibitors has been demonstrated to be a direct effect on myocardial ECs (Mikaelian et al. 2010).

Combretastatins, a class of natural stilbenoid phenols, were the first microtubule-binding agents identified to have vascular disrupting properties and are the leading agents in the colchicinoid class, if not the entire category of VDAs. They are competitive inhibitors of the binding of colchicine to tubulin (Lin et al. 1988). In particular, combretastatin A-4 (CA4) was first isolated from the bark of the South African bush willow, *Combretum caffrum* (Pettit et al. 1987, 1989) (Fig. 2.2). Indeed, over a dozen combretastatins have been isolated and reported from *C. caffrum* and related plants. CA4 displayed potent cytotoxic activities against a wide range of human cancer cell lines, including multidrug resistant (MDR) cancer cell lines, with IC_{50} values consistently in low nanomolar to subnanomolar range. CA4 is not recognized by P-glycoprotein, often cited as responsible for MDR, and therefore CA4 is less apt to result in drug resistance (Atalay et al. 2006). Besides their effects on tumor cells themselves, CA4 and derivatives target ECs and cause disruption of the endothelial cytoskeleton by binding to the colchicine binding site of the β- subunit of tubulin, thereby inhibiting the polymerization of tubulin to microtubules. CA4 has been demonstrated to be more potent, anti-proliferative and pro-cytotoxic to ECs than would be expected for a tubulin inhibitor and at doses significantly lower than the maximum tolerated dose (MTD) in mice. In fact, in vitro studies using human umbilical vein endothelial cells (HUVEC) by several groups demonstrated that CA4 is cytotoxic if the cells are proliferating, resulting in cell rounding, retraction, membrane blebbing, and substratum detachment, but not if they are quiescent (Dark et al. 1997; Galbraith et al. 2001). CA4 has also been shown to inhibit active angiogenesis in a HUVEC tube formation assay (Grosios et al. 1999). The perhaps unexpected safety profile in mice may be due to the resistance of quiescent endothelia to the effects of combretastatin—likely due to the discriminating dependence of nascent endothelia on the tubulin cytoskeleton (Galbraith et al. 2001; Young and Chaplin 2004). In support of this, a major change that occurs after EC exposure to CA4 is an alteration in shape and morphology, which is profoundly less than that which occurs in quiescent ECs, offering combretastatin, and tubulin inhibitors with similar mechanism of action, an exquisite level of selectivity for ECs in the pathological milieu (Galbraith et al. 2001).

Significant selectivity of colchicinoid compounds for vascular targeting and disruption has been demonstrated in preclinical and clinical studies. Of the combretastatin derivatives, CA4P a simple water soluble prodrug derivative (active when dephosphorylated to CA4 which can then bind tubulin), was the first and perhaps most extensively to be studied (Banerjee et al. 2008) (Fig. 2.2). CA4P induces microtubule disruption by targeting neovascular ECs in the highly proliferative tumor environment which denude from the vessel wall, thereby disrupting tumor blood flow and increasing vascular permeability, resulting in central hemorrhagic necrosis (Galbraith et al. 2001; Tozer et al. 2001). CA4P effects in vivo are rapid and severe, resulting in an acute reduction in tumor blood flow within minutes of drug administration. Significantly, the precise mechanism of action has been defined, in that CA4P interrupts VE-cadherin signaling and induces the regression of susceptible emerging tumor neovessels thereby selectively disrupting tumor vasculature. CA4P actions on VE-cadherin and associated beta-catenin/AKT signaling pathways, critical for endothelial cell survival, results in increased vascular permeability and endothelial cell death (Vincent et al. 2005). However, reduction in tumor growth is not attainable with a single dose and requires repeated administration of the combretastatin over time. Notably synergism is attainable by combination treatments with anti-VE-cadherin neutralizing antibodies (Vincent et al. 2005).

Although CA4P was shown in Phase I studies to selectively reduce tumor blood flow at well-tolerated doses, it unfortunately does not result in tumor regressions or substantial tumor growth delay as a single agent—so more selective targeting, combination therapies (if toxicities can be mediated), and/or improved colchicinoids may be required (Banerjee et al. 2008). In fact, CA4P was recently shown to improve overall survival in a phase II/III trial in anaplastic thyroid cancer in combination with paclitaxel and carboplatin. Currently, CA4P is in studies for NSCLC and platinum resistant ovarian carcinoma, and in combination with anti-VEGF bevacizumab for persistent ovarian epithelial, fallopian tube, or primary peritoneal carcinoma.

This is particularly interesting as combining VTAs with already established angiogenesis inhibitory agents may provide the one-two punch necessary to prevent tumor regrowth.

Naturally found congeners of CA4 include combretastatin A-1 (CA1), a simple hydroxyl derivative of CA4 which binds to the colchicine binding site at high affinity (Fig. 2.2). The diphosphate prodrug of CA1, referred to as CA1P or Oxi4503, is currently in clinical development (Fig. 2.2). Preclinical studies with CA1P in mice demonstrated profound damage to tumor microvasculature with minimal uninvolved organ vascular bed damage, however complete tumor eradication was not achieved—suggesting that combination therapies with other chemotherapeutic modalities might achieve complete tumor eradication (Chan et al. 2007). CA1P is currently in Phase I clinical trials for relapsed and refractory acute myelogenous leukemia (AML) and myelodysplastic syndrome (MDS). Furthermore, several synthesized derivatives of combretastatin have been pursued and are in advanced clinical development. In particular the combretastatin-serine AVE8062 is currently being evaluated in clinical cancer trials (Fig. 2.2). AVE8062 is a water soluble synthetic combretastatin analog with nanomolar potency inducing G2/M arrest and apoptosis in mouse ECs in vitro and in vivo, which is enhanced by combination therapy with docetaxel (Kim et al. 2007). Phase I and II studies of CA1P and CA4P (and several other colchinoids—both natural products and synthetics) reported cardiovascular toxicities. Cardiac events were dose-limiting in phase I trials with VDA monotherapy and combination therapy (Subbiah et al. 2011). These studies suggest that monitoring of cardiac biomarkers could play a valuable role in future trials of VDAs.

In 1994, Judah Folkman's laboratory demonstrated that a previously identified anti-angiogenic metabolite of estradiol, 2-methoxyestradiol (2MeO-E2, Panzem) (Fig. 2.2), inhibited angiogenesis in the chick chorioallantoic membrane (CAM) via binding to the colchicine site of tubulin (D'Amato et al. 1994). 2-MeO-E2's anti-angiogenic activity lies within its ability to reduce endothelial cell proliferation, induce EC apoptosis, and reduce the transcription and expression, nuclear accumulation, and transcriptional activity of HIF-1α—thereby demonstrating that 2MeO-E2 is a true anti-neoangiogenic compound (Mabjeesh et al. 2003; Verenich and Gerk 2010). This agent also inhibits tumor cell growth by binding to tubulin, resulting in antimitotic activity, and by inducing caspase activation, resulting in cell cycle arrest in the G2 phase, DNA fragmentation, and apoptosis. Phase I and II clinical trials revealed that orally administered 2MeO-E2 is well tolerated by patients with only grade 2 and 3 toxicities observed (Verenich and Gerk 2010). However, 2MeO-E2 faces many challenges due to poor solubility, no MTD determined in early clinical studies, only nanogram per milliliter levels detected in plasma due to presystemic metabolism despite large doses, and in vitro cytotoxicity effects on tumor cells in the micromolar to submicromolar range (Verenich and Gerk 2010).

In addition to toxicities outlined above, one of the main concerns with using colchinoids/combretastatins for vascular targeting as single agents is the potential for reactive angiogenesis from the residual viable rim left after the central necrosis events have taken place (Horsman and Siemann 2006). There are two working non-mutually exclusive hypotheses for what may contribute to the remaining viable rim after colchinoid/combretastatin treatment: (1) functional and structural tumor endothelial heterogeneity (interstitial pressure, vessel caliber, vascular compensation) resulting in drug susceptibility of the inner vasculature vs. outer vasculature; (2) sustenance of the outer rim tumor cells from the surrounding normal vasculature, largely resistant to VDAs (Tozer et al. 2005). One could argue that neither colchinoids nor any tubulin inhibitors were designed to be single therapy agents and are much more likely to be effective when administered in intermittent doses and in synergy with other therapeutics (Young and Chaplin 2004). To combat the reactive rim rebound angiogenesis, several groups suggested and evaluated the efficacy of combination therapies, including nonconventional therapies, to inhibit the regrowth of the tumor from the reactive angiogenic rim—including markedly successful combinations with cisplatin, cyclophosphamide, radiation and a host of other standards of care (Dark et al. 1997; Li et al. 1998; Chaplin et al. 1999; Horsman et al. 2000; Siemann et al. 2002; Young and Chaplin 2004; Madlambayan et al. 2010). The differential sensitivity for alternative mechanism of action chemotherapies and radiation vs. combretastatins allows for a unique opportunity for dual population targeting (central hypoxic tumor cells and tubulin inhibitor resistant tumor rim), resulting in additive or synergistic long-term outcomes. Notably, other groups have successfully used combretastatin-mediated vascular targeting to enhance the retention of therapeutic antibodies within the tumor microenvironment (Lankester et al. 2007). In addition, because therapeutic antibody targeting strategies often have greatest affect at the tumor rim due to tumor penetration, it makes sense to combine these strategies with VDAs, such as colchinoids. It is of note that vascular and tumoral heterogeneity can confound the response of the patient or preclinical animal to CA4P—not the least of which are tumor size, vascular perfusion, interstitial fluid pressure, and vascular permeability. Using biomarker and clinical strategies for appropriate drug administration should improve the long-term benefits of colchinoid class tubulin inhibitors in the clinic, resulting in personally designed dosing regimens with less toxicity and improved efficacy indices—ultimately saving patient lives.

2.4 *Vinca* Alkaloids and Peptide Tubulin Inhibitors as Vascular Disrupting Agents

Vinca alkaloids and peptide tubulin inhibitors are distinct classes of natural products which share similar function as microtubule depolymerizing agents. The earliest reported *Vinca* alkaloids were natural products derived from the periwinkle *Vinca rosea*, later renamed *Catharanthus roseus*. Vinblastine and later vincristine were the first natural alkaloids discovered from extracts of these plants (Kruczynski and Hill 2001) (Fig. 2.3). Medicinal chemistry efforts successfully generated multiple analogs via semisynthetic approaches, several of which underwent clinical evaluation, including vinorelbine, vindesine, and vinflunine (Fig. 2.3). The peptide tubulin

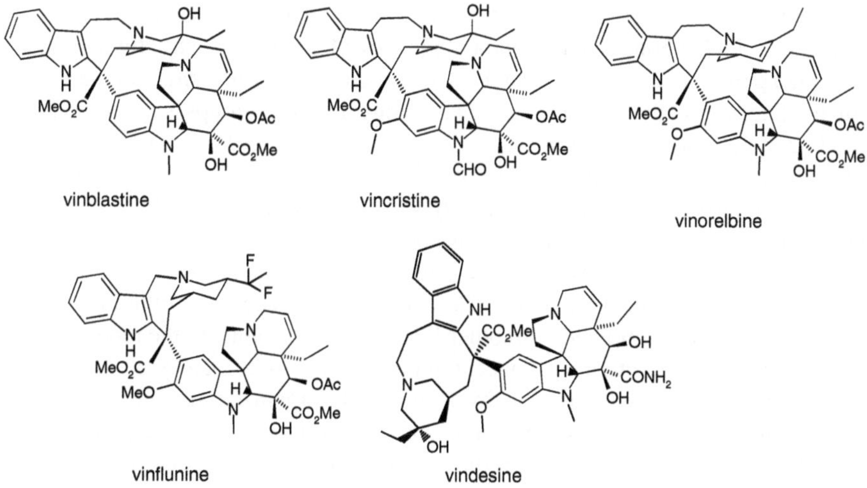

Fig. 2.3 Vinca alkaloids

inhibitors are a highly diverse class of natural products, most of which are composed of nonnatural amino acids. Representatives of this class include dolastatins, auristatins, hemiasterlins, cryptophycins, and others. Synthetic efforts have also produced multiple analogs with modified tubulin binding, cellular potency, and in vivo efficacy. *Vinca* alkaloids and peptide tubulin inhibitors share an overlapping binding site on β-tubulin that is distinct from the colchicine and taxane binding sites. In general, *Vincas* noncompetitively inhibit the binding of peptides such as dolastatin and analogs to tubulin (Bai et al. 1990). Both classes of tubulin inhibitors destabilize microtubules and shift tubulin into the soluble, heterodimeric form. During normal physiological processes, microtubules undergo treadmilling, the net addition of subunits at the growing "+" end of the microtubule and loss at the "–" end, and dynamic instability, the switching at microtubule ends between phases of slow and rapid growth. *Vincas* and peptides bind to microtubule ends and suppress both dynamic instability and treadmilling. Several members of both *Vinca* and peptide classes have demonstrated activity against both epithelial and endothelial cells. We highlight vinflunine and soblidotin, one member of each class with enhanced anti-vascular activity compared to their related analogs.

2.4.1 *Vinflunine*

In order to identify potentially more active *Vinca* analogs with an improved safety profile, chemists and biologists at the Pierre Fabre research group designed, synthesized, and analyzed novel *Vinca* analogs, ultimately advancing vinflunine (Kruczynski and Hill 2001) (Fig. 2.3). Vinflunine represents a modification of vinorelbine at the 20′ position with the addition of two fluorine atoms and reduction of the 3′,4′ double bond in the catharanthine moiety. Despite the structural similarities among the four major *Vinca* analogs, there are notable differences in their tubulin binding and cellular properties. Data from multiple groups suggest that the binding of vinflunine to tubulin is actually weaker than other *Vincas*, on the order of vincristine > vinblastine > vinorelbine > vinflunine. Moreover, drug competition studies demonstrated that vinflunine only weakly inhibited the binding of radiolabeled vinblastine or vinorelbine and did not inhibit vincristine (Kruczynski et al. 1998). Vinflunine also showed unique effects on microtubule dynamics compared with vinblastine (Ngan et al. 2000). These distinct differences in vinflunine's interaction with tubulin compared with other *Vincas* may contribute to the reported anti-vascular effects of vinflunine at concentrations that are lower than cytotoxic doses.

Tubulin mediates cell division by forming the organized spindle network of microtubules that control chromosomal segregation into daughter cells. However, microtubules also play critical roles during interphase, including cytoskeletal structure, cell migration, adhesion site dynamics, and protein transport. These activities are particularly important during neovascularization. Therefore, drugs interfering with tubulin function have pleiotropic effects on cell physiology, in addition to the cytotoxicity promoted by mitotic catastrophe. ECs may be more sensitive to the non-mitotic effects of tubulin inhibitors due to their dependence on microtubules for a dynamic cytoskeleton, cell migration, and other functions.

The Braguer lab elegantly observed the effects of vinflunine on microtubule dynamics in human dermal microvascular endothelial cells (HMVEC), allowing direct assessment of vinflunine's properties in an endothelial cell system (Honore et al. 2008). In general, tubulin depolymerizing agents cause a decrease in the amount of stabilized or "paused" microtubules. Using time-lapse video microscopy, the number of microtubule pauses were measured in HMVEC and found to be approximately 50% in untreated cells. Vinflunine doses as low as 10 pM reduced microtubule pauses to ~20%, and also shortened the duration of the pauses. These effects were apparent between 0.01 and 2 nM. Therefore, at very low concentrations, even below the doses typically causing cytotoxicity, vinflunine destabilizes microtubules and prevents them from participating in normal interphase functions. A direct comparison among *Vincas* on microtubule dynamics conducted by another group confirmed that vinflunine neither reduces the rate of microtubule shortening nor the duration of the "pause" state, in contrast to vinblastine (Ngan et al. 2000). This reduced affinity of vinflunine with tubulin likely results in the observed 4–7-fold reduced microtubule treadmilling rates compared with vinorelbine and vinblastine.

In HMVEC, vinflunine also decreased the duration of microtubule end interaction with adhesion sites from 2.5 min in untreated cells to 38 min in the presence of 1 nM vinflunine (Honore et al. 2008). In addition, vinflunine altered actin dynamics by increasing the size and duration of stress fibers between focal adhesions. Microtubule-associated proteins such as EB1 typically mediate tubulin dynamics by binding to the growing tubule at the (+) end. Vinflunine at 1 nM eliminated the proper localization of transfected GFP-EB1 from microtubules in HMVEC. A direct correlation was also observed between vinflunine concentrations that alter microtubule dynamics and the phenotype of cell migration. HMVEC speed was decreased from a normal rate of 0.4 to 0.2 μm/min in the presence of 0.01–1 nM vinflunine, doses which are commensurate with the destabilization of microtubules.

The anti-proliferative effects of vinflunine were compared with other *Vinca* analogs in a panel of tumor cell lines after 2 or 3 day treatment (Kruczynski et al. 1998). In general, vinflunine was 2–44-fold less potent than vinblastine, vincristine, or vinorelbine in eight cell lines, and >10- to >1,000-fold less potent in two cell lines. The Kruczynski lab later reported the effects of vinflunine on ECs (Kruczynski et al. 2006). One hour treatment with vinflunine caused the dissociation of newly formed endothelial cell tubes at doses of 100 nM and greater. Statistically significant inhibition of HUVEC migration was observed at 100 nM vinflunine. Much higher concentrations of vinflunine were required to induce endothelial cell death during acute exposure, ranging from 8 μM upon 4 h exposure and >100 μM upon 1 h exposure. Similar results were reported for HMVECs (Pourroy et al. 2006). Low concentrations of vinflunine increased microtubule dynamics in HMVEC and were sufficient to inhibit Matrigel tube formation and cell speed without altering mitotic spindle function. This suggests that vinflunine selectively affects interphase microtubule functions that are necessary for anti-vascular effects at concentrations that do not induce mitotic arrest or cytotoxicity.

The effects of vinflunine on tumor vasculature were assessed in the poorly differentiated and rapidly growing colon adenocarcinoma model, MAC15A (Holwell et al. 2001). At the single-administration maximally tolerated dose of 50 mg/kg, vinflunine caused >50% tumor growth inhibition 1 week following a single injection. Assessment of vascular damage in tumors was conducted by infusion with Hoechst dye following drug treatment. Microscopy revealed staining only of ECs and confirmed >50% vascular shutdown in tumors 4 h after treatment with vinflunine at 10–40 mg/kg. Approximately 80% inhibition of functional vasculature was evident 24 h after a single dose at concentrations below the MTD.

Vinflunine also inhibited growth factor induced angiogenesis when bFGF-embedded Matrigel plugs were implanted subcutaneously in mice (Kruczynski et al. 2006). At vinflunine doses of 1.25 mg/kg and greater, 2–3-fold decreases in hemoglobin content were observed in the plugs, suggesting impaired vascularization. In the same study, LS174T tumor cells were injected into mice spleens to assess the effect of vinflunine on experimentally induced metastasis. Vinflunine treatment reduced the number of liver metastases at 1.25 mg/kg, or 16-fold lower than the MTD. In contrast, colchicine or vinblastine produced similar effects only at doses near MTD. Vinflunine was also efficacious across multiple studies in xenografts derived from a variety of human tumor cell lines, including lung NCI-H69 and LX1, renal RXF944LX, colon TC37, prostate PC3, breast MX1, and pancreas PAXF546 (Hill et al. 1999).

Several clinical studies were conducted to evaluate the safety and anticancer efficacy of vinflunine. The predominate adverse events for vinflunine therapy across all trials were grade 3 or 4 neutropenia (approximately 50% of patients) and constipation, fatigue, and/or anemia (approximately 10–20%). As a result of an acceptable safety profile in various phase I trials, vinflunine was advanced to proof-of-concept studies. Phase II trials were conducted in patients with multiple tumor types including renal cell carcinoma, ovarian cancer, and melanoma, with the most impressive responses observed in lung, breast, and bladder cancers (Yun-San Yip et al. 2008). In a phase II study conducted in 63 patients with NSCLC, an 8.3% partial response rate was observed and responses persisted for an average of 7 months. In another phase II study, metastatic breast cancer patients who failed taxane and anthracycline therapies showed a 30% ORR and 14.3 month median overall survival. Another study with 51 bladder cancer patients who previously failed platinum therapies showed 18% partial response rate and 67% disease control; 14% of patients previously treated with a vinblastine-containing cocktail responded (Culine et al. 2006). Overall, single-agent vinflunine showed higher response rates in clinical trials compared with trials evaluating paclitaxel, docetaxel, oxaliplatin, lapatinib, bortezomib, or pemetrexed (Yun-San Yip et al. 2008). Several trials were also conducted to evaluate combination of vinflunine with platins, gemcitabine, or capecitabine. Combination studies of vinflunine with anti-Her2 antibody trastuzumab produced relatively high 62.5% and 73.7% response rates, at 280 or 320 mg/m^2 vinflunine respectively.

A phase III trial was conducted in transitional cell carcinoma of the urinary tract (TCCU) to follow-up on the preliminary phase I and II activity in this subtype of bladder cancer (Bellmunt et al. 2009). A 2 month survival advantage was observed in patients treated with vinflunine plus best supportive care (6.9 months) vs. BSC alone (4.6 month) and a significant 23% improvement in overall survival. These data suggest that vinflunine can benefit patients for second-line treatment of TCCU.

Specific vascular endpoints do not appear to have been evaluated or reported systematically across the majority of clinical trials, so it is difficult to determine if anti-vascular activity is a major contributor of the clinical efficacy. Also, it remains unknown whether the reduced affinity of vinflunine for tubulin ultimately imparts an improved clinical profile compared with first generation *Vincas*. Interestingly, the tubulin binding properties of the *Vincas*, more so than the cytotoxicity potencies, corresponded to the doses utilized in the clinic. As noted above, tubulin binding affinities were vincristine > vinblastine > vinorelbine > vinflunine. In contrast, the cellular potency for vincristine, vinblastine, vinorelbine in a panel of tumor cell lines were within fivefold of each other, while vinflunine was typically >10-fold higher. Weekly dosing regimens for these drugs were approximately 0.4–1.4 mg/m^2 for vincristine, 4–20 for vinblastine, 25–35 for vinorelbine, and 280–320 mg/m^2 for vinflunine. These activity differences among the highly related *Vinca* alkaloids further emphasizes that minimal but selective structural changes in analogs can significantly alter binding to the therapeutic target and translate to relevant pharmacological differences. Chemotherapeutic agents are typically administered to patients at maximally tolerated doses. Hence, it may be difficult to tease out the subtle biological differences in tubulin binding and anti-vascular effects observed at lower concentrations for a drug like vinflunine. At high doses, cytotoxicity in normal proliferating cells, such as hematopoietic and gastrointestinal compartments, can reduce the therapeutic index. Rational and well-titrated combinations of novel anti-vascular agents plus debulking cytotoxics remain an important focus in oncology.

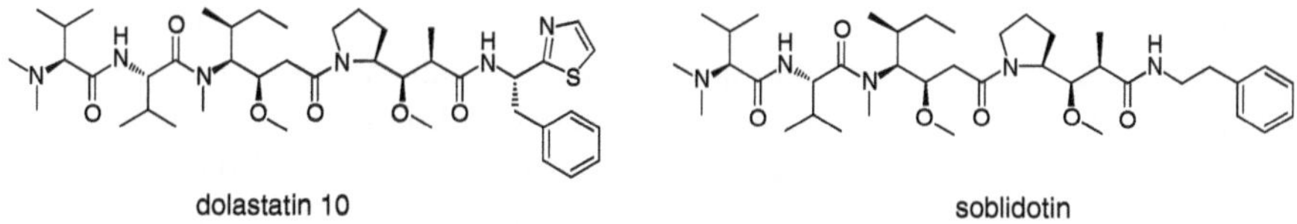

Fig. 2.4 Peptide tubulin inhibitors

2.4.2 *Soblidotin*

Soblidotin, also known as auristatin PE or TZT-1027, is an analog of dolastatin-10 (Kobayashi et al. 1997) (Fig. 2.4). Dolastatin-10 was originally extracted in 1987 from the Indian ocean mollusk *Dolabella auricularia.* Soblidotin demonstrated anti-vascular activity that was more pronounced than dolastatin and related anti-microtubule agents. Soblidotin inhibits purified tubulin in a cell-free system with an IC_{50} of approximately 2.2 μM, within range of other tubulin destabilizing agents. Soblidotin was found to bind to tubulin at both high (K_d=0.2 nM) and low (K_d= 10 μM) affinity sites, and to non-competitively inhibit the binding of vinblastine from tubulin (Natsume et al. 2000). The anti-vascular activity of soblidotin was first observed as altered cell contraction and membrane blebbing in HUVEC upon acute exposure that was more pronounced than vincristine (Otani et al. 2000). Soblidotin induces mitotic arrest and apoptosis in cultured tumor and endothelial cells and is cytotoxic at subnanomolar concentrations. The effect of drug exposure time and cytotoxicity was reported for soblidotin and other anticancer agents in murine colon carcinoma C26 cells and HUVEC. Soblidotin was highly potent against both cell types, with an IC_{50} of 0.014 ng/mL against C26 after 3 day treatment; soblidotin was threefold more potent against HUVEC at approximately 0.0046 ng/mL (Watanabe et al. 2007). Interestingly, soblidotin retained ultra-potency after only 24 h treatment in HUVEC (0.098 ng/mL) while other chemotherapeutics were 10–100,000-fold less potent. Short-term exposure of all drugs to C26 cancer cells was ineffective (IC_{50}>1,000 ng/mL). These findings were extended to the assessment of vascular permeability with HUVEC cells, where the ability of inhibitors to damage the junctions between ECs was assessed by using FITC-dextran (Watanabe et al. 2006, 2007). Soblidotin increased permeability by approximately 12-fold, compared with 9- and 4-fold changes for vinblastine and vincristine, respectively, and no effect for docetaxel, 5-fluorouracil, and cisplatin (Watanabe et al. 2007). These data suggest differential effects of soblidotin on ECs compared with other chemotherapeutics, including inhibitors of tubulin and DNA function.

The effect of soblidotin on HUVEC vascular tube formation was explored using cells plated on extracellular basement membrane (Watanabe et al. 2007). Drug added at the time of HUVEC plating impairs the development of newly formed tubules and when soblidotin was added 20 h after cell plating, it disrupted existing tubules with an IC_{50} range of approximately 0.1–1 ng/mL. The effect of soblidotin on angiogenesis was assessed using a chick embryo CAM assay (Watanabe et al. 2007). After 2 days of exposure to soblidotin, >80% inhibition of neovascularization was observed. Transcriptional profiling was also conducted in a lung carcinoma cell line treated with various anti-tubulin agents, and the observed soblidotin gene expression pattern was distinct from that of vinblastine, vincristine, paclitaxel, and docetaxel (Shimoyama et al. 2006). Interestingly, soblidotin increased the expression of the MMP inhibitor, TIMP3, which is known to interfere with VEGF binding to its receptor. Proteomic or RNA profiling of soblidotin-treated ECs may provide additional support for the role of the peptide on anti-vascular activity.

The in vivo efficacy of soblidotin was evaluated in a VEGF-dependent model where human small cell carcinoma cell line SBC-3 was transfected with VEGF cDNA and implanted into nude mice (Natsume et al. 2003). SBC-3/VEGF tumors showed increased neovascularization and faster growth rate compared with control cells. Nearly complete regressions were observed against both control tumors and SBC-3/VEGF tumors treated with soblidotin. This activity was more pronounced than combretastatin-A4P, vincristine, docetaxel, or cisplatin at or near their maximally tolerated doses. Histological analyses showed enhanced erythrocyte accumulation, leakage, and scattering from tumor vasculature of SBC-3/VEGF cells compared with SBC-3/neo cells 3–48 h post-administration of soblidotin. The authors suggest that soblidotin preferentially induced thrombi, extravasation and necrosis in a VEGF-dependent manner. Combretastatin-A4P was ineffective in these models, suggesting that a combination of both cytotoxic and anti-vascular activity is required for maximal effect. Soblidotin also reduced tumor perfusion in C26 tumor-bearing mice, caused erythrocyte leakage, and potentiated vascular permeability induced by VEGF (Watanabe et al. 2006).

Multiple clinical trials have been conducted with soblidotin. Separate Phase I studies were performed in patients with various advanced solid tumors, as well as in a cohort of non-small cell lung cancer (NSCLC) (Schoffski et al. 2004; Greystoke et al. 2006; Tamura et al. 2007). In the NSCLC trial (Horti et al. 2008), 49 patients treated by dose-escalation had dose-limiting toxicities of neutropenia, fever, myalgia, and neuropathic pain. One complete response, 3 partial responses, and 20 stable disease were reported.

A Phase II trial was conducted in 32 NSCLC patients previously treated with platinum-based therapy, and the majority had also been treated with microtubule inhibitor taxanes or vinblastine (Riely et al. 2007). Predominate adverse events included leukemia and neutropenia, which are commonly seen with tubulin inhibitors at MTD. However, no objective responses were observed and median disease progression was 1.5 months. No data was reported specifically on anti-vascular markers. A Phase II study was also conducted in patients with metastatic soft-tissue sarcomas; 21% of patients showed stable disease but no partial responses were observed (Patel et al. 2006). No further Phase II development work has been conducted on soblidotin as a free drug beyond these trials.

2.5 Conclusions and Future Directions

A variety of natural products exert pronounced anti-angiogenic effects in preclinical models, however, their clinical development was frequently limited by the onset of side effects, including damage in normal tissues. A key question to overcome the current limitations of VDAs is how to select for natural products with optimal selectivity for tumor vasculature, sparing normal, resting vasculature.

The balance between efficacy and safety is evident in clinical studies with ultra-potent anti-microtubule agents such as soblidotin, dolastatin, and related molecules. Clearly it is necessary to minimize toxicities associated with tubulin inhibitors against rapidly proliferating cell compartments, such as hematological and gastrointestinal tissues.

A class of therapeutics targeting tumor vasculature, also referred to as VTAs, that interfere selectively with key regulators of tumor angiogenesis, including agents blocking VEGF-A and VEGFR-2 signaling, showed promising antitumor activities in preclinical studies and in clinical trials (Ferrara et al. 2007). However, the increase in overall survival by targeted therapeutics is frequently limited, and these agents are most commonly administered in combination with standard of care cytotoxic regimens. For example, the use of bevacizumab is approved only when combined with cytotoxic or cytokine therapy (with the exception of patients with GBM) and patients with metastatic disease are refractory or acquire resistance to VEGF inhibitors (Jain et al. 2006; Bergers and Hanahan 2008). Therefore, novel compounds or therapeutic modalities targeting tumor vasculature with improved efficacy and safety characteristics are needed to increase the response to anti-angiogenic treatment.

One potential strategy to achieve this goal is to combine the unique potency and selectivity of VDAs including tubulin inhibitors and flavonoids with the selectivity of large molecules binding to antigens expressed on tumor vasculature. This approach is showing success, in particular with auristatin-based tubulin inhibitors, as evidenced by the clinical approval of brentuximab vedotin (SGN-35) for Hodgkin lymphoma. The delivery of soblidotin and other peptide inhibitors to anti-vascular antigen targets via tumor endothelial-specific antibodies may enhance the therapeutic index and clinical activity of such agents.

Additional therapeutic modalities are currently developed in oncology which may achieve this goal, including nanoparticles targeted to the tumor vasculature. It will be interesting to test the potential of VDAs in the context of ADCs or nanoparticles and to study their ability to interfere with tumor angiogenesis and to block tumor growth.

References

Asahina A, Tada Y et al (2001) Colchicine and griseofulvin inhibit VCAM-1 expression on human vascular endothelial cells—evidence for the association of VCAM-1 expression with microtubules. J Dermatol Sci 25(1):1–9

Atalay C, Deliloglu Gurhan I et al (2006) Multidrug resistance in locally advanced breast cancer. Tumour Biol 27(6):309–318

Baguley BC, Siemann DW (2010) Temporal aspects of the action of ASA404 (vadimezan; DMXAA). Expert Opin Investig Drugs 19(11):1413–1425
Baguley BC, Holdaway KM et al (1991) Inhibition of growth of colon 38 adenocarcinoma by vinblastine and colchicine: evidence for a vascular mechanism. Eur J Cancer 27(4):482–487
Bai RL, Pettit GR et al (1990) Binding of dolastatin 10 to tubulin at a distinct site for peptide antimitotic agents near the exchangeable nucleotide and vinca alkaloid sites. J Biol Chem 265(28):17141–17149
Banerjee S, Wang Z et al (2008) Efficacy of selected natural products as therapeutic agents against cancer. J Nat Prod 71(3):492–496
Bellmunt J, Theodore C et al (2009) Phase III trial of vinflunine plus best supportive care compared with best supportive care alone after a platinum-containing regimen in patients with advanced transitional cell carcinoma of the urothelial tract. J Clin Oncol 27(27):4454–4461
Bergers G, Hanahan D (2008) Modes of resistance to anti-angiogenic therapy. Nat Rev Cancer 8(8):592–603
Bibby MC, Double JA (1993) Flavone acetic acid—from laboratory to clinic and back. Anticancer Drugs 4(1):3–17
Bibby MC, Double JA et al (1987) Factors involved in the anti-cancer activity of the investigational agents LM985 (flavone acetic acid ester) and LM975 (flavone acetic acid). Br J Cancer 55(2):159–163
Bibby MC, Double JA et al (1989) Reduction of tumor blood flow by flavone acetic acid: a possible component of therapy. J Natl Cancer Inst 81(3):216–220
Brooks AC, Kanthou C et al (2003) The vascular targeting agent combretastatin A-4-phosphate induces neutrophil recruitment to endothelial cells in vitro. Anticancer Res 23(4):3199–3206
Carmeliet P, Jain RK (2011a) Molecular mechanisms and clinical applications of angiogenesis. Nature 473(7347):298–307
Carmeliet P, Jain RK (2011b) Principles and mechanisms of vessel normalization for cancer and other angiogenic diseases. Nat Rev Drug Discov 10(6):417–427
Chabot GG, Branellec D et al (1993) Tumour necrosis factor-alpha plasma levels after flavone acetic acid administration in man and mouse. Eur J Cancer 29A(5):729–733
Chan LS, Malcontenti-Wilson C et al (2007) Effect of vascular targeting agent Oxi4503 on tumor cell kinetics in a mouse model of colorectal liver metastasis. Anticancer Res 27(4B): 2317–2323
Chaplin DJ, Pettit GR et al (1999) Anti-vascular approaches to solid tumour therapy: evaluation of combretastatin A4 phosphate. Anticancer Res 19(1A):189–195
Ching LM, Goldsmith D et al (1999) Induction of intratumoral tumor necrosis factor (TNF) synthesis and hemorrhagic necrosis by 5,6-dimethylxanthenone-4-acetic acid (DMXAA) in TNF knockout mice. Cancer Res 59(14):3304–3307
Ching LM, Cao Z et al (2002) Induction of endothelial cell apoptosis by the antivascular agent 5,6-dimethylxanthenone-4-acetic acid. Br J Cancer 86(12):1937–1942
Ching LM, Zwain S et al (2004) Relationship between tumour endothelial cell apoptosis and tumour blood flow shutdown following treatment with the antivascular agent DMXAA in mice. Br J Cancer 90(4):906–910
Chung F, Liu J et al (2008) Consequences of increased vascular permeability induced by treatment of mice with 5,6-dimethylxanthenone-4-acetic acid (DMXAA) and thalidomide. Cancer Chemother Pharmacol 61(3):497–502
Clauss M, Sunderkotter C et al (2001) A permissive role for tumor necrosis factor in vascular endothelial growth factor-induced vascular permeability. Blood 97(5):1321–1329
Cronstein BN, Molad Y et al (1995) Colchicine alters the quantitative and qualitative display of selectins on endothelial cells and neutrophils. J Clin Invest 96(2):994–1002
Culine S, Theodore C et al (2006) A phase II study of vinflunine in bladder cancer patients progressing after first-line platinum-containing regimen. Br J Cancer 94(10):1395–1401
D'Amato RJ, Lin CM et al (1994) 2-Methoxyestradiol, an endogenous mammalian metabolite, inhibits tubulin polymerization by interacting at the colchicine site. Proc Natl Acad Sci U S A 91(9):3964–3968

Dark GG, Hill SA et al (1997) Combretastatin A-4, an agent that displays potent and selective toxicity toward tumor vasculature. Cancer Res 57(10):1829–1834
Darland DC, D'Amore PA (1999) Blood vessel maturation: vascular development comes of age. J Clin Invest 103(2):157–158
Denekamp J (1982) Endothelial cell proliferation as a novel approach to targeting tumour therapy. Br J Cancer 45(1):136–139
Denekamp J, Hobson B (1982) Endothelial-cell proliferation in experimental tumours. Br J Cancer 46(5):711–720
Double JA, Bibby MC et al (1986) Pharmacokinetics and anti-tumour activity of LM985 in mice bearing transplantable adenocarcinomas of the colon. Br J Cancer 54(4):595–600
Ferrara N, Mass RD et al (2007) Targeting VEGF-A to treat cancer and age-related macular degeneration. Annu Rev Med 58:491–504
Finkelstein Y, Aks SE et al (2010) Colchicine poisoning: the dark side of an ancient drug. Clin Toxicol (Phila) 48(5):407–414
Folkman J (1985) Tumor angiogenesis. Adv Cancer Res 43:175–203
Galbraith SM, Chaplin DJ et al (2001) Effects of combretastatin A4 phosphate on endothelial cell morphology in vitro and relationship to tumour vascular targeting activity in vivo. Anticancer Res 21(1A):93–102
Greystoke A, Blagden S et al (2006) A phase I study of intravenous TZT-1027 administered on day 1 and day 8 of a three-weekly cycle in combination with carboplatin given on day 1 alone in patients with advanced solid tumours. Ann Oncol 17(8):1313–1319
Grosios K, Holwell SE et al (1999) In vivo and in vitro evaluation of combretastatin A-4 and its sodium phosphate prodrug. Br J Cancer 81(8):1318–1327
Harris SR, Thorgeirsson UP (1997) Flavone acetic acid stimulates nitric oxide and peroxynitrite production in subcutaneous mouse tumors. Biochem Biophys Res Commun 235(3):509–514
Havlin KA, Kuhn JG et al (1991) Phase I clinical and pharmacokinetic trial of flavone acetic acid. J Natl Cancer Inst 83(2):124–128
Hill BT, Fiebig HH et al (1999) Superior in vivo experimental antitumour activity of vinflunine, relative to vinorelbine, in a panel of human tumour xenografts. Eur J Cancer 35(3):512–520
Holwell SE, Hill BT et al (2001) Anti-vascular effects of vinflunine in the MAC 15A transplantable adenocarcinoma model. Br J Cancer 84(2):290–295
Honore S, Pagano A et al (2008) Antiangiogenic vinflunine affects EB1 localization and microtubule targeting to adhesion sites. Mol Cancer Ther 7(7):2080–2089
Horsman MR, Siemann DW (2006) Pathophysiologic effects of vascular-targeting agents and the implications for combination with conventional therapies. Cancer Res 66(24):11520–11539
Horsman MR, Murata R et al (2000) Combretastatins novel vascular targeting drugs for improving anti-cancer therapy. Combretastatins and conventional therapy. Adv Exp Med Biol 476:311–323
Horti J, Juhasz E et al (2008) Phase I study of TZT-1027, a novel synthetic dolastatin 10 derivative, for the treatment of patients with non-small cell lung cancer. Cancer Chemother Pharmacol 62(1):173–180
Jain RK (2003) Molecular regulation of vessel maturation. Nat Med 9(6):685–693
Jain RK (2005) Normalization of tumor vasculature: an emerging concept in antiangiogenic therapy. Science 307(5706):58–62
Jain RK, Duda DG et al (2006) Lessons from phase III clinical trials on anti-VEGF therapy for cancer. Nat Clin Pract Oncol 3(1):24–40
Joseph WR, Cao Z et al (1999) Stimulation of tumors to synthesize tumor necrosis factor-alpha in situ using 5,6-dimethylxanthenone-4-acetic acid: a novel approach to cancer therapy. Cancer Res 59(3):633–638
Kanthou C, Tozer GM (2002) The tumor vascular targeting agent combretastatin A-4-phosphate induces reorganization of the actin cytoskeleton and early membrane blebbing in human endothelial cells. Blood 99(6):2060–2069
Kim TJ, Ravoori M et al (2007) Antitumor and antivascular effects of AVE8062 in ovarian carcinoma. Cancer Res 67(19):9337–9345

Kobayashi M, Natsume T et al (1997) Antitumor activity of TZT-1027, a novel dolastatin 10 derivative. Jpn J Cancer Res 88(3):316–327

Kruczynski A, Hill BT (2001) Vinflunine, the latest Vinca alkaloid in clinical development. A review of its preclinical anticancer properties. Crit Rev Oncol Hematol 40(2):159–173

Kruczynski A, Barret JM et al (1998) Antimitotic and tubulin-interacting properties of vinflunine, a novel fluorinated Vinca alkaloid. Biochem Pharmacol 55(5):635–648

Kruczynski A, Poli M et al (2006) Anti-angiogenic, vascular-disrupting and anti-metastatic activities of vinflunine, the latest vinca alkaloid in clinical development. Eur J Cancer 42(16):2821–2832

Lankester KJ, Maxwell RJ et al (2007) Combretastatin A-4-phosphate effectively increases tumor retention of the therapeutic antibody, 131I-A5B7, even at doses that are sub-optimal for vascular shut-down. Int J Oncol 30(2):453–460

Lara PN Jr, Douillard JY et al (2011) Randomized phase III placebo-controlled trial of carboplatin and paclitaxel with or without the vascular disrupting agent vadimezan (ASA404) in advanced non-small-cell lung cancer. J Clin Oncol 29(22):2965–2971

Lash CJ, Li AE et al (1998) Enhancement of the anti-tumour effects of the antivascular agent 5,6-dimethylxanthenone-4-acetic acid (DMXAA) by combination with 5-hydroxytryptamine and bioreductive drugs. Br J Cancer 78(4):439–445

Li L, Rojiani A et al (1998) Targeting the tumor vasculature with combretastatin A-4 disodium phosphate: effects on radiation therapy. Int J Radiat Oncol Biol Phys 42(4):899–903

Lin CM, Singh SB et al (1988) Interactions of tubulin with potent natural and synthetic analogs of the antimitotic agent combretastatin: a structure-activity study. Mol Pharmacol 34(2):200–208

LoRusso PM, Gadgeel SM et al (2008) Phase I clinical evaluation of ZD6126, a novel vascular-targeting agent, in patients with solid tumors. Invest New Drugs 26(2):159–167

Ludford RJ (1948) Factors determining the action of colchicine on tumour growth. Br J Cancer 2(1):75–86

Mabjeesh NJ, Escuin D et al (2003) 2ME2 inhibits tumor growth and angiogenesis by disrupting microtubules and dysregulating HIF. Cancer Cell 3(4):363–375

Madlambayan GJ, Meacham AM et al (2010) Leukemia regression by vascular disruption and antiangiogenic therapy. Blood 116(9):1539–1547

McKeage MJ, Von Pawel J et al (2008) Randomised phase II study of ASA404 combined with carboplatin and paclitaxel in previously untreated advanced non-small cell lung cancer. Br J Cancer 99(12):2006–2012

McKeage MJ, Reck M et al (2009) Phase II study of ASA404 (vadimezan, 5,6-dimethylxanthenone-4-acetic acid/DMXAA) 1800 mg/m(2) combined with carboplatin and paclitaxel in previously untreated advanced non-small cell lung cancer. Lung Cancer 65(2):192–197

Micheletti G, Poli M et al (2003) Vascular-targeting activity of ZD6126, a novel tubulin-binding agent. Cancer Res 63(7):1534–1537

Mikaelian I, Buness A et al (2010) Primary endothelial damage is the mechanism of cardiotoxicity of tubulin-binding drugs. Toxicol Sci 117(1):144–151

Natsume T, Watanabe J et al (2000) Characterization of the interaction of TZT-1027, a potent antitumor agent, with tubulin. Jpn J Cancer Res 91(7):737–747

Natsume T, Watanabe J et al (2003) Antitumor activity of TZT-1027 (Soblidotin) against vascular endothelial growth factor-secreting human lung cancer in vivo. Cancer Sci 94(9):826–833

Neri D, Bicknell R (2005) Tumour vascular targeting. Nat Rev Cancer 5(6):436–446

Ngan VK, Bellman K et al (2000) Novel actions of the antitumor drugs vinflunine and vinorelbine on microtubules. Cancer Res 60(18):5045–5051

Otani M, Natsume T et al (2000) TZT-1027, an antimicrotubule agent, attacks tumor vasculature and induces tumor cell death. Jpn J Cancer Res 91(8):837–844

Parkins CS, Holder AL et al (2000) Determinants of anti-vascular action by combretastatin A-4 phosphate: role of nitric oxide. Br J Cancer 83(6):811–816

Patel S, Keohan ML et al (2006) Phase II study of intravenous TZT-1027 in patients with advanced or metastatic soft-tissue sarcomas with prior exposure to anthracycline-based chemotherapy. Cancer 107(12):2881–2887

Pettit GR, Cragg GM et al (1987) Antineoplastic agents, 122. Constituents of Combretum caffrum. J Nat Prod 50(3):386–391

Pettit GR, Singh SB et al (1989) Isolation and structure of the strong cell growth and tubulin inhibitor combretastatin A-4. Experientia 45(2):209–211

Pili R, Rosenthal MA et al (2010) Phase II study on the addition of ASA404 (vadimezan; 5,6-dimethylxanthenone-4-acetic acid) to docetaxel in CRMPC. Clin Cancer Res 16(10): 2906–2914

Plowman J, Narayanan VL et al (1986) Flavone acetic acid: a novel agent with preclinical antitumor activity against colon adenocarcinoma 38 in mice. Cancer Treat Rep 70(5):631–635

Pluda JM (1997) Tumor-associated angiogenesis: mechanisms, clinical implications, and therapeutic strategies. Semin Oncol 24(2):203–218

Pourroy B, Honore S et al (2006) Antiangiogenic concentrations of vinflunine increase the interphase microtubule dynamics and decrease the motility of endothelial cells. Cancer Res 66(6):3256–3263

Pratt CB, Relling MV et al (1991) Phase I study of flavone acetic acid (NSC 347512, LM975) in patients with pediatric malignant solid tumors. Am J Clin Oncol 14(6):483–486

Prise VE, Honess DJ et al (2002) The vascular response of tumor and normal tissues in the rat to the vascular targeting agent, combretastatin A-4-phosphate, at clinically relevant doses. Int J Oncol 21(4):717–726

Rewcastle GW, Atwell GJ et al (1989) Potential antitumor agents. 58. Synthesis and structure-activity relationships of substituted xanthenone-4-acetic acids active against the colon 38 tumor in vivo. J Med Chem 32(4):793–799

Riely GJ, Gadgeel S et al (2007) A phase 2 study of TZT-1027, administered weekly to patients with advanced non-small cell lung cancer following treatment with platinum-based chemotherapy. Lung Cancer 55(2):181–185

Schoffski P, Thate B et al (2004) Phase I and pharmacokinetic study of TZT-1027, a novel synthetic dolastatin 10 derivative, administered as a 1-hour intravenous infusion every 3 weeks in patients with advanced refractory cancer. Ann Oncol 15(4):671–679

Shimoyama T, Hamano T et al (2006) Reference profiling of the genomic response induced by an antimicrotubule agent, TZT-1027 (Soblidotin), in vitro. Pharmacogenomics J 6(6):388–396

Siegenthaler P, Kaye SB et al (1992) Phase II trial with flavone acetic acid (NSC.347512, LM975) in patients with non-small cell lung cancer. Ann Oncol 3(2):169–170

Siemann DW, Mercer E et al (2002) Vascular targeting agents enhance chemotherapeutic agent activities in solid tumor therapy. Int J Cancer 99(1):1–6

Siim BG, Lee AE et al (2003) Marked potentiation of the antitumour activity of chemotherapeutic drugs by the antivascular agent 5,6-dimethylxanthenone-4-acetic acid (DMXAA). Cancer Chemother Pharmacol 51(1):43–52

Stafford SJ, Schwimer J et al (2005) Colchicine and 2-methoxyestradiol inhibit human angiogenesis. J Surg Res 125(1):104–108

Subbiah IM, Lenihan DJ et al (2011) Cardiovascular toxicity profiles of vascular-disrupting agents. Oncologist 16(8):1120–1130

Tamura K, Nakagawa K et al (2007) Phase I study of TZT-1027, a novel synthetic dolastatin 10 derivative and inhibitor of tubulin polymerization, which was administered to patients with advanced solid tumors on days 1 and 8 in 3-week courses. Cancer Chemother Pharmacol 60(2):285–293

Thomsen LL, Ching LM et al (1990) Evidence for the production of nitric oxide by activated macrophages treated with the antitumor agents flavone-8-acetic acid and xanthenone-4-acetic acid. Cancer Res 50(21):6966–6970

Thomsen LL, Ching LM et al (1991) Tumor-dependent increased plasma nitrate concentrations as an indication of the antitumor effect of flavone-8-acetic acid and analogues in mice. Cancer Res 51(1):77–81

Thomsen LL, Ching LM et al (1992) Nitric oxide production in endotoxin-resistant C3H/HeJ mice stimulated with flavone-8-acetic acid and xanthenone-4-acetic acid analogues. Biochem Pharmacol 43(11):2401–2406

Tozer GM, Prise VE et al (2001) Mechanisms associated with tumor vascular shut-down induced by combretastatin A-4 phosphate: intravital microscopy and measurement of vascular permeability. Cancer Res 61(17):6413–6422

Tozer GM, Kanthou C et al (2005) Disrupting tumour blood vessels. Nat Rev Cancer 5(6):423–435

Venditti JM, Wesley RA et al (1984) Current NCI preclinical antitumor screening in vivo: results of tumor panel screening, 1976–1982, and future directions. Adv Pharmacol Chemother 20:1–20

Verenich S, Gerk PM (2010) Therapeutic promises of 2-methoxyestradiol and its drug disposition challenges. Mol Pharm 7(6):2030–2039

Vincent L, Kermani P et al (2005) Combretastatin A4 phosphate induces rapid regression of tumor neovessels and growth through interference with vascular endothelial-cadherin signaling. J Clin Invest 115(11):2992–3006

Wahl O, Oswald M et al (2011) Inhibition of tumor angiogenesis by antibodies, synthetic small molecules and natural products. Curr Med Chem 18(21):3136–3155

Watanabe J, Natsume T et al (2006) Antivascular effects of TZT-1027 (Soblidotin) on murine Colon26 adenocarcinoma. Cancer Sci 97(12):1410–1416

Watanabe J, Endo Y et al (2007) Antiangiogenic activity of TZT-1027 (soblidotin) on chick chorioallantoic membrane and human umbilical vein endothelial cells. In Vivo 21(2):297–304

Young SL, Chaplin DJ (2004) Combretastatin A4 phosphate: background and current clinical status. Expert Opin Investig Drugs 13(9):1171–1182

Yun-San Yip A, Yuen-Yuen Ong E et al (2008) Vinflunine: clinical perspectives of an emerging anticancer agent. Expert Opin Investig Drugs 17(4):583–591

Zhao L, Ching LM et al (2002) The antitumour activity of 5,6-dimethylxanthenone-4-acetic acid (DMXAA) in TNF receptor-1 knockout mice. Br J Cancer 87(4):465–470

Zhao L, Ching LM et al (2005) Mechanisms of tumor vascular shutdown induced by 5,6-dimethylxanthenone-4-acetic acid (DMXAA): increased tumor vascular permeability. Int J Cancer 116(2):322–326

Zwi LJ, Baguley BC et al (1989) Blood flow failure as a major determinant in the antitumor action of flavone acetic acid. J Natl Cancer Inst 81(13):1005–1013

Zwi LJ, Baguley BC et al (1994) Correlation between immune and vascular activities of xanthenone acetic acid antitumor agents. Oncol Res 6(2):79–85

Chapter 3
Discodermolide: Past, Present, and Future

Ross E. Longley

Abstract (+)-Discodermolide, a polyhydroxylated lactone isolated from the marine sponge *Discodermia dissoluta*, is a promising antitumor agent that continues to be the subject of intensive chemical, biological, and pharmaceutical research since its discovery two decades ago. Although structurally distinct from taxol, discodermolide shares a common mechanism of action as a potent tubulin polymerizer, resulting in blockage of cells in the G_2/M phase of the cell cycle, aberrant microtubule function, and cell death. Following its licensing from Harbor Branch Oceanographic Institution in 1998 by Novartis AG as an antitumor agent, discodermolide was afforded an unprecedented priority status for synthesis on an industrial scale, borrowing and modifying in some cases, academic synthetic schemes to produce quantities sufficient for clinical trial. From a co-discoverer's viewpoint, the story of discodermolide will be described here, starting with its discovery, its progression through preclinical evaluation, highlights from the synthetic contributions of academia and industry, a brief SAR walk around the molecule, the results of the first clinical trial and the more recent journey from the actual compound, through its various analogs and finally to its evolution to some exciting novel "hybrid" molecules.

3.1 Discovery and Initial Studies

The discovery of discodermolide came about as part of an overall program in marine natural products drug discovery in the late 80s and early 90s at Harbor Branch Oceanographic Institution in Fort Pierce, Florida. The activities of that program

R.E. Longley, Ph.D. (✉)
Lake Erie College of Osteopathic Medicine-Bradenton,
5000 Lakewood Ranch Boulevard, Bradenton, FL 34211, USA
e-mail: rlongley@lecom.edu

F.E. Koehn (ed.), *Natural Products and Cancer Drug Discovery*, Cancer Drug Discovery and Development, DOI 10.1007/978-1-4614-4654-5_3,

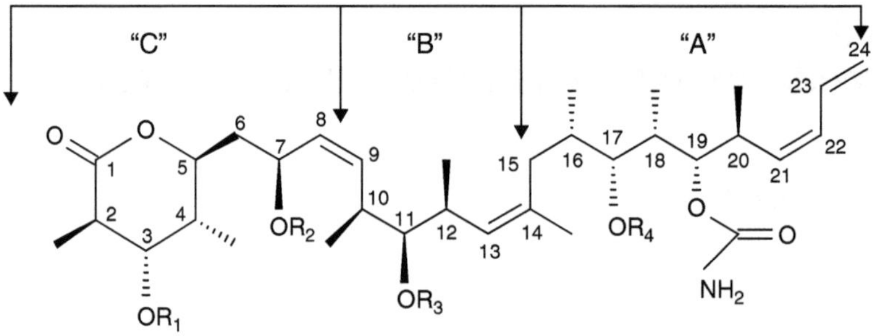

Discodermolide R = H

Fig. 3.1 Structure of discodermolide where R = H. The molecule can be additionally divided into three regions A, B, and C which represent fragments which can be obtained synthetically and then coupled together to form the complete molecule

were geared toward the discovery of potentially clinically useful compounds with antitumor, antimicrobial, and immunomodulatory activities, through a systematic screening of extracts from the marine environment and purification of the active components following a bioassay guided approach. The isolation and structural determination of discodermolide was first reported by Gunasekera et al. (1990). The structure of discodermolide was described as a polyhydroxylated lactone which was isolated from a methanol/toluene extract of a Caribbean marine sponge, *Discodermia dissoluta* which was collected by scuba in March, 1987 at Lucay, Grand Bahama Island, at a depth of 33 m and was immediately frozen for later analyses. The compound was isolated using silica gel and reverse-phase chromatography and the structure was determined primarily using NMR spectroscopy and verified by X-ray crystallographic analysis (Fig. 3.1). The extract of the sponge originally demonstrated inhibition of the in vitro proliferation of P388 leukemia cells, inhibition of the growth of *Candida albicans* yeast, and suppression of the two-way mixed lymphocyte reaction (MLR). However, upon further purification, anti-proliferative properties towards fungal organisms disappeared. The pure compound displayed potent antitumor activity towards P388 cells with an IC_{50} of 0.5 μg/mL and suppression of the two-way mixed lymphocyte response of both mice and humans (0.5 μg/mL and 5.0 μg/mL, respectively). The compound appeared not to be overly toxic to normal murine splenocytes. The carbon skeleton of discodermolide was reported as novel and the relative stereochemistry revealed a hairpin structure in the middle of the carbon chain (Fig. 3.2). Initial isolation schemes revealed an extremely low yield of the compound, with only 7.0 mg of pure compound isolated from the initial 434 g of frozen sponge (0.002%).

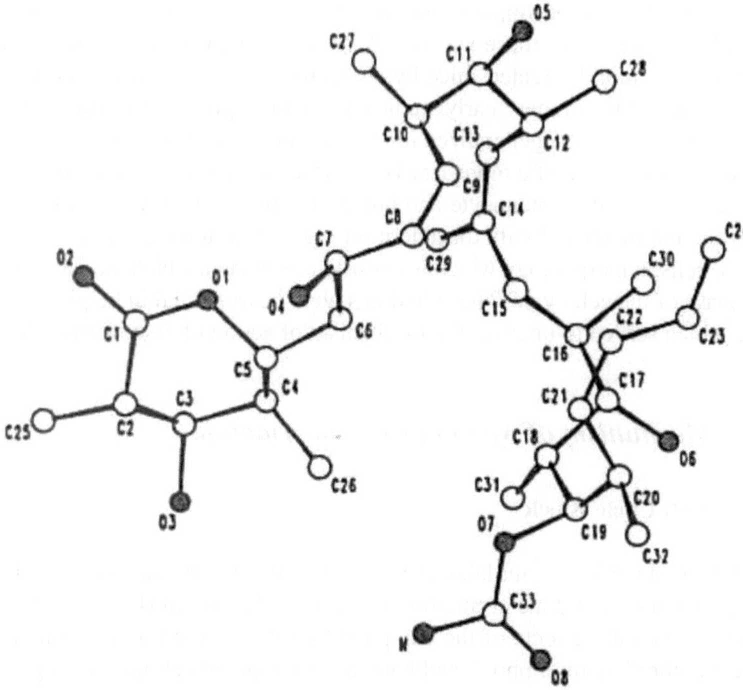

Fig. 3.2 URANUS drawing showing numbering scheme and hairpin structure (reprinted with permission from Gunasekera et al. (1990). Copyright 1990 American Chemical Society)

3.1.1 Early In Vitro and In Vivo Studies with Discodermolide

The initial biological activity of discodermolide was further characterized by Longley et al. (1991a) to possess a wide range of in vitro immunosuppressive activities, including inhibition of mitogenic stimulation of murine and human lymphocytes, suppression of PMA-ionomycin stimulation, and interleukin-2 receptor expression of purified T-cells of both murine and human origin. The compound appeared not to be overly cytotoxic in vitro to normal cells. The first in vivo immunosuppressive studies with discodermolide were reported in our companion paper (Longley et al. 1991b) which described discodermolide's ability to suppress the graft-versus-host splenomegaly response of BALB/c => CB6F$_1$ (BALB/c×C57BL/6J)F$_1$ grafted mice when administered as daily, i.p. injections (5.0, 2.5, and 1.25 mg/kg) for 7 days. Some toxicity and morbidity was associated with the two higher doses, but significant suppression was observed at nontoxic doses of 1.25 mg/kg and two additional tested lower doses of 0.625 and 0.313 mg/kg. Splenocytes obtained from these discodermolide-treated animals were found to be deficient in their abilities to respond to mitogens in vitro and an impaired killing of YAC-1 tumor cells by natural killer (NK) cells isolated from these animals. Antibody responses following immunization of discodermolide-treated mice by sheep red blood cells, however, remained entirely intact. Thus, in these early studies, it became apparent that discodermolide was exhibiting more of a generalized immunosuppressive effect on immune responsiveness rather than one of a more T or B-cell specific type. In addition, microscopic observations of mouse splenocyte and human lymphocytes stimulated with PMA-ionomycin and incubated with discodermolide resulted in the appearance of large, blast-like cells; a morphology which is consistent with cells which are blocked at the G_2/M phase of the cell cycle. These findings were instrumental in helping to design studies which served to uncover the mechanism of action of discodermolide.

3.1.2 Mechanism of Action of Discodermolide

3.1.2.1 Cell Cycle Block

One of the first studies in our laboratory which pointed to the mechanism of action of discodermolide was done in collaboration with colleagues at Hoffmann-La Roche which described the effects of the compound on the proliferation of a number of murine lymphoid, nonlymphoid, and human nonlymphoid cell lines (Longley et al. 1993). Our results showed that IC_{50} values for the inhibition of proliferation of discodermolide-treated cell lines were all very similar, with ranges of 37–84 nM. These results, combined with our previous observations of mitotically blocked “blast cells,” led us to examine the effects of discodermolide on the progression of cells through the cell cycle, using flow cytometry. Murine DO11.10T hybridoma cells were used as cell cycle targets for discodermolide. Cells were incubated with varying concentrations of discodermolide for 3, 5, and 24 h, stained with propidium iodide and analyzed on a flow cytometer. Our findings indicated that in as little 3 h incubation of cells with 1.0 μg/mL of discodermolide, increasing populations of cells entering S and G_2/M phases were observed with more than 50% of the cells accumulating in G_2/M phase by 24 h. Forward vs. side scatter patterns of discodermolide blocked cells confirmed a size range of approximately twofold, with a DNA content of 2 N compared to untreated cells, thus confirming our initial observations of the formation of large blast-like cells in discodermolide-treated cultures.

3.1.2.2 Tubulin Target and Discodermolide Binding Site

In collaborative studies with Billy Day’s laboratory at the University of Pittsburg (Kowalski et al. 1995), discodermolide was shown to stabilize cellular microtubules and polymers of tubulin and microtubule-associated protein (MAPs). In comparison

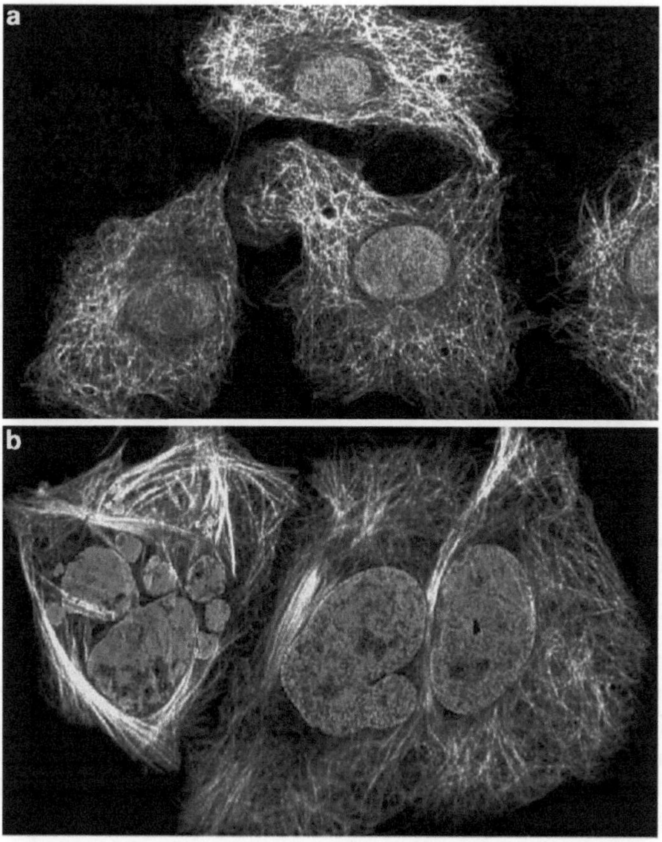

Fig. 3.3 Effects of discodermolide on A549 human lung cancer cells. A549 human lung cancer cells were incubated with and without 100 nM discodermolide for 24 h fixed in ethanol, stained with fluorescent anti-alpha tubulin and examined with confocal microscopy. (**a**) 0.05% ethanol control. Note regular, mesh-like microtubule network, and oval shaped nuclei. (**b**) 100 nM discodermolide. Spectacular rearrangement of microtubules into thick bundles and fragmented nuclei (used with kind permission from Harbor Branch Oceanographic Institution, Fort Pierce, FL)

to other microtubule polymerizing agents, epothilone A/B and taxol, discodermolide was observed to be the most potent of these agents. In a subsequent paper (ter Haar et al. 1996), the Day laboratory confirmed our original findings by the observation of a G_2/M block for discodermolide-treated Burkett lymphoma cells. They also observed that 10 nM discodermolide treatment of MCF-7 breast cancer cells resulted in a spectacular rearrangement of microtubule filaments into clusters or "bundles," similar to that which was routinely required 1.0 μM of taxol. These observations extended to our own in which we observed a similar phenomenon with A549 human lung adenocarcinoma cancer cells (Fig. 3.3). In vitro incubation of purified tubulin plus MAPs with discodermolide resulted in almost total polymerization of tubulin at 0°C, conditions in which taxol at an identical concentration was inactive.

EC_{50} values for induction of tubulin assembly were 3.2 μm for discodermolide vs. 23 μm for taxol. Similar findings of G_2/M block were reported by Hung et al. (1994) in Stuart Schreiber's laboratory, with some of the first synthetically obtained discodermolide. The (+) enantiomer of synthetic discodermolide was found to have potent anti-proliferative and G_2/M blocking activity in several cell lines in the nanomolar range. Interestingly, the synthetic (−) enantiomer was 6–19-fold less potent compared to the (+) enantiomer and its mechanism of action appeared to be linked with accumulation of cells in S-phase rather than G_2/M.

In subsequent studies, the Day group (Kowalski et al. 1997) found discodermolide to be strikingly potent in inhibiting the binding of [^{3}H] taxol to tubulin polymer, with an apparent K_i value of 0.4 μM. The morphology of the tubulin polymers formed with discodermolide was clearly different than that observed for taxol i.e., discodermolide polymers appeared much shorter in length. The Schreiber group (Hung et al. 1996a) confirmed the discodermolide targeting of tubulin and showed that discodermolide bound one molecule per tubulin dimer, with a higher affinity than taxol. The binding of microtubules by discodermolide and taxol were mutually exclusive, indicating either overlapping or different tubulin binding sites. More recent studies by the Horwitz lab (Khrapunovich-Baine et al. 2009) indicate that while taxol enjoys major interactions with the M-loop of tubulin, discodermolide is oriented away from this loop and toward the N-terminal H1-S2 loop. Discodermolide appears to stabilize microtubules primarily through tubulin interdimer contacts on the R-tubulin side, and to a lesser extent on interprotofilament contacts between adjacent β-tubulin subunits. These complementary stabilizing effects of discodermolide and taxol may help to understand the synergy of these two drugs observed in vivo. Several groups continue to investigate the 3D structure of discodermolide, relating the solution structure (Canales et al. 2008; Sanchez-Pedregal et al. 2006) with that of the original proposed solid state conformation (Gunasekera et al. 1990; Smith et al. 2000, 2001) in relation to its conformational binding to tubulin Surprisingly, the solid state and solution conformations are similar, and more recent studies by Jogalekar et al. (2010) suggest a new binding mode (pose) of discodermolide for β-tubulin which is more consistent with some of SAR data involving certain lactone and diene substitutions which will be discussed later.

3.1.2.3 Activity Against Drug Resistant Tumor Cell Lines

Discodermolide was shown to have potent cytotoxic activity against various multidrug resistant cell lines (Kowalski et al. 1997). P-glycoprotein over-expressing human colon and ovarian carcinoma cell lines which were 25–89-fold more resistant relative to their parental lines were sensitive to discodermolide, with IC_{50} values in the range of 70 nM for SWR620AD-300 MDR resistant colon carcinoma cells (compared to 260 nM for paclitaxel) and 580 nM for A2789AD MDR resistant ovarian carcinoma cells (compared to 3,900 nM for paclitaxel). For ovarian carcinoma cells made resistant to paclitaxel due to mutations in their β-tubulin, IC_{50}

values for discodermolide were 7.2 nM for 1A9PTX10 cells (compared to 32 nM for paclitaxel) and 13 nM for 1A9PTX22 cells (compared to 41 nM for paclitaxel). Susan Horwitz's laboratory (Martello et al. 2000) was the first to show that discodermolide could combine synergistically with taxol to kill A549-T12 human lung carcinoma cells which had been made resistant to taxol, but were sensitive to discodermolide. This same cell line also required low concentrations (2–6 nM) of taxol for growth. The microtubule active compounds such as the epothilones and elu-etherobin could substitute for taxol in sustaining the growth of these cells, whereas discodermolide could not. Thus, discodermolide's mechanism of action seemed to differ somewhat from that of taxol, and could explain the synergistic cytotoxic effects observed with discodermolide/taxol combinations. Further studies in Mary Ann Jordan's laboratory (Honore et al. 2004) confirmed and extended these findings in normal A549 cells. Discodermolide and taxol synergistically blocked cell cycle progression at G_2/M interface (41%) compared to each drug alone (9.6% and 16% for discodermolide and taxol, respectively), and they synergistically enhanced cell death due to apoptosis. The combination of discodermolide and taxol at their respective anti-proliferative IC_{50} values (7 and 2 nM) for discodermolide and taxol however, altered most parameters of microtubule dynamic instability in a synergistic fashion, except for a time-based rescue frequency. The significance of these findings lie in the ability of low concentrations of discodermolide to synergize with paclitaxel to effect microtubule dynamics directly and to contribute to a synergistic G_2/M arrest, inhibition of proliferation, and induction of apoptosis.

One of the first in vivo studies to test the synergistic effects of discodermolide and paclitaxel was described by Huang et al. (2006) in the Horwitz laboratory. Discodermolide was used singly or in combination with taxol in the treatment of nude mice bearing human ovarian SKOV-3 carcinoma cell xenografts. Engrafted mice which were given a single tail vein injection of taxol (20 mg/kg) or similarly treated with one dose of discodermolide (5 mg/kg) exhibited only minimal tumor growth suppression 6 days following treatment. Taxol at the maximum tolerated dose (MTD), in nude mice (60 mg/kg i.v. administered as 20 mg/kg qd X3), resulted in tumor growth suppression but not regression. Engrafted mice which were given a combined taxol/discodermolide dose showed 40% tumor regression. At 20 days posttreatment, the tumor volumes of the combined treatment group were significantly smaller compared to those of the single-agent groups. Additionally, the combination treatment was well tolerated as evidenced by no weight loss or observable toxicities compared to the taxol MTD group, where transient body weight loss occurred (10%). Fibroblast growth factor-supplemented Matrigel plugs obtained from animals treated with discodermolide/taxol combinations and separately had significantly decreased vascularization in the combination-treated animals compared with single-agent. The significance of these findings is multifold; the in vivo results validate the previously described in vitro synergism of discodermolide and taxol and point to the potential of combination therapy using these two drugs. In addition, they give us insight on adjunctive mechanism of action of discodermolide here, i.e., as a potent antiangiogenic agent.

3.1.2.4 Other Mechanisms of Action of Discodermolide

A further definition of discodermolide's mechanism of action was investigated by Klein et al. (2005) in the Horwitz laboratory. They reported a type of "accelerated senescence" of A549 cells which had been grown for more than a year in 8 nM of discodermolide, before switching to a higher concentration of the drug. These cells appeared to stop proliferating and exhibited a flattened, enlarged and granular morphology. These cells expressed β-galactosidase, up regulation of plasminogen activator inhibitor (PAI-1), and induction of three important proteins of the MAPK signaling pathway, p66Shc, a mediator of oxidase stress and Erk1 and Erk2, kinases which phosphorylate cytoskeletal proteins and transcription factors, resulting ultimately in regulation of cell growth. Interestingly, incubation of these cells with equimolar concentrations of taxol had either no effect or diminished the expression of levels of these proteins. This was the first study which demonstrated that discodermolide could accelerate cellular senescence suggesting yet another mechanism by which discodermolide suppresses tumor cell growth.

In yet another potential mechanism of action of discodermolide, Bröker et al. (2002) and Huisman et al. (2002) in the Giaccone group had previously reported that paclitaxel, epothilone B, and discodermolide triggered cell death primarily via caspase-independent routes in non small cell lung carcinoma (NSCLC) cell lines Neither expression of the caspase-8 inhibitor CrmA nor preincubation with the broad-spectrum caspase blocker zVAD-fmk decreased the cytotoxic effects of these compounds. Further studies by Bröker et al. (2004) showed that discodermolide could trigger disruption of lysosomes leading to release and subsequent activation of the lysosomal protease cathepsin B, representing an early stage of the cell death process in this type of cell. Inhibition of cathepsin B, and not of caspases or other proteases such as calpains or cathepsin D, strongly protected against cell death, indicating that cathepsin B was essential for the caspase-independent cell death triggered by paclitaxel, epothilone B, and discodermolide. A most interesting mechanism of action of discodermolide was reported by Smith et al. (2005d) involving the potential for use of a discodermolide analog compound in bacteriolytic therapy (Dang et al. 2001) of tumors. A single intravenous injection of (+)-2,3-anhydrodiscodermolide combined with genetically modified *Clostridium novyi*-NT spores caused rapid and complete regressions of tumors in mice bearing HCT116 colorectal cancer xenografts. Treatment with each agent alone was not effective in causing tumor regression. The combination of anaerobic growth of *C. novyi*-NT spores in the hypoxic regions of the tumor and the cell growth inhibitory action of (+)-2,3-anhydrodiscodermolide in the vascularized compartments of the tumor were postulated to be the mechanism of action of tumor regression.

3.2 Supply

Discodermolide was originally isolated in very limited quantities from the sponge, *D. dissoluta* (Gunasekera et al. 1990). The original 7 mg of the compound was obtained from an initial quantity of 434 g of the frozen sponge (0.002%) and was utilized for some of the first in vitro studies. Additional milligram amounts were isolated from subsequent recollections of the sponge by the Harbor Branch Group, and were used for initial in vivo studies. It was clear from the start that harvesting large quantities of sponge to provide isolated amounts of the pure compound sufficient for anticipated clinical trials was both economically and ecologically prohibitive. Alternative sources of the compound were therefore sought, which led to the consideration of two avenues of investigation; (1) isolation of a potential microorganism(s) from the source sponge which theoretically could be responsible for production of the compound or (2) access via synthetic methods.

3.2.1 Microbial Sources of Discodermolide

Sponges contain a wide variety of microorganisms which live in a commensal type relationship. Some of these microorganisms isolated from their respective sponges have been shown to produce the very metabolites which were once thought to be solely associated with the sponge source (Bewley et al. 1996; Davidson et al. 2001; Unson and Faulkner 1993). Early attempts were made to isolate microorganisms from *D. dissoluta* sponge material, both from freshly collected and frozen specimens. Several morphological types were cultured and partially identified, but none were shown to produce discodermolide (personal observations and unpublished data). Shirmer et al. (2005) through the screening of sponge metagenomic libraries, described a sponge-specific, filamentous bacteria found in whole tissue samples of *D. dissoluta*, which belonged to the genus *Entotheonella*, This organism appeared to constitute the majority of the filamentous population within the sponge. Brück et al. (2008), in collaboration with Pomponi's group at Harbor Branch, used centrifugal sponge cell fractionation of dissociated tissue from *Discodermia* sp. and was able to identify enriched populations of *Entotheonella* sp. by the use of 16sRNA genomic analyses which was confirmed through the use of fluorescent in situ hybridization. The organism appeared to locate in the mesohyl of the sponge. While the tantalizing possibility still exists that there may be a sponge-associated microorganism which could be isolated from *Discodermia* sp. sponge, cultured and fermented to produce discodermolide, none has been reported thus far.

3.2.2 Synthesis of Discodermolide

A very thorough and excellent review of the synthesis of discodermolide has been recently described in a paper by Florence et al. (2008) in the Paterson group, so those details will not be presented here. However, a brief mention of the history of the contributions of each of the groups involved in the syntheses, the relative yields, number of steps, and number of grams provided (where appropriate) will be made here. The first synthesis of the negative antipode of discodermolide was carried out by Schreiber and coworkers (Nerenberg et al. 1993) and helped to establish the first absolute configuration. Synthesis of the natural (+) antipode followed shortly thereafter (Hung et al. 1994), and as mentioned earlier, the two antipodes exhibited quite different activities in relation to cell cycle block, with the natural (+) antipode blocking cells at the G_2/M phase of the cell cycle, while the unnatural (–) antipode blocking cells in the S phase. The Schreiber synthesis was accomplished with an overall yield of 4.3% over the 23 steps in the longest linear sequence and was the first to establish the absolute stereo configuration of discodermolide which allowed for subsequent preparation of a number of discodermolide analogs and initial structure activity relationship studies. The laboratory of Amos Smith was similarly successful in providing a total synthesis of unnatural (–)-discodermolide with his first attempts resulting in a 2.2% overall yield obtained over 28 steps in the longest linear sequence (Smith et al. 1995). Synthesis of natural (+)-discodermolide was subsequently reported by his group (Smith et al. 1999). The overall synthetic scheme improved the yield to 6% over 24 steps in the longest linear sequence and is most notable for providing the greatest amount of discodermolide at that time (1.04 g) which provided quantities sufficient for more in depth in vitro and in vivo biological studies. A further improved synthesis which helped to negate a difficult high-pressure step, a clear roadblock to scale-up synthesis, was reported by Smith et al. (2003), however, yield suffered, with a 1.9% overall yield and 24 steps in the longest linear sequence. A fourth scheme was then initiated by Smith et al. (2005c) which described modifications which increased the overall efficiency and convergence of the approach. This resulted in an impressive increase in overall yield of 9.0% with only 17 steps in the longest linear sequence.

Myles and coworkers first reported the synthesis of unnatural (–) discodermolide in 2003 (Harried et al. 2003), followed by synthesis of natural (+) discodermolide in 2003 (Myles 2003). The Myles synthesis relied on a convergent strategy and included a diastereoselective alkylation of a ketone enolate to establish the key C15–C16 bond. Overall yield was 1.1% with 22 steps in the longest linear sequence.

Marshall and Johns (1998) described a total synthesis of (+)-discodermolide in 1998, which featured the assembly of three key stereotriad polypropionate subunits using allenylmetal addition methodology. The overall yield using this strategy was 2.2% in 29 steps in the longest linear sequence.

The Paterson group reported their first initial total synthesis of (+)-discodermolide in 2000 (Paterson et al. 2000). The basis of their synthetic strategy was the use of stereo controlled aldol reactions to join three key subunits. This was their first attempt at total synthesis with an impressive yield of 10.3% over 23 steps (longest linear sequence). A second strategy, aimed at reducing the total number of steps and to reduce the need to use all chiral reagents was subsequently reported (Paterson et al. 2003) and resulted in a lower overall yield (5.1%), but greatly reduced the total number of steps to complete the discodermolide molecule from 58 to 35 steps. A third generation synthesis by the Paterson group was later described (Paterson and Lyothier 2004, 2005), in which a revised strategy, utilizing a modified Still-Gennari HWE coupling of advanced subunits which served to reduce the overall linear sequence with increased yields and contributed to potential application of a more feasible industrial scale-up plan. The overall yield was an impressive 11.1% with 21 steps (longest linear sequence).

Novartis licensed discodermolide from Harbor Branch Oceanographic Institution in 1998 for clinical development as an anticancer drug and quickly realized that sufficient and sustainable supplies of the drug could not be easily attained by isolation of compound from collected specimens, nor was there a fermentable microorganism isolated from source sponge which had been indentified to produce the compound. A decision was made to engage the Novartis Process Chemistry Group, led by Stuart Mickel, to tackle the complete synthesis of discodermolide in order to obtain at least 50 g of discodermolide for the anticipated clinical trials. A combination of several key synthetic schemes taken from both the Paterson and Smith approaches resulted in the production of 64 g of API discodermolide. Details of this plan are reported in five papers (Mickel et al. 2004a, b, c, d, e).

Arefolov and Panek (2005) offered a stereo-controlled synthesis of (+)-discodermolide by introducing the absolute stereochemistry of the C1–C6, C7–C14, and C15–C24 subunits by introducing asymmetric crotylation methodology. Overall, the Panek's synthesis resulted in an overall yield of 2.1% based on 27 steps (longest linear synthesis). The Betzer/Ardisson total synthesis of (+) discodermolide (de Lemos et al. 2007) utilized some of the strategies of Marshall and Johns (1998), Smith et al. (1995) and Arefolov and Panek (2005). The resultant yield was 1.6% from 21 steps (longest linear sequence).

3.2.3 Structure Activity Relationships of Discodermolide

An excellent review of the structure activity relationships of discodermolide analogs was recently described by Shaw (2008). We will attempt to summarize and present some examples in the literature which have helped to "map" the structure activity of the discodermolide molecule. The original work by Gunasekara (1990) presented the X-ray crystal structure of discodermolide which first suggested a hairpin feature of the molecule comprising the C8 through C14 region, although the absolute configuration was not determined (Fig. 3.2). The tetraacetate derivative of discodermolide prepared in this study and additional mono, di- and tri-acetates were later evaluated for cytotoxic activity against P388 murine leukemia cells (Table 3.1) (Gunasekera et al. 2001). We found that acetylation at C3 and/or C7 increased cytotoxic activity towards P388 cells compared to parent discodermolide. Interestingly, acetylation at C11 and/or C17 decreased the cytotoxic activity of discodermolide, even in the presence of "enhancing" C3 and/or C7 acetylation, indicating the importance of this region of the molecule for activity. In our subsequent paper (Gunasekera et al. 2002a) we found similar effects of acetylation of the compound on cytotoxicity towards A549 human lung cancer cells, again pointing out the crucial importance of the C11 and C17 region of the molecule. We also examined the ability of these acetylated discodermolide analogs to induce the in vitro polymerization of purified bovine brain tubulin compared to taxol and

Table 3.1 Cytotoxicity of discodermolide and discodermolide analogs in P388 murine leukemia and A549 human lung adenocarcinoma cells as determined by MTT metabolism following 72 h exposure to each test agent

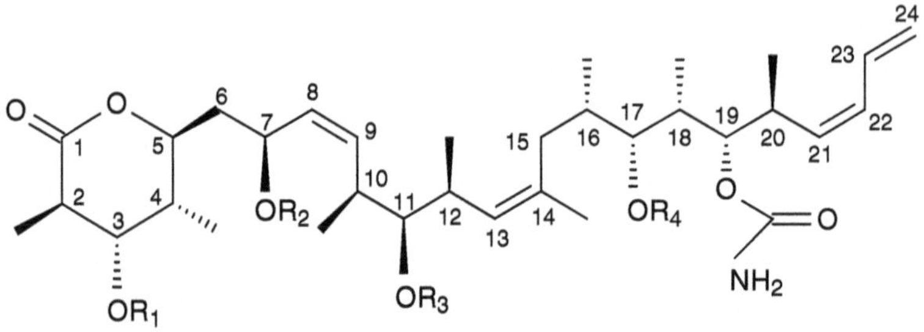

	R_1	R_2	R_3	R_4	P388 (nM)	A549 (nM)
Discodermolide	H	H	H	H	35.0	3.5
D-3-OAc	$COCH_3$	H	H	H	3.6	3.8
D-7-OAc	H	$COCH_3$	H	H	3.9	0.8
D-3,7-OAc	$COCH_3$	$COCH_3$	H	H	0.7	0.8
D-3,11-OAc	$COCH_3$	H	$COCH_3$	H	103.0	164.0
D-3,17-OAc	$COCH_3$	H	H	$COCH_3$	1109.0	524.0
D-3,7,11-OAc	$COCH_3$	$COCH_3$	$COCH_3$	H	166.0	545.0
D-3,7,17-OAc	$COCH_3$	$COCH_3$	H	$COCH_3$	6925.0	1000.0
D-3,7,11,17-OAc	$COCH_3$	$COCH_3$	$COCH_3$	$COCH_3$	ND	1000.0

Values expressed as IC_{50} (nM). R_1–R_4 indicate group substitutions (reprinted with permission from Gunasekera et al. (2002a), Copyright 2002 American Chemical Society)

discodermolide, under conditions of varying temperature and time (Fig. 3.4) (Isbrucker et al. 2001). The kinetics of discodermolide include rapid polymerization of tubulin (within seconds of addition) and maximum polymerization within 20 min at a temperature of approximately 10°C. Taxol, on the other hand, induces tubulin polymerization at a temperature of approximately 20°C, 15 min following addition at 4°C, thereafter reaching to a maximum polymerization value at 35°C, which is approximately one-half that induced by discodermolide. The surprising finding was that none of the analogs induced the maximal tubulin polymerization response observed with parent discodermolide, however, the D-3,7 analog reached a maximum response which was approximately 35% greater than that observed for taxol (but less than discodermolide). The D-7 and D-3 analogs values were slightly less than for taxol. The kinetics of the three discodermolide analogs revealed a slow polymerization of tubulin (compared to discodermolide) with maximal polymerization at approximately the same temperature as taxol. Our results suggested that these three analogs appeared to have polymerization profiles more like taxol than discodermolide. More importantly, these studies were some of the first to show the critical importance of the C-11 through C-17 region of the discodermolide molecule in helping to support and maintain the biological activity of the compound. Additional analogs

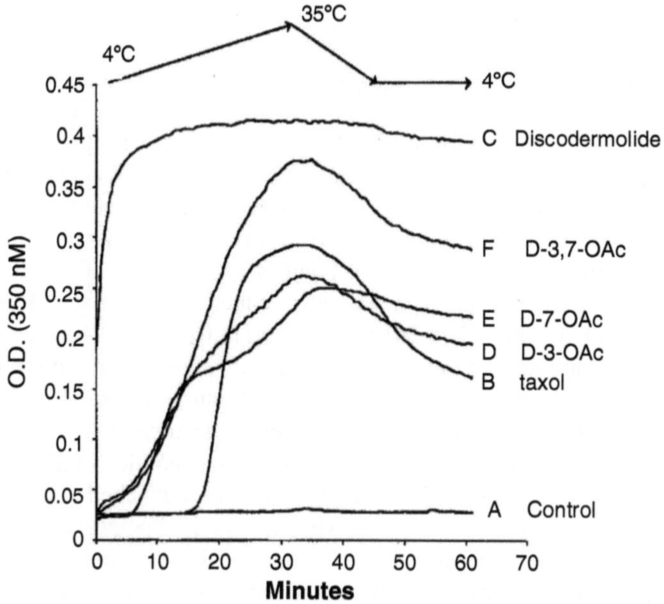

Fig. 3.4 Tubulin Polymerization Profiles of Discodermolide and Analogs. One mg/mL purified bovine brain tubulin in G-PEM buffer induced by 10 mM of compounds B through F and vehicle control A. *Arrows* above graph indicate temperature of cuvette holder throughout the assay (with kind permission from Springer Science+Business Media: Isbrucker et al. 2001). Copyright Springer-Verlag 2001

of discodermolide have been prepared by others through semi-synthesis, or as by-products in total synthetic processes. In addition, academia and industry have reported direct synthetic schemes to generate analogs for further study.

3.2.3.1 Fragments A, B, and C: A Walk Down the Molecule

The discodermolide molecule can be divided into three general fragments with somewhat equal complexity (A, B, and C), which are then coupled together to form the complete compound (Fig. 3.1). Synthetic intermediates and modifications to each of these fragments have provided a wealth of information on structure activity relationships. A walk down the molecule through a brief review of some of the work from several groups will reveal some interesting insights on the SAR of the discodermolide molecule. Unfortunately, space does not allow a complete review of all the literature in this area, so only a few research highlights from various groups will be described here.

3.2.3.2 The A Fragment

Some of the earlier work by several groups with the A fragment and attachments to the terminal diene, either as linkers (Hung et al. 1996b) or as fluorescent probes (Smith et al. 2005e) resulted in a discodermolide analog with potency similar to parent compound, suggesting that space existed for the attachment of larger groups within this domain. Additionally, the Novartis group showed that while reduction of the terminal olefin did not impact the anti-proliferative activity of the compound, further reduction of the second double bond or replacement of the diene with a phenyl or benzyl ether resulted in dramatic reduction in potency (Kinder et al. 2002; Palmero et al. 2002). The carbamate group in fragment A has been analyzed as well for structure activity relationships, and we were the first to report a naturally occurring decarbamoyl analog, 19 desaminocarbonyldiscodermolide (Gunasekera et al. 2002b), isolated from the source sponge, which showed only a slight (3–5-fold) drop in potency compared to discodermolide. Smith et al. (2005a) synthesized analogs in which alkyl or aryl groups were attached to the nitrogen of the carbamate. These analogs, not surprisingly, exhibited potencies similar to discodermolide, however, interestingly, the activity of these analogs towards discodermolide sensitive cell lines which expressed the P-glycoprotein pump were lost.

The 17-hydroxyl portion of the molecule in the A fragment appears to play an important role in maintaining activity, as we (Gunasekera et al. 2001, 2002a) and others (Hung et al. 1994, 1996a) have shown that modifications in this area leads to loss of biological activity presumably through altered stereochemistry.

3.2.3.3 The AB Linkage

Modifications to the AB linkage of discodermolide by several groups have yet to yield compounds which are more potent than discodermolide. With the previous report by Schreiber's group that the activity of the 16-demethyldiscodermolide analog was similar to that of parent discodermolide (Hung et al. 1996b), the Smith group prepared and thoroughly investigated a series of 14-demethyl analogs (Smith et al. 2005a, b). The summation of these studies resulted in the preparation of a number of analogs whose activity either approached that of the parent compound or was multifold less. An interesting finding was that for those analogs which were equipotent for activity in discodermolide sensitive cell lines, the activity dropped off considerably when tested in MDR cell lines, indicating a role for the 14-methyl group in substrate specificity for the MDR efflux pump.

3.2.3.4 The B Fragment and BC Linkage

There have been only a few series of analogs prepared in the region of the B fragment and BC linkage. Our previous work with acetylated discodermolide analogs which include the 11-hydroxyl group have shown the importance of this B region of the molecule in maintaining potent cytotoxic properties, presumably through the preservation of crucial stereochemistry (Gunasekera et al. 2001, 2002b). In addition, we analyzed several 9,13-cyclopentane analogs for cytotoxic activity and were found to be active only in the micromolar range (Gunasekera et al. 2002b). Martello et al. (2001), additionally showed that the 8,9-*trans* analog prepared during the course of the Smith group discodermolide synthesis demonstrated 100-fold less potency compared to discodermolide, again supporting the idea of the importance of this region in maintaining proper positioning and stereochemical alignment of the molecule in order to retain potent cytotoxic activity.

3.2.3.5 The C Fragment

The C fragment region of the molecule, containing the lactone ring with the 3- and 7-hydroxyls has yielded the greatest number of analogs generated and thus yielded a wealth of information regarding the structure activity relationship of discodermolide. In some of our first SAR studies, we reported that acylation of the 3-hydroxyl group, to our surprise, resulted in a compound which was almost tenfold more potent than discodermolide (Gunasekera et al. 2001). Similarly and consistent with our findings was the report by Sundermann et al. (2005) that alkylation of the 7-hydroxyl to form the methoxymethyl or methyl analog resulted in compounds with much increased activity compared to discodermolide. Regarding changes in the lactone ring, our studies (Gunasekera et al. 2002a) as well as those of Martello et al. (2001) have shown that the 2,3 anhydrodiscodermolide was slightly more potent or slightly less potent, depending on the cell line tested, compared to discodermolide. The 3-hydroxyl appears not to be required for activity and is also carried through to other components surrounding the lactone ring (Shaw et al. 2005, 2006). However, as we observed with the isolation of the naturally occurring, 2-epi-discodermolide, the inversion of the stereocenter revealed a potency which was tenfold less compared to discodermolide, suggesting a change in the conformation of the ring structure which disrupts a crucial interaction of discodermolide with tubulin.

There have been many attempts to make simplified analogs by altering and/or replacing the lactone ring, some of which have resulted in compounds which come within threefold potency of discodermolide, but none reported so far have succeeded in surpassing cytotoxicity exhibited by discodermolide (reviewed by Shaw 2008).

3.3 Clinical Trial of Discodermolide

Discodermolide was licensed from Harbor Branch Oceanographic Institution to Novartis AG in 1998. The clinical trials that ensued were designed to evaluate the discodermolide (XAA296A) in patients with solid tumors. The results of the first

Phase I trial of the compound were reported at the American Society for Clinical Oncologists meeting by Mita et al. (2004). A summary of the results follows. The compound was administered IV as a fixed infusion rate of 0.77 mg/mL/min once every 3 weeks. Twenty six patients (17 male, 9 females) were treated with the following dose levels (mg/m^2): 0.6 (3 patients), 1.2 (3 patients), 2.4 (4 patients), 4.8 (4 patients), 9.6 (4 patients), 14.4 (6 patients), and 19.2 (2 patients). The median age of the patients was 59.5 years with a range of 19–79. Nine of the 26 patients had prior taxane therapy. There were no dose limiting toxicities (DLT) reported nor were there signs of neuropathy or neutropenia in the study at that time. One patient with appendix carcinoma experienced stable disease for four cycles. Discodermolide demonstrated a nonlinear PK characterized by a second broad peak at the terminal phase.

To date, there have been several references to pulmonary toxicity reported for discodermolide which occurred later in the original clinical trial, although this data has not been published. A full paper on the clinical trial of discodermolide is currently being prepared by Mita (personal communication), and will hopefully describe this pulmonary toxicity in greater detail.

3.4 Future Studies with Discodermolide

Despite Novarits having dropped its plans for further development of discodermolide as an antitumor agent, the outlook for the compound and its analogs remains promising. New and more potent discodermolide analogs are yet to be discovered which share the mechanism of action of the parent compound, or which possess novel and totally unexpected modes of action are anticipated as well as those with fewer toxic side effects. The effectiveness of a discodermolide analog in the bacteriolytic therapy of experimental cancers is one such example (Smith et al. 2005d). Targeting of the bacterial tubulin "homolog," FtsZ, an essential cell division protein in bacteria, by taxanes as a type of novel antibacterial therapy (Huang et al. 2006a, b) may additionally lend itself to the investigation of discodermolide and its analogs for a similar purpose. A recent report by Fan et al. (2009) in Billy Day's laboratory which describes for the first time, the metabolism of discodermolide may also lead to the development of new discodermolide analogs which may possess greater activity with fewer side effects. Our discovery of the mechanism of action of the marine natural product, dictyostatin, as a potent microtubule stabilizer (Isbrucker et al. 2003) and its structural similarity to discodermolide have led other groups to synthesize the compound (Paterson et al. 2004; Shin et al. 2004; O'Neil and Philips 2006) and make possible the preparation of discodermolide-dictyostatin "hybrids" which have shown enhanced anti-proliferative activity towards tumor cells in vitro, compared to dictyostatin or discodermolide alone (Shin et al. 2002, 2007; Paterson and Gardner 2007; Paterson et al. 2008, 2010). Finally, the original immunosuppressive properties of discodermolide which we first described may be a productive avenue for a second look at this compound as a new immunosuppressive agent.

Whether discodermolide itself, its analogs or a hybrid molecule make it to the clinic will continue to rely on the hard work ingenuity and genius of chemists and biologists who continue to work with the molecule as well as those who are yet to come. It has been my great fortune to be associated with the initial discovery and early work with discodermolide and I and my colleagues will continue to work and watch with great amazement as the story of this interesting molecule continues to unfold.

References

Arefolov A, Panek JS (2005) Crotylsilane reagents in the synthesis of complex polyketide natural products: total synthesis of (+)-discodermolide. J Am Chem Soc 127:5596–5603

Bewley CA, Holland ND, Faulkner DJ (1996) Two classes of metabolites from *Theonella swinhoei* are localized in distinct populations of bacterial symbionts. Experientia 52:716–722

Bröker LE, Huisman C, Ferreira CG, Rodriguez JA, Kruyt FAE, Giaccone G (2002) Late activation of apoptotic pathways plays a negligible role in mediating the cytotoxic effects of discodermolide and epothilone B in non-small cell lung cancer cells. Cancer Res 62:4081–4088

Bröker LE, Huisman C, Span SW, Rodriguez JA, Kruyt FE, Giaccone G (2004) Cathepsin B mediates caspase-independent cell death induced by microtubule stabilizing agents in non-small cell lung cancer cells. Cancer Res 64:27–30

Brück WM, Sennett SH, Pomponi SA, Willenz P, McCarthy PJ (2008) Identification of the bacterial symbiont Entotheonella sp. in the mesohyl of the marine sponge Discodermia sp. ISME J 3:335–339

Canales A, Matesanz R, Gardner NM, Andru JM, Paterson I, Diaz JF, Jiménez-Barbero J (2008) The bound conformation of microtubule-stabilizing agents: NMR insights into the bioactive 3D structure of discodermolide and dictyostatin. Chem Eur J 14:7557–7569

Dang LH, Bettegowda C, Huso DL, Kinzler KW, Vogelstein B (2001) Combination bacteriolytic therapy for the treatment of experimental tumors. Proc Natl Acad Sci U S A 98:15155–15160

Davidson SK, Allen SW, Lim GE, Anderson CM, Haygood MG (2001) Evidence for the biosynthesis of bryostatins by the bacterial symbiont "Candidatus Endobugula sertula" of the bryozoan *Bugula neritina*. Appl Environ Microbiol 67:4531–4537

de Lemos E, Poree FH, Commercon A, Betzer JF, Pancrazi A, Ardisson J (2007) Alpha-oxygenated crotyltitanium and dyotropic rearrangement in the total synthesis of discodermolide. Angew Chem Int Ed Engl 46:1917–1921

Fan Y, Schreiber M, Day BW (2009) Human liver microsomal metabolism of (+)-discodermolide. J Nat Prod 72:1748–1754

Florence GJ, Gardner NM, Paterson I (2008) Development of practical syntheses of the marine anticancer agents discodermolide and dictyostatin. Nat Prod Rep 25:342–375

Gunasekera SP, Gunasekera M, Longley RE (1990) Discodermolide, a new bioactive polyhydroxylated lactone from the marine sponge, *Discodermia dissoluta*. J Org Chem 55:4912–4915

Gunasekera SP, Longley RE, Isbrucker RA (2001) Acetylated analogues of the microtubule-stabilizing agent discodermolide: preparation and biological activity. J Nat Prod 64:171–174

Gunasekera SP, Longley RE, Isbrucker RA (2002a) Semisynthetic analogues of the microtubule-stabilizing agent discodermolide: preparation and biological activity. J Nat Prod 65:1830–1837

Gunasekera SP, Paul GK, Longley RE, Isbrucker RA, Pomponi SA (2002b) Five new discodermolide analogues from the marine sponge Discodermia species. J Nat Prod 65:1643–1648

Harried SS, Lee CP, Yang G, Lee TI, Myles DC (2003) Total synthesis of the potent microtubule-stabilizing agent (+)-discodermolide. J Org Chem 22(68):6646–6660

Honore S, Kamath K, Braguer D, Horwitz SB, Wilson L, Briand C, Jordan MA (2004) Synergistic suppression of microtubule dynamics by discodermolide and paclitaxel in non-small cell lung carcinoma cells. Cancer Res 64:4957–4964

Huang Q, Kirikae F, Kirikae T, Pepe A, Amin A, Respicio L, Slayden RA, Tonge PJ, Ojima I (2006a) Targeting FtsZ for anti-tuberculosis drug discovery: non-cytotoxic taxanes as novel anti-tuberculosis agents. J Med Chem 49:463–466

Huang GS, Barcons-Lope L, Freeze SB, Smith AB III, Goldberg GL, Horwitz SB, McDaid HM (2006b) Potentiation of taxol efficacy by discodermolide in ovarian carcinoma xenograft-bearing mice. Clin Cancer Res 12(1):298–304

Huisman C, Ferreira CG, Bröker LE, Rodriguez JA, Smit EF, Postmus PE, Kruyt FAE, Giaccone G (2002) Paclitaxel triggers cell death primarily via caspase-independent routes in the non-small cell lung cancer cell line NCI-H460. Clin Cancer Res 8:596–606

Hung DT, Nerenberg JB, Schreiber SL (1994) Distinct binding and cellular properties of synthetic (+) and (−) discodermolides. Chem Biol 1:67–71

Hung DT, Chen J, Schreiber SL (1996a) (+)-Discodermolide binds to microtubules in stoichiometric ratio to tubulin dimers, blocks taxol binding and results in mitotic arrest. Chem Biol 3:287–293

Hung DT, Nerenberg JB, Schreiber SL (1996b) Syntheses of discodermolides useful for investigating microtubule binding and stabilization. J Am Chem Soc 118:11054–11080

Isbrucker RA, Gunasekera SP, Longley RE (2001) Structure-activity relationship studies of discodermolide and its semisynthetic acetylated analogs on microtubule function and cytotoxicity. Cancer Chemother Pharmacol 48:29–36

Isbrucker RA, Cummins J, Pomponi SA, Longley RE, Wright AE (2003) Tubulin polymerizing activity of dictyostatin-1, a polyketide of marine sponge origin. Biochem Pharmacol 66:75–82

Jogalekar AS, Kriel FH, Shi Q, Cornett B, Cicero D, Snyder JP (2010) The discodermolide hairpin structure flows from conformationally stable modular motifs. J Med Chem 53:155–165

Khrapunovich-Baine M, Menon V, Verdier-Pinard P, Smith AB III, Angeletti RH, Fiser A, Horwitz SB, Xiao H (2009) Distinct pose of discodermolide in taxol binding pocket drives a complementary mode of microtubule stabilization. Biochemistry 48:11664–11677

Kinder FR Jr, Bair KW, Chen W, Florence, G, Francavilla C, Geng P, Gunasekera S, Guo Q, Lassota PT, Longley RE, Palermo MG, Paterson I, Pomponi S, Ramsey TM, Rogers L, Sabio M, Sereinig N, Sorensen E, Wang RM, Wright A (2002) Synthesis and antitumor activity of analogues of the novel microtubule stabilizing agent discodermolide. Abstracts of papers American Chemical Society, 224th ACS national meeting, Boston, MA, MEDI-236

Klein LE, Freeze BS, Smith AB III, Horwitz SB (2005) The microtubule stabilizing agent discodermolide is a potent inducer of accelerated cell senescence. Cell Cycle 4:501–507

Kowalski RJ, ter Haar E, Longley RE, Gunasekera SP, Lin CM, Day BW, Hamel E (1995) Comparison of novel microtubule polymerizing agents, discodermolide and epothilone A/B, with taxol. Mol Biol Cell 6:368a

Kowalski RJ, Giannakakou P, Gunasekera SP, Longley RE, Day BW (1997) The microtubule stabilizing agent discodermolide competitively inhibits the binding of paclitaxel (taxol) to tubulin polymers, enhances tubulin nucleation reactions more potently than paclitaxel and inhibits the growth of paclitaxel-resistant cells. Mol Pharmacol 52:613–622

Longley RE, Caddigan D, Harmody D, Gunasekera M, Gunasekera SP (1991a) Discodermolide—a new, marine-derived immunosuppressive compound. I. In vitro studies. Transplantation 52:650–656

Longley RE, Caddigan D, Harmody D, Gunasekera M, Gunasekera SP (1991b) Discodermolide—a new, marine-derived immunosuppressive compound. II. In vivo studies. Transplantation 52:656–661

Longley RE, Gunasekera SP, Faherty D, Mclane J, Dumont F (1993) Immunosuppression by discodermolide. Ann N Y Acad Sci 696:94–107

Marshall JA, Johns BA (1998) Total synthesis of (+)-discodermolide. J Org Chem 63:7885–7892

Martello LA, McDaid HM, Regl DL, Yang CH, Meng D, Pettus TRR, Kaufman MD, Arimoto H, Danishefsky SJ, Smith AB, Horwitz SB (2000) Taxol and discodermolide represent a synergistic drug combination in human carcinoma cell lines. Clin Cancer Res 6:1978–1987

Martello LA, LaMarche MJ, He L, Beauchamp TJ, Smith AB, Horwitz SB (2001) The relationship between taxol and (+)-discodermolide: synthetic analogs and modeling studies. Chem Biol 8:843–855

Mickel SJ, Sedelmeier GH, Niederer D, Daeffler R, Osmani A, Schreiner K, Seeger-Weibel M, Berod B, Schaer K, Gamboni R (2004a) Large-scale synthesis of the anti-cancer marine natural product (+)-discodermolide. Part 1: synthetic strategy and preparation of a common precursor. Org Process Res Dev 8:92–100

Mickel SJ, Sedelmeier GH, Niederer D, Schuerch F, Grimler D, Koch G, Daeffler R, Osmani A, Hirni A, Schaer K, Gamboni R (2004b) Large-scale synthesis of the anti-cancer marine natural product (+)-discodermolide. Part 2: synthesis of fragment C1-6 and C9-14. Org Process Res Dev 8:101–106

Mickel SJ, Sedelmeier GH, Niederer D, Schuerch F, Koch G, Kuesters E, Daeffler R, Osmani A, Seeger-Weibel S, Schmid E, Hirni A, Schaer K, Gamboni R (2004c) Large-scale synthesis of the anti-cancer marine natural product (+)-discodermolide. Part 3: synthesis of fragment C15-21. Org Process Res Dev 8:107–112

Mickel SJ, Sedelmeier GH, Niederer D, Schuerch F, Seger M, Schreiner K, Daeffler R, Osmani A, Bixel D, Loiseleur O, Cercus J, Stettler H, Schaer K, Gamboni R (2004d) Large-scale synthesis of the anti-cancer marine natural product (+)-discodermolide. Part 4: preparation of fragment C7-24. Org Process Res Dev 8:113–121

Mickel SJ, Niederer D, Daeffler R, Osmani A, Kuesters E, Schmid E, Schaer K, Gamboni R (2004e) Large-scale synthesis of the anti-cancer marine natural product (+)-discodermolide. Part 5: linkage of fragments C1-6 and C7-24 and finale. Org Process Res Dev 8:122–130

Mita A, Lockhart AC, Chen TL, Bochinski K, Curtright J, Cooper W, Hammond L, Rothenberg M, Rowinsky E, Sharma S (2004) A phase I pharmacokinetic (PK) trial of XAA296A (discodermolide) administered every 3 wks to adult patients with advanced solid malignancies. J Clin Oncol 22:2025

Myles DC (2003) Novel biologically active natural and unnatural products. Curr Opin Biotechnol 14:627–633

Nerenberg JB, Hung DT, Somers PK, Schreiber SL (1993) Total synthesis of the immunosuppressive agent (−)-discodermolide. J Am Chem Soc 115:12621–12622

O'Neil GW, Philips AJ (2006) Total synthesis of (−)-dictyostatin. J Am Chem Soc 128: 5340–5341

Palmero MG, Bair W, Chen W, Guo Q, Lassota PT, Sorrensen E, Wang RM, Kinder FR Jr (2002) Abstracts of papers, 224th meeting of the American Chemical Society. American Chemical Society, Boston, MA, Abstract ORGN 791

Paterson I, Gardner NM (2007) Design, synthesis and biological evaluation of a macrocyclic discodermolide/dictyostatin hybrid. Chem Commun 1:49–51

Paterson I, Lyothier I (2004) Total synthesis of (+)-discodermolide: an improved endgame exploiting a Still–Gennari-type olefination with a C1–C8 β-ketophosphonate fragment. Org Lett 6:4933–4936

Paterson I, Lyothier I (2005) Development of a third-generation total synthesis of (+)-discodermolide: an expedient Still–Gennari-type fragment coupling utilizing an advanced β-ketophosphonate. J Org Chem 70:5494–5507

Paterson I, Florence GJ, Gerlach K, Scott JP (2000) Total synthesis of the antimicrotubule agent (+)-discodermolide using boron-mediated aldol reactions of chiral ketones. Angew Chem Int Ed Engl 39:377–380

Paterson I, Delgado O, Florence GJ, Lyothier I, Scott JP, Sereinig N (2003) 1,6-Asymmetric induction in boron-mediated aldol reactions: application to a practical total synthesis of (+)-discodermolide. Org Lett 5:35–38

Paterson I, Britton R, Delgado O, Meyer A, Poullennec KG (2004) Total synthesis and configurational assignment of (−)-dictyostatin, a microtubule-stabilizing macrolide of marine sponge origin. Angew Chem Int Ed Engl 43:4629–4633

Paterson I, Naylor G, Wright AE (2008) Total synthesis of a potent hybrid of the anticancer natural products dictyostatin and discodermolide. Chem Commun 38:4628–4630

Paterson I, Naylor GJ, Fujita T, Guzmán E, Wright AE (2010) Total synthesis of a library of designed hybrids of the microtubule-stabilising anticancer agents taxol, discodermolide and dictyostatin. Chem Commun 46:261–263

Sanchez-Pedregal VS, Kubicek K, Meiler J, Lyothier I, Paterson I, Carlomagno T (2006) The tubulin-bound conformation of discodermolide derived by NMR studies in solution supports a common pharmacophore model for epothilone and discodermolide. Angew Chem Int Ed Engl 45:7388–7394

Shaw SJ (2008) The structure activity relationship of discodermolide analogues. Mini Rev Med Chem 8:276–284

Shaw SJ, Sundermann KF, Burlingame MA, Myles DC, Freeze BS, Xian M, Brouard I, Smith AB III (2005) Toward understanding how the lactone moiety of discodermolide affects activity. J Am Chem Soc 127:6532–6533

Shaw SJ, Sundermann KF, Burlingame MA, Shang D, Petryka J, Myles DC (2006) A series of 23,24-dihydrodiscodermolide analogues with simplified lactone regions. Bioorg Med Chem Lett 16:1961–1964

Shin Y, Choy N, Turner TR, Balachandran R, Madiraju C, Day BW, Curran DP (2002) Discodermolide/dictyostatin hybrids: synthesis and biological evaluation. Org Lett 4:4443–4446

Shin Y, Fournier J-H, Fukui Y, Brückner AM, Curran DP (2004) Total synthesis of dictyostatin 1 provides its relative and absolute configuration. Angew Chem Int Ed Engl 43:4634–4637

Shin Y, Fournier J-H, Bruckner A, Madiraju C, Balachandran R, Raccor BS, Edler MC, Hamel E, Sikorski RP, Vogt A, Day BW, Curran DP (2007) Synthesis and biological evaluation of (−)-dictyostatin and stereoisomers. Tetrahedron 63:8537–8562

Shirmer A, Gadkari R, Reeves CD, Ibrahim F, DeLonge EF, Hutchinson CR (2005) Metagenomic analysis reveals diverse polyketide synthase gene clusters in microorganisms associated with the marine sponge *Discodermia dissoluta*. Appl Environ Microbiol 71:4840–4849

Smith AB III, Qiu YP, Jones DR, Kobayashi K (1995) Total synthesis of (−)-discodermolide. J Am Chem Soc 117:12011–12012

Smith AB III, Kaufman MD, Beaucham TJ, LaMarche MJ, Arimoto H (1999) Gram-scale synthesis of (+)-discodermolide. Org Lett 1:1823–1826

Smith AB, Beauchamp TJ, LaMarche MJ, Kaufman MD, Qiu Y, Arimoto H, Jones DR, Kobayashi K (2000) Evolution of gram-scale synthesis of (+) discodermolide. J Am Chem Soc 122:8654–8664

Smith AB, LaMarche M, Falcone-Hindley M (2001) Solution structure of (+)-discodermolide. Org Lett 3:695–698

Smith AB, Freeze BS, Brouard I, Hirose T (2003) A practical improvement, enhancing the large-scale synthesis of (+)-discodermolide: a third-generation approach. Org Lett 5:4405–4408

Smith AB III, Freeze BS, Lamarche MJ, Hirose T, Brouard I, Rucker PV, Xian M, Sundermann KF, Shaw SJ, Burlingame MA, Horwitz SB, Myles DC (2005a) Design, synthesis, and evaluation of carbamate-substituted analogues of (+)-discodermolide. Org Lett 7:311–314

Smith AB III, Freeze BS, LaMarche MJ, Hirose T, Brouard I, Xian M, Sundermann KF, Shaw SJ, Burlingame MA, Horwitz SB, Myles DC (2005b) Design, synthesis, and evaluation of analogues of (+)-14-normethyldiscodermolide. Org Lett 7:315–318

Smith AB III, Freeze BS, Xian M, Hirose T (2005c) Total synthesis of (+)-discodermolide: a highly convergent fourth-generation approach. Org Lett 7:1825–1828

Smith AB III, Freeze BS, LaMarche MJ, Sager J, Kinzler KW, Vogelstein B (2005d) Discodermolide analogues as the chemical component of combination bacteriolytic therapy. Bioorg Med Chem Lett 15:3623–3626

Smith AB III, Rucker PV, Brouard I, Freeze BS, Xia S, Horwitz SB (2005e) Design, synthesis, and biological evaluation of potent discodermolide fluorescent and photoaffinity molecular probes. Org Lett 7:5199–5202

Sundermann KF, Shaw SJ, Burlingame MA, Myles DC, Freeze SB, Xian M, Brouard J, Smith AB III (2005) Abstracts of papers, 229th meeting of the American Chemical Society. American Chemical Society, San Diego, CA, Abstract MEDI 417

ter Haar E, Kowalski RJ, Hamel E, Lin CM, Longley RE, Gunasekera SP, Rosenkranz HS, Day BW (1996) Discodermolide, a cytotoxic marine agent that stabilizes microtubules more potently than taxol. Biochemistry 35:243–250

Unson MD, Faulkner DJ (1993) Cyanobacterial symbiont biosynthesis of chlorinated metabolites from *Dysidea herbacea* (Porifera). Experientia 49:349–353

Chapter 4
HDAC Inhibitors and Other Histone Modifying Natural Products as Emerging Anticancer Agents

Lilibeth A. Salvador and Hendrik Luesch

Abstract Histone deacetylase (HDAC) inhibitors represent a new class of anticancer agents that target dysregulated acetylation of histone lysines, an epigenetic rather than a genetic event. Certain HDACs are overexpressed and hyperactive in cancer cells, and suppression of these enzymes' activities provides superior selectivity over more traditional anticancer agents. To date, two HDAC inhibitors—vorinostat and romidepsin—have reached the market, with romidepsin being an actual natural product and vorinostat closely related to the natural product HDAC inhibitor trichostatin A. Over the past 15 years, several secondary metabolites with high structural diversity from microorganisms, marine sponges, and cyanobacteria have been discovered to possess HDAC inhibitory activity and are currently at the clinical and preclinical stages. In this chapter, we recapitulate the discovery of natural product HDAC inhibitors, enumerate the challenges in their development and provide insights in the continuing role of natural products in the discovery of HDAC inhibitors as well as new modulators for other clinically-relevant epigenetic events, including histone methylation.

4.1 Introduction

Nature has provided several of the most potent anticancer agents with different biological targets, including paclitaxel and Vinca alkaloids that affect microtubule assembly, the topoisomerase inhibitors camptothecin and etoposide, and the DNA

L.A. Salvador • H. Luesch (✉)
Department of Medicinal Chemistry, University of Florida,
1600 SW Archer Road, Gainesville, FL 32610, USA
e-mail: luesch@cop.ufl.edu

F.E. Koehn (ed.), *Natural Products and Cancer Drug Discovery*, Cancer Drug Discovery and Development, DOI 10.1007/978-1-4614-4654-5_4,

interacting agents mitomycin C and ET743. Natural products have been a validated starting point for the development of cancer chemotherapeutic agents, with ~70% of those introduced from 1981–2010 either natural products or natural products-derived (Newman and Cragg 2012). Natural products are again leading the way for the development of a new generation of anticancer agents which target epigenetic events at the levels of transcriptional regulation. Epigenetic modifications act independent of the DNA sequence but are nonetheless capable of affecting gene expression, thereby contributing to genomic instability (Jones and Baylin 2007; Sharma et al. 2010). These modifications are amenable to chemotherapeutic interventions because of their reversibility that is regulated by selected enzymes. Modulation of enzymes involved in epigenetic modifications thus presents a new method of combating cancer. Among the known modifications, histone acetylation has been successfully targeted by secondary metabolites through inactivation of the key enzymes involved in this process—histone deacetylases (HDACs). More than 75% of human cancer cell lines have higher expression of class I HDACs compared to normal cells and 5–40% of various solid tumors showed overexpression depending on the tissue (Nakagawa et al. 2007). In the past four years, two HDAC inhibitors have gained approval for clinical use, vorinostat and romidepsin, a natural product-like compound and a natural product itself, respectively, further supporting that natural products continue to have a major impact in the field and provide first-in-class lead molecules. In this review, we describe the discovery and development of HDAC inhibitors, their contributions to our knowledge about transcriptional regulation, the challenges in their development, and the continuing role of natural products in the era of epigenetic drugs.

4.2 Covalent Modifications to Nucleosomes

Nucleosomes are the building blocks of chromatin and each unit is composed of ~146 bp of DNA coiled to a histone octameric core. Units are joined together by linker DNA, and 14 contact points between histones and DNA of each nucleosome provide an extremely stable structure under physiological conditions (Luger et al. 1997; Kornberg and Lorch 1999). The protein core of nucleosomes consists of four different histones (H2A, H2B, H3, and H4) that are characterized by unstructured amino terminal domains that appear as tails of variable length and are rich in basic amino acids which are particularly susceptible to posttranslational modifications (Luger et al. 1997). To date, there are eight distinct types of histone posttranslational modifications at ~65 amino acid residues mostly on histone tails. As for basic amino acids, Lys can undergo acetylation, methylation, sumoylation, and ubiquitination, while Arg can be methylated or deimidated (Kouzarides 2007; Li et al. 2007). Other posttranslational modifications include Pro isomerization and phosphorylation of Ser and Thr. Covalent modifications primarily affect the interactions between histone and DNA as well as histone–histone interactions; thus it influences higher order chromatin structure and consequently access to genetic information without

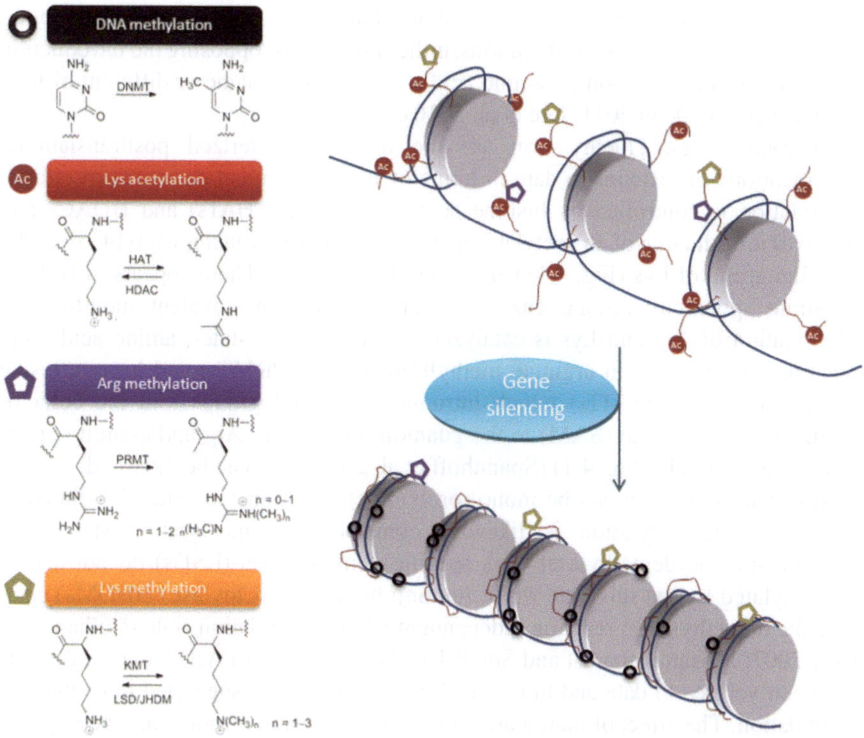

Fig. 4.1 Dynamic chromatin regulation is achieved through posttranslational modification of histone amino acids and DNA bases. Transcriptional activation is marked by histone lysine acetylation and methylation of histone H3K4, while repression is characterized by DNA methylation at promoter region, lack of acetylation, and methylation at histone H3K9 (Sharma et al. 2010; Liang et al. 2004)

alteration of the primary DNA sequence (Li et al. 2007). Acetylation has been regarded to favor an open chromatin structure and activation of transcription due to neutralized charged interactions between histones and DNA (Fig. 4.1). The relationship of acetylation to transcription has been proven to be more than due to histone–DNA interactions, such that these covalent modifications also affect the binding of transcription factors to the chromatin (Kouzarides 2007). Bromodomain-containing non-histone proteins recognize acetylated residues, while methylation and phosphorylation are marks for chromo-like domain of the Royal family, nonrelated PHD domains and domains of 14-3-3 proteins, respectively (Kouzarides 2007). These histone modifications occur in specific combinations likened to a "histone code" that determines the downstream cellular effects (Oliver and Denu 2011). Although stable, these modifications are dynamically regulated and changing within the cell. Global levels of posttranslational modifications are dictated by enzymes that introduce and remove these covalent changes as well as crosstalk among histone posttranslational modifications, either favoring or opposing the introduction of a covalent modification on amino acid residues on the same or different histone tail (Oliver and Denu 2011; Lee et al. 2010).

Acetylation and methylation are the most characterized posttranslational modifications of histones to date in terms of mechanism and downstream effects. Acetylation is controlled by histone acetyltransferases (HATs) and HDACs and involves introduction of an acetyl group derived from the cofactor acetyl-CoA to the ε-amino group of Lys (Fig. 4.1) (Smith and Denu 2009). There appears to be little substrate specificity among enzymes that mediate this covalent modification. Methylation of Arg and Lys is catalyzed by methylation state-, amino acid-, and position-specific protein arginine methyltransferases (PRMTs) and histone lysine methyltransferases (KMTs) which introduce a methyl group from the cofactor *S*-adenosyl methionine (SAM) to the guanidino moiety of Arg and ε-amino group of Lys, respectively (Fig. 4.1) (Spannhoff et al. 2009). Lys can be mono-, di-, or trimethylated, while Arg can be mono- or symmetrically/unsymmetrically dimethylated. Lys demethylation is likewise controlled by methylation state- and position-specific demethylases. Lys specific demethylases (LSDs) do not act on trimethylated Lys as substrate while Jumonji histone demethylases (JHDMs) catalyze the demethylation reaction independent of the methylation state (Culhane and Cole 2007; Mosammaparast and Shi 2010). The existence of Arg demethylases has not been verified to date and turnover of methylated Arg is suggested to be through deimidation. The effect of methylation on gene expression is more varied compared to acetylation and depends on the amino acid residue, methylation state, and the region of the gene to which it belongs (Fig. 4.2) (Li et al. 2007). Methylation does not alter the charge of Lys and Arg on histone tails but affects the pK_a, lipophilicity, and hydrophobic interactions. Transcription-repressive methylation marks occur on H3K9, H3K27, H4K20 while activation is favored by methylation of H3K4, H3K36, and H3K39 (Fig. 4.2) (Li et al. 2007). The specificity of histone methylases and demethylases as well as the varied transcriptional effects of methylation is postulated to selectively regulate a small subset of genes.

Aside from histones, methylation can also occur on DNA, predominantly on cytosine (Fig. 4.1). This covalent modification introduces a methyl group on the C5 position of cytosine through DNA methyltransferases (DNMTs) using SAM as the cofactor (Cheng and Blumenthal 2008; Jurkowska et al. 2011). DNA methylation occurs mostly on CpG islands having GC content of >55% on promoter regions of genes (Jones and Baylin 2002). This heritable modification has been extensively documented to favor transcriptional repression by modulating the binding of regulatory proteins to DNA (Jones and Baylin 2002). Crosstalk between histone modifications and DNA methylation also occur with histone methyltransferases and HDACs associating with DNMTs to regulate gene expression and the chromatin structure (Jones and Baylin 2002, 2007; Sharma et al. 2010).

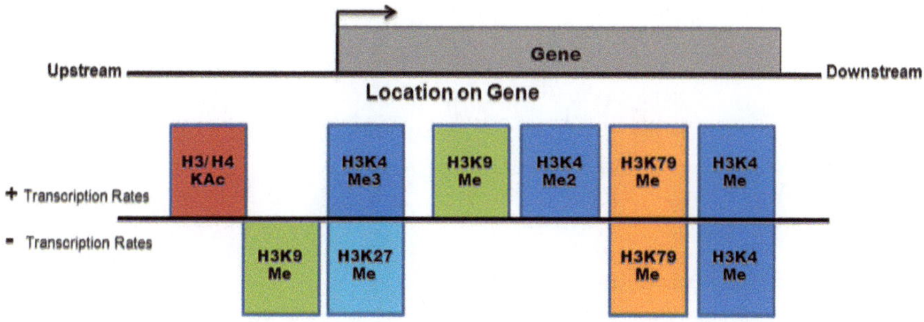

Fig. 4.2 Histone posttranslational modifications and their effects on transcription rates depending on location (Li et al. 2007)

4.2.1 Mechanism of Histone Acetylation

Three distinct structural families of HATs—GNAT, MYST, and p300/CBP—control the acetylation of histones as well as non-histone proteins and polyamines such as spermidines. The majority of these enzymes has an acetyl-CoA binding site and a bromodomain, but otherwise is largely structurally unrelated (Hodawadekar and Marmorstein 2007). Although HATs utilize a common acetyl-CoA dependent reaction mechanism for acetylation, the kinetics varies between families (Fig. 4.3). The GNAT family catalyzes a single-step acetyl transfer and forms a ternary complex with the lysine substrate and acetyl-CoA (Smith and Denu 2009). Structural information on the catalytic domain of GCN5, a member of the GNAT family, indicated an active site Glu173 serving as a base to deprotonate the ε-amino group of Lys and the latter serving as the nucleophile to attack the carbonyl carbon of acetyl-CoA (Lin et al. 1999; Tanner et al. 1999). The MYST family undergoes a sequential ternary complex (ordered Bi-Bi) kinetic mechanism of acetyl transfer where the active site Glu residue also deprotonates the Lys substrate after cofactor and peptide binding (Berndsen et al. 2007). The p300/CBP family on the other hand undergoes a Theorell-Chance mechanism of acetyl transfer, where the ternary complex possesses a short lifetime (Liu et al. 2008).

There are 18 characterized HDACs to date that can be broadly classified into classical HDACs (classes I, II, IV), which utilize a metal-dependent deacetylation mechanism, and the structurally and mechanistically distinct class III sirtuins (SIRTs) which require NAD^+ for the reaction (Gregoretti et al. 2004). Mammalian class I HDACs (HDAC1, 2, 3, and 8) show homology with the yeast HDAC Rpd and are localized mainly in the nucleus (Gregoretti et al. 2004; De Ruijter et al. 2003; Ficner 2009). Class II HDACs (HDAC4–7, 9, 10) are closely related to the yeast HDAC Hda1 and localize both in the nucleus and cytoplasm. This class is further subdivided into class IIa (HDAC 4, 5, 7) and class IIb (HDAC6 and 10), differentiated by a functionally important N-terminal domain and two putative catalytic domains, respectively (Ficner 2009). HDAC11 is the sole member of class IV

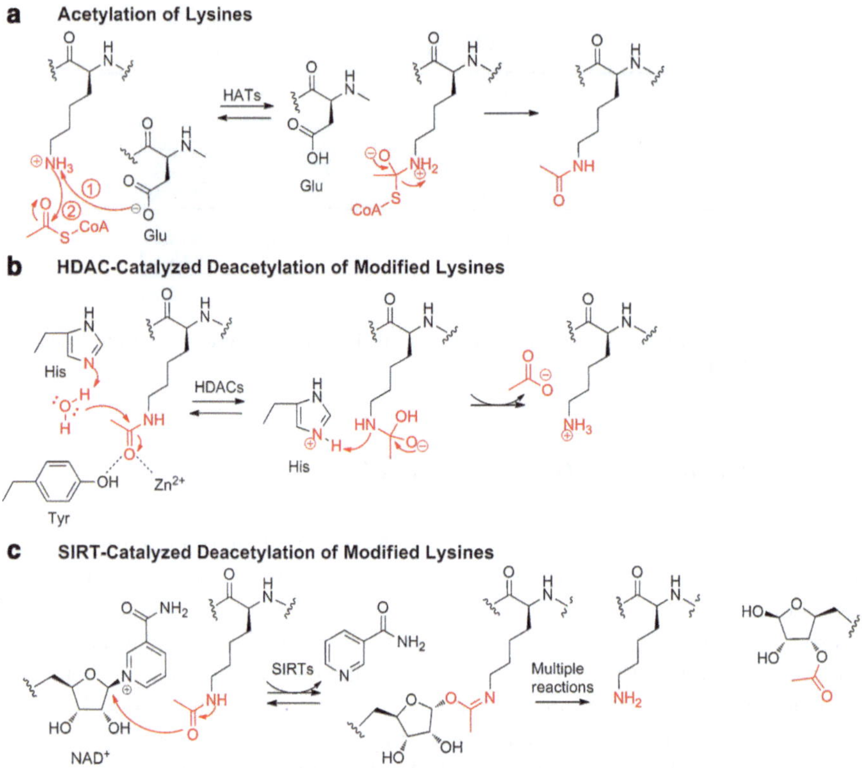

Fig. 4.3 Mechanism of lysine acetylation catalyzed by HATs and deacetylation of ε-acetyl-lysines via HDAC and SIRT-mediated reactions. (**a**) Acetylation of terminal amino group of Lys requires the cofactor acetyl-CoA. The reaction requires an initial deprotonation of the amino group by an active site Glu residue and subsequently nucleophilic attack to the acetyl group of the cofactor. (**b**) HDAC catalyzed deacetylations are Zn^{2+}-dependent reactions that liberate free acetate and Lys. (**c**) SIRTs utilize a distinct mechanism of deacetylation compared to HDACs and require NAD^+. Instead of liberating acetate, SIRTs conjugate the acetate group with ADP-ribose to generate 3′-*O*-acetyl-ADP ribose and Lys

and is localized in the nucleus but coprecipitates with the mainly cytoplasmic HDAC6. Classical HDACs consists of ~390 amino acids that share an identical α/β domain and have a homologous catalytic site with a characteristic 11 Å tubular channel that accommodates the acetylated lysine substrate, a penta-coordinated Zn^{2+} ion that serves as the catalytic center and a His–Asp charge-relay system (His143, His144, Asp176 and Asp183, based on HDAC8 numbering) which acts as a general acid–base pair (Ficner 2009; Somoza et al. 2004). The His–Asp dyads convert water into a nucleophile and also protonate the ε-amino group of the Lys substrate after deacetylation (Fig. 4.3) (Ficner 2009). The acetyl oxygen of the substrate is bonded and polarized by Tyr306 and the Zn^{2+} ion, thereby rendering the carbonyl carbon susceptible to nucleophilic attack by water, forming a tetrahedral intermediate that collapses to liberate acetate and lysine. Mammalian sirtuins

(SIRT1–7) are closely related to the yeast Sir2p deacetylase and, compared to the mainly nuclear and cytosolic HDACs, have a more variable cellular localization. Aside from SIRT1 and SIRT6, which are found in the nucleus, other members of this enzyme class are distributed in the cytosol (SIRT2), nucleoli (SIRT7), and mitochondria (SIRT3–5) (Balcerczyk and Pirola 2010). Except for SIRT1–3, all other members of this enzyme family have weak deacetylase activity, and instead function mainly as an ADP-ribosyltransferase. SIRTs, like canonical HDACs, also have varied protein substrates that are not limited to histones; however, only SIRT1 and SIRT2 deacetylate modified lysines of histones H3 and H4 (Balcerczyk and Pirola 2010). SIRTs utilize Zn^{2+} mainly for structural stability rather than catalysis, and instead require a 1:1 stoichiometry between the cofactor NAD^+ and the acetyl-Lys substrate (Dittenhafer-Reed et al. 2011; Sanders et al. 2010). These enzymes are composed of ~275 amino acids that form a large Rossman fold domain characteristic for NAD^+ binding proteins and a small Zn^{2+}-binding domain that are connected by several loops which form the NAD^+ and acetyl-Lys binding sites (Sanders et al. 2010). The substrate and cofactor bind on opposite sides of the active site. Substrate binding creates significant conformational changes that bring the Zn^{2+}-binding and Rossman domains closer to facilitate correct orientation of the reactive centers (Sanders et al. 2010). While canonical HDACs liberate acetate from the deacetylation reaction via nucleophilic attack of water to the electrophilic carbonyl carbon, SIRTs transfer the acetate group to NAD^+ after elimination of nicotinamide and utilize the carbonyl oxygen as the main reactive center for nucleophilic attack to generate the α-1′-*O*-alkylamidate intermediate (Fig. 4.3) (Dittenhafer-Reed et al. 2011; Sanders et al. 2010). Both SN1 and SN2 mechanisms have been proposed for the formation of this intermediate. Following a series of reactions, this intermediate degrades to liberate Lys and 3′-*O*-acetyl-ADP ribose (Dittenhafer-Reed et al. 2011; Sanders et al. 2010).

4.2.2 Mechanism of Histone Methylation

PRMTs are classified according to the product of the methylation reaction with type I (PRMT1–4, 6, 8) giving monomethyl- and asymmetric dimethyl-Arg, while type II (PRMT5, 7, 9) yields only symmetric dimethyl-Arg (Sippl and Jung 2009; Di Lorenzo and Bedford 2011). PRMT1 accounts for ~85% of cellular mammalian PRMT activity and has been demonstrated to be critical for embryo development and neural differentiation (Di Lorenzo and Bedford 2011). It is responsible for the transcription activation mark at H4R3Me2a and the methylation of non-histone proteins involved in RNA processing and transport (Di Lorenzo and Bedford 2011; Zhang and Cheng 2003). It is characterized by a β-barrel that is unique to PRMTs, distinct SAM and substrate binding pockets, and a dimerization arm (Zhang and Cheng 2003). Homodimerization is required for enzymatic activity and postulated to improve cofactor binding and would also favor processive methylation reactions to yield the dimethylated Arg product (Zhang and Cheng 2003). HKMTs, like

PRMTs, are also characterized by distinct cofactor and substrate peptide binding sites and, except for the DOT1 family of HKMT, contain a conserved SET domain (Min et al. 2003; Kwon et al. 2003; Zhang et al. 2003a; Xiao et al. 2003). The substrate and SAM binding pockets are located on opposite sides of the SET domain. The size and amino acid residues in the enzyme active site determine the methylation state of the substrate lysine residue (Zhang et al. 2003a). PRMTs and KMTs utilize a similar mechanism for the methylation reaction, where the cofactor SAM is held in place by hydrogen bonding and nonpolar interactions of the ribose and adenine with amino acids of the enzyme (Xiao et al. 2003). This aligns the CH_3 group of SAM for an SN2 reaction with the terminal amino or guanidino group of the substrate Lys or Arg, respectively (Smith and Denu 2009; Xiao et al. 2003).

Demethylation of modified Lys is carried out either in a flavin adenine dinucleotide (FAD) or α-ketoglutarate dependent mechanism by LSD and JHDM family of demethylases, respectively (Fig. 4.4). These enzymes, while utilizing distinct reaction mechanisms, both yield a hemiaminal intermediate that degrades to liberate the demethylated product and formaldehyde. LSDs are members of the monoamine oxidase superfamily which reduce FAD to $FADH_2$ (Fig. 4.4), but are structurally distinct from other members of this superfamily by possessing a Tower domain that associates with the cofactor protein CoRest to render it catalytically active (Yang et al. 2007). The oxidation–reduction step involved in the LSD-mediated demethylation reaction requires a free electron pair, and hence this enzyme can only mediate the demethylation of mono- and dimethylated Lys residues. JHDM demethylases, on the other hand, are metalloenzymes that bear a catalytic Fe^{2+} ion and require α-ketoglutarate as a cofactor and are distinct from LSDs in their ability to demethylate trimethylated Lys (Anand and Marmorstein 2007). The key step to the JHDM-mediated reaction is the formation of an Fe^{IV}-oxo intermediate from the reaction of the catalytic Fe^{2+} with molecular oxygen and α-ketoglutarate (Fig. 4.4) (Anand and Marmorstein 2007). This intermediate oxidizes the CH_3 moiety of the methylated Lys substrate to yield the hemiaminal intermediate (Smith and Denu 2009; Anand and Marmorstein 2007).

4.3 HDAC Inhibitors as Chemotherapeutic Agents

4.3.1 General Pharmacophore

Compared to other histone modifying enzymes, inhibitors of HDACs are the most represented and extensively studied to date (Fig. 4.5); the class III SIRTs are largely unaffected by these agents. The cocrystal structure of two structurally related hydroxamate-based inhibitors, vorinostat and trichostatin A, bound to the histone deacetylase-like protein (HDLP) of *Aquifex aeolicus*, provided initial insights into the binding and essential features of an HDAC inhibitor: a Zn^{2+} binding moiety, an

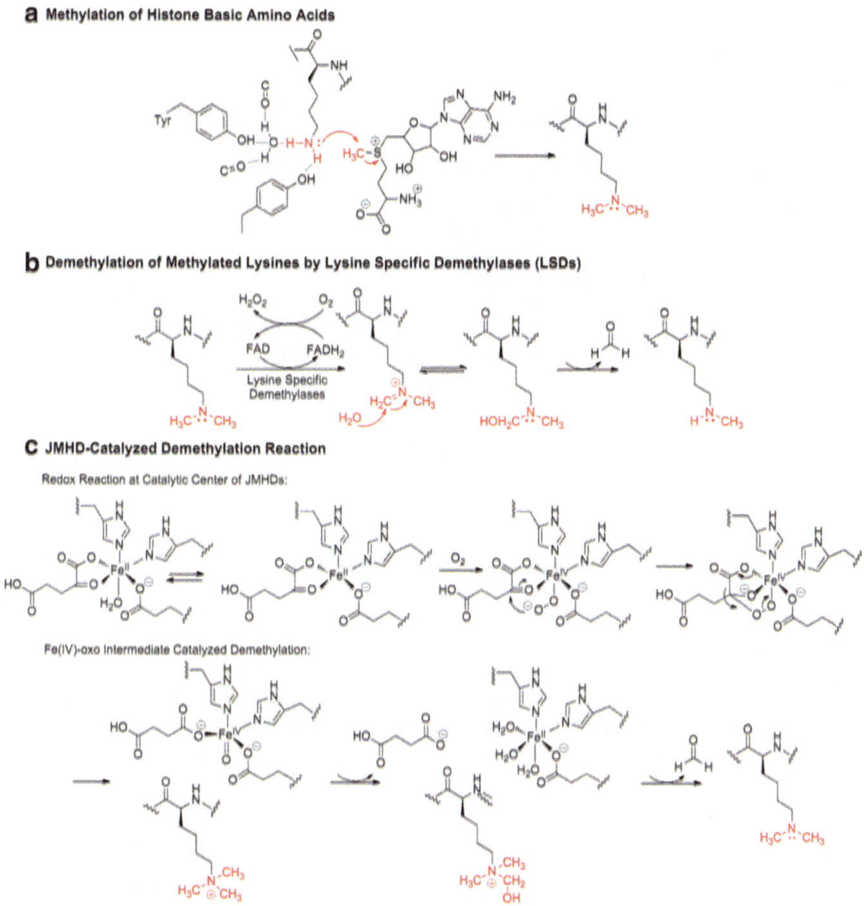

Fig. 4.4 Mechanism of histone lysine methylation and demethylation. (**a**) Histone methyltransferases employ a common mechanism, utilizing the nucleophilic terminal amino or guanidino group to abstract the methyl group of SAM. (**b**) Mono- or dimethylated Lys can be modified by lysine specific demethylases through a FAD-dependent reaction, removing a CH_3 group that is converted to formic acid. (**c**) Demethylation of trimethylated Lys is catalyzed by JHDM enzymes through an Fe^{2+}-mediated reaction, distinct from LSD-catalyzed reactions

alkyl chain which mimics the aliphatic chain of acetylated Lys, and a cap group that interacts with amino acid residues on the rim of the enzyme's tubular channel (Fig. 4.6) (Finnin et al. 1999).

The hydroxamate moiety of suberoylanilidine hydroxamic acid (SAHA) and trichostatin A acts as a bidentate ligand of the catalytic Zn^{2+} ion, utilizing both the hydroxy and carbonyl groups, and also forms hydrogen bonding interactions with several amino acid residues in the active site (Finnin et al. 1999). Other documented

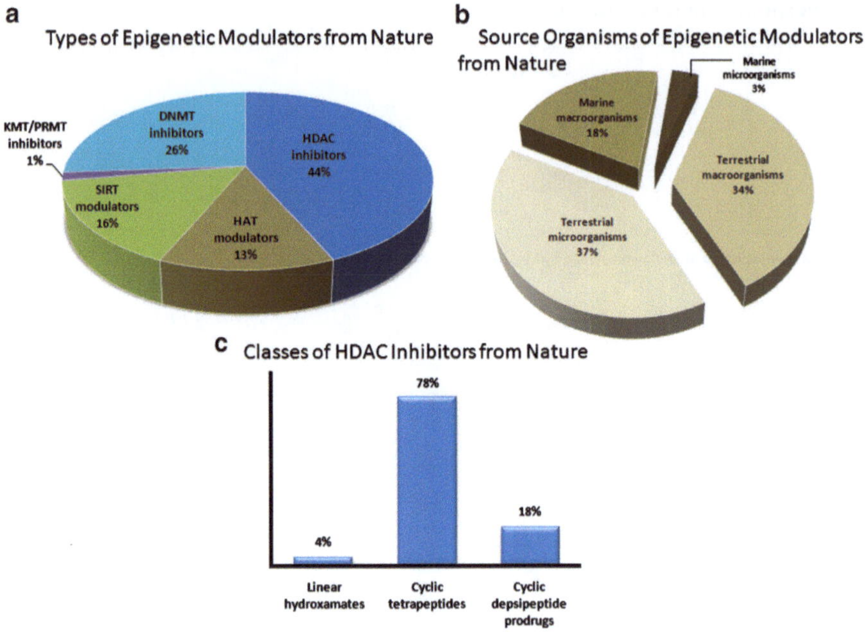

Fig. 4.5 Distribution of epigenetic modulators from Nature. (**a**) HDAC inhibitors represent the majority of natural epigenetic modulators. (**b**) Source organisms of these modulators are also well-distributed, including terrestrial as well as marine macroorganisms and microorganisms. (**c**) Among the natural HDAC inhibitors, cyclic tetrapeptides and depsipeptides are the predominant class of compounds and feature a variety of "warhead" and cap moieties

HDAC inhibitor "warheads" include mercapto (thiol), epoxide, carboxylic acid, amide, hydroxy, and keto functionalities. An alkyl chain length of four to five carbons is ideal for the best fit in the tubular channel and maximum interaction between the catalytic center and "warhead" moiety. Cocrystallization of trichostatin A and SAHA with HDAC8 showed similar interactions between the inhibitor and enzyme as in HDLP (Vannini et al. 2004). Significant positional change was observed for the side chain of Lys33, causing conformational distortion of amino acid residues that make up the binding pocket of HDAC8 (Vannini et al. 2004). The cap moiety (dimethyl aniline) of trichostatin A makes stabilizing π–π interactions with Y100 to cause an ordered L2 loop (Glu98–Pro103). Based on sequence alignment, different HDAC isoforms have a highly homologous enzyme active site, while less conserved residues make up the rim of the substrate binding site that can be exploited for isoform selectivity (Somoza et al. 2004; Vannini et al. 2004). Trichostatin A utilizes its dimethyl aniline moiety for this purpose, which can only interact with regions close to the enzyme active site due to limited pharmacophore size of the aromatic ring. In contrast, the cyclic depsipeptide HDAC inhibitors such as romidepsin and largazole possess a highly functionalized cap moiety comprised of multiple peptide–polyketide-derived units that can interact with divergent residues of the enzyme, leading to pronounced class I HDAC selectivity (Bowers et al. 2009;

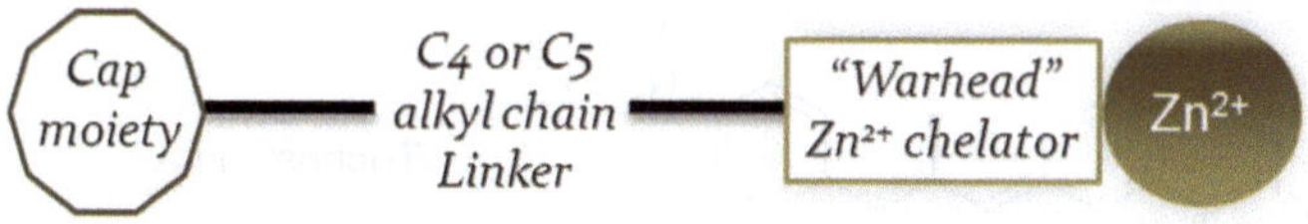

Fig. 4.6 Essential features of an HDAC inhibitor include a Zn^{2+}-binding moiety attached to a C_4 or C_5 alkyl chain. This is capped by structural elements that may confer HDAC isoform selectivity

Hong and Luesch 2012). The recently published cocrystal structure for HDAC8 and largazole thiol (the bioactive species of largazole) rationalized the importance of the 3*S*-hydroxy-7-mercapto-4-heptenoic acid moiety and the depsipeptide ring structure (Cole et al. 2011). The mercapto (thiol) moiety of largazole thiol was shown to exist as the thiolate anion which interacts with the catalytic Zn^{2+} ion in a nearly perfect tetrahedral arrangement (Cole et al. 2011). This conformation of the thiolate–Zn^{2+} coordination is stabilized by the fused thiazole–thiazoline system, which rigidifies the depsipeptide ring structure. While the conformation of the depsipeptide core does not change upon HDAC8 binding, the enzyme, however, showed notable conformational changes between Leu98 and Phe109, specifically with Tyr100 and Asp101, not observed with small and less bulky HDAC inhibitors cocrystallized with HDAC8 (Cole et al. 2011). Consequently, these changes also alter the conformation of the L1 loop (Leu31–Lys36) of HDAC8 (Cole et al. 2011).

4.3.2 *Clinical HDAC Inhibitors*

4.3.2.1 Zolinza® (Vorinostat)

Vorinostat (Fig. 4.7) was approved in October 2006 as third line treatment for cutaneous T-cell lymphoma (CTCL). The development of vorinostat started with the observation that dimethyl sulfoxide induces growth arrest and erythroid differentiation. Several series of polar, small molecules were synthesized and assessed for their ability to differentiate Friend leukemia cells (Marks and Breslow 2007; Reuben et al. 1976; Breslow et al. 1991; Richon et al. 1996; Tanaka et al. 1975). Hexamethylene bisacetamide was the first member of this class of compounds to enter clinical trials for myelodysplastic syndrome and acute myeloid leukemia (AML), which did not progress due to short-lived cellular effects (Marks and Breslow 2007; Young et al. 1988). To increase the potency, a hydroxamic acid moiety and a bulky phenyl group were introduced together with an optimized alkyl chain length joining these two terminal functionalities to maximize the interaction with the target of this series of compounds, yielding SAHA (Richon et al. 1996). The structural similarity of SAHA to the natural product HDAC inhibitor trichostatin A (Fig. 4.7) isolated from *Streptomyces hygroscopicus* suggested the same molecular target and mechanism of action for the former (Finnin et al. 1999; Marks and Breslow 2007). Both trichostatin A and SAHA are nonselective HDAC inhibitors.

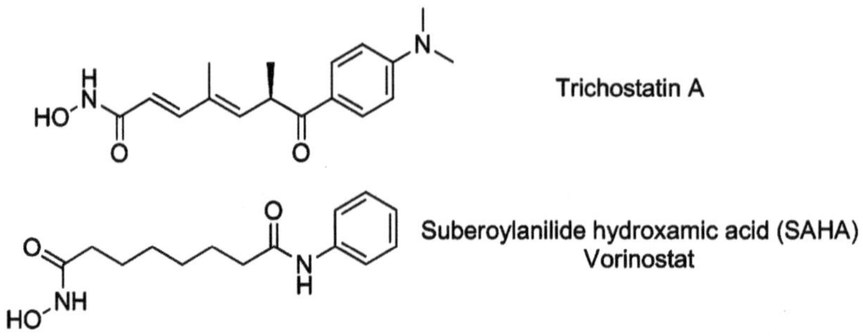

Fig. 4.7 The hydroxamate-based HDAC inhibitors trichostatin A and vorinostat

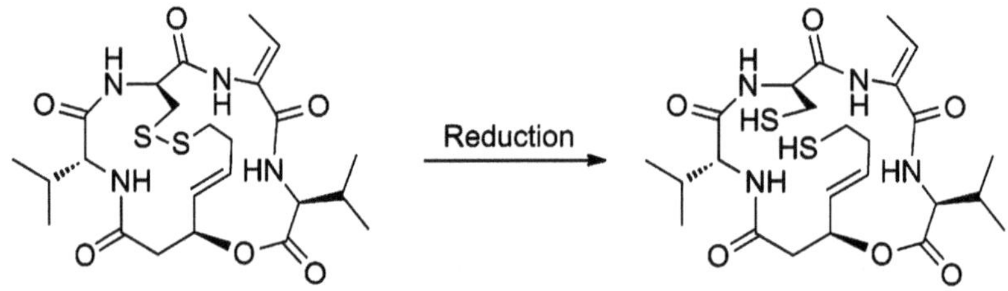

FK228/Romidepsin redFK228

Fig. 4.8 Cyclic depsipeptide FK228 and the mechanism of its metabolic activation

4.3.2.2 Istodax® (Romidepsin)

Romidepsin (Fig. 4.8) was approved in November 2009 for use in CTCL patients who have received at least one prior systemic therapy. Formerly known as FR901228, FK228, and depsipeptide, romidepsin is a natural product HDAC inhibitor purified from the fermentation broth of *Chromobacterium violaceum* (Ueda et al. 1994a). It is characterized by a bicyclic depsipeptide structure and has a mixed peptide–polyketide-derived unit (3*S*-hydroxy-7-mercapto-4-heptenoic acid) fused through a disulfide bridge with D-Cys (Shigematsu et al. 1994). This compound was initially shown to revert the *Ras*-transformed NIH3T3 phenotype to normal, and initial in vitro cytotoxicity screening showed potent inhibitory activity against 11 cancer cell lines with IC_{50} range of 0.3–2.4 ng/mL while having weak cytotoxic activity on human and mouse fibroblasts (IC_{50}>500 ng/mL) (Ueda et al. 1994a, b). The mechanism of action of FK228 was elucidated based on the similarity of its cellular effects and transcriptional activation of SV40 promoter in comparison to trichostatin A (Nakajima et al. 1998). Despite the structural unrelatedness, FK228, like trichostatin A, caused accumulation of acetylated histones and induced G1 and G2/M cell cycle arrest of M-8 cells (Nakajima et al. 1998). The HDAC inhibitory activity of FK228 is dependent on its metabolic activation and requires glutathione-dependent reduction of the disulfide moiety to liberate the active species redFK228, bearing two free sulfhydryl groups (Furumai et al. 2002). Molecular docking studies using the HDLP crystal structure demonstrated that one of the sulfhydryl groups can access the catalytic Zn^{2+} ion via a water molecule (Furumai et al. 2002). The importance of the glutathione-mediated cellular reduction was demonstrated in yeast mutants with glutathione synthesis defects, which showed partial to complete resistance to FK228 treatment and suppression of histone H4 acetylation (Furumai et al. 2002). The prodrug character of FK228 improves its cellular permeability and stability, while redFK228 was less stable, based on comparison of the half-lives in serum and growth media of these two species. FK228 has a half-life of >12 h and 4.7 h, while redFK228 has a half-life of 0.54 h and <0.3 h in growth medium and serum, respectively (Furumai et al. 2002).

4.3.3 Preclinical Natural HDAC Inhibitors

The majority of histone modifying natural products are HDAC inhibitors, and the more recently discovered HDAC inhibitors can be classified mostly into tetrapeptides and depsipeptides (Fig. 4.5). Cyclic tetrapeptide HDAC inhibitors (Fig. 4.9) which include trapoxins (Itazaki et al. 1990), microsporins (Gu et al. 2007), apicidins (Singh et al. 1996, 2001), FR235222 (Mori et al. 2003a, b, c) and azumamides (Nakao et al. 2006) are characterized by a 12- or 13-membered cyclic peptide core structure consisting of three nonpolar and/or aromatic amino acid residues and a fourth nonproteinogenic unit that is a C_9/C_{10} modified aliphatic chain with a terminal functional group as an epoxide, hydroxy-, or keto-group. The structurally rigid peptide ring interacts with hydrophobic pockets of HDACs while the terminal functionality acts as the Zn^{2+} binding moiety. The presence of aromatic amino acids can provide additional π-π interactions between the cyclic peptide backbone and the enzyme (Di Micco et al. 2008; Maulucci et al. 2007). Azumamides A–E from the marine sponge *Mycale izuensis* are different from other cyclic tetrapeptide HDAC inhibitors: they are comprised of three unnatural α-amino acids with D-configuration and bear a β-amino acid derived unit acid either as 3*R*-amino-2*S*-methyl-5-nonenoic-1,9-diacid (Amnda) or the corresponding amide, giving a 13-membered ring system instead of the usual 12-membered ring (Nakao et al. 2006). These metabolites also do not contain a Pro or pipecolic moiety that is common among other cyclic tetrapeptides. Trapoxins from *Helicoma ambiens* and the related antibiotics HC-toxin, chlamydocin (Closse and Huguenin 1974), WF-3161 (Umehara et al. 1983), and Cyl-2 (Hirota et al. 1973) utilize a distinct inhibition mechanism compared to other HDAC inhibitors; these cyclic tetrapeptides are characterized by a terminal epoxide ring that can undergo a nucleophilic reaction with an active site residue, presumably a Cys, to form an irreversible enzyme–inhibitor complex (Kijima et al. 1993). The phytotoxic compound depudecin, which bears an epoxide moiety, was demonstrated to be a covalent HDAC inhibitor as well (Shimada et al. 1995).

Cyclic depsipeptides including spiruchostatins A and B (Masuoka et al. 2001) and FR901375 (Chen et al. 2003) (Fig. 4.10) are structurally related to FK228 and hence are also prodrug HDAC inhibitors that require glutathione-dependent

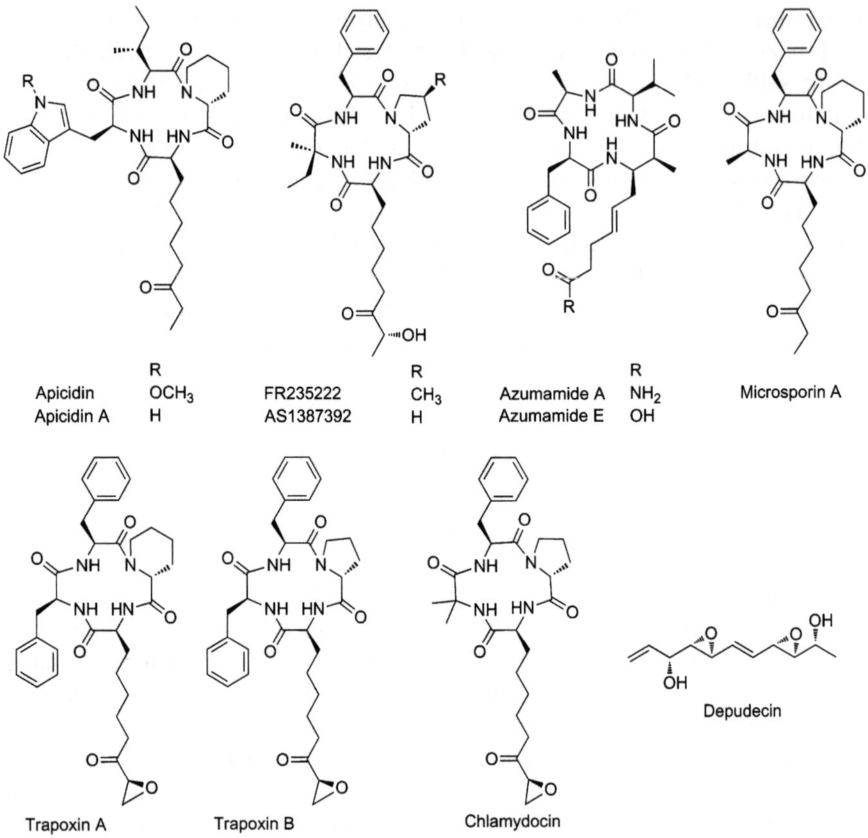

Fig. 4.9 Cyclic tetrapeptide HDAC inhibitors interact with the catalytic Zn^{2+} ion through a variety of functionalities such as a terminal keto, hydroxy or amide group. Covalent HDAC inhibitors (trapoxins, chlamydocin, and depudecin) are characterized by a terminal epoxide moiety. The electrophilic epoxide ring is susceptible to an attack by the nucleophilic thiol of Cys located in the enzyme active site

reduction to expose the biologically relevant 3*S*-hydroxy-7-mercapto-4-heptenoic acid portion (Furumai et al. 2002). The linear HDAC inhibitor psammaplin A (Fig. 4.10) is a prodrug and requires activation by glutathione as well (Pina et al. 2003; Kim et al. 2007). Largazole (Fig. 4.10), a cyclic depsipeptide isolated from a marine cyanobacterium *Symploca* sp., possesses a unique protecting group as an octanoyl thioester rather than a disulfide; hence the sulfur atom bears a different oxidation state (Taori et al. 2008; Ying et al. 2008; Hong and Luesch 2012). Through protein-assisted hydrolysis, the metabolic activation of largazole also renders 3*S*-hydroxy-7-mercapto-4-heptenoic acid as in other cyclic depsipeptide HDAC inhibitors (Ying et al. 2008; Liu et al. 2010). The thioester moiety is rapidly metabolized in serum and cell lysates, with the bioactive largazole thiol being stable in its free form or as a covalent adduct (Liu et al. 2010).

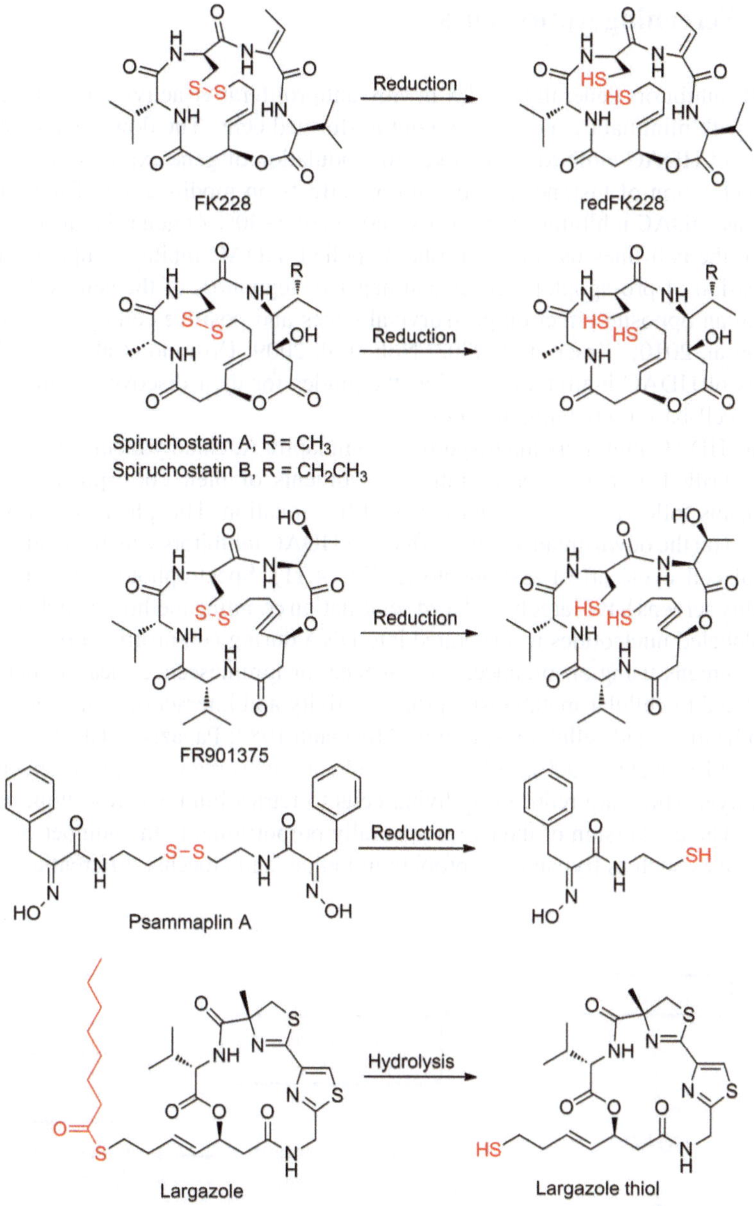

Fig. 4.10 Prodrug HDAC inhibitors require metabolic activation through a glutathione-mediated reduction or protein-assisted hydrolysis

4.4 Screening Approaches

HDAC inhibitors generally display in vitro antiproliferative activity on transformed cells, with minimal cytotoxicity to nontransformed cells. The downstream cellular effects of HDAC inhibitors are linked to modulation of gene expression via lysine (de)acetylation of histones and pleiotropic effects on modification of non-histone proteins. HDAC inhibitors alter the expression of 5–40% of genes in vitro, depending on the cell lines used and threshold applied. HDAC inhibitors upregulate the expression of pro-apoptotic genes and negative regulators of the cell cycle, while having an opposite effect on pro-survival genes and positive cell cycle regulators (Liu et al. 2010; Glaser et al. 2003; Noh et al. 2009; Hoshino et al. 2007). These effects of HDAC inhibitors served as the guides for their discovery using mostly whole cell-based screening methods.

The HDAC inhibitors microsporins, psammaplin A, chlamydocin, and largazole were purified as the antiproliferative constituents of their corresponding source organisms following a cytotoxicity-guided fractionation. This phenotypic response is based on the downstream cellular effects of HDAC inhibitors which include induction of cell cycle arrest and apoptosis (Fig. 4.11). Spectrophotometry-based cell viability screens have largely replaced scintillation counting methods which measure radiolabeled nucleotides incorporated into DNA during cell proliferation. Cell viability screens using absorbance, fluorescence, or luminescence measurements are correlated to cellular metabolism, protein activity and interactions, membrane permeabilization, and cellular respiration (Mosmann 1983; Papazisis et al. 1997; Voigt 2005; O'Brien et al. 2000). Colorimetric cell viability screens use formazan or resazurin dye, which are reduced by living cells to tetrazolium and resorufin, respectively. The conversion of the dye is generally proportional to the number of viable cells. Dyes such as trypan blue, propidium iodide, and Hoechst differentiate viable

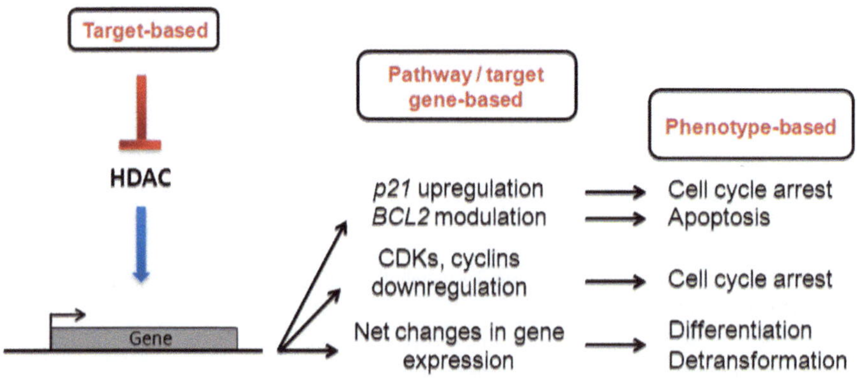

Fig. 4.11 Screening approaches for HDAC inhibitors have mostly been based on phenotypic screens as a result of changes in gene expression. HDAC inhibitors induce cell cycle arrest, cell death, and differentiation, while reverting oncogenic cell transformation

and nonviable cells based on their ability to stain the nuclei or cytoplasm (Strober 2001; Altman et al. 1999). Cell staining and sorting of cell populations has been used to quantitatively assess growth arrest at specific points in the cell cycle as well as apoptosis.

Phenotypic assays based on the reversion of the transformed phenotype following oncogene transfection of mouse fibroblast cells also yielded several of the natural HDAC inhibitors (Fig. 4.11). FK228, trichostatin A, depudecin, and trapoxins were shown to revert the transformed phenotype to normal. NIH3T3 cells were utilized due to their high sensitivity to transformation following activation of the signal transduction pathway. *Ras*-, *v-sis*, and *src*-oncogenes were commonly used in transformation assays (Fry et al. 1988). Transformed cells are observed to undergo changes in cell morphology, loss of density-dependent growth regulation, and possess increased growth rate, all of which are linked to altered gene expression, which in turn can be modulated with HDAC inhibitors. Compared to the normal flat morphology of NIH3T3 cells, transformed cells form swirling clusters (foci) and are highly refractile in appearance (Clark et al. 1995). Morphological changes and foci formation are monitored in either live cells or fixed cells stained with crystal violet (Clark et al. 1995).

Screening for positive regulators of gene expression of *p21* and plasminogen activator inhibitor-1 using luciferase-based reporter assays yielded trichostatin A and spiruchostatins as hit compounds, respectively (Masuoka et al. 2001; Abe et al. 1994; Uekia et al. 2001; Nie et al. 2001). The activation of *p21* expression is among the validated downstream cellular effects of HDAC inhibitors and its expression has been used as a biomarker for effective HDAC inhibition (Fig. 4.11) (Wilson et al. 2006). The p21 protein, also known as cyclin-dependent kinase inhibitor 1, is a negative regulator of the cell cycle and controls the progression to S phase by direct inhibition of cyclin E/CDK2 and cyclin D/CDK4/6 complexes. Plasminogen activator inhibitor-1 is induced via TGF-β signaling and has been linked to tumor suppression; hence, chemical entities capable of phenocopying this effect could represent a new class of anticancer agents (Abe et al. 1994). Recently, psammaplin A was obtained as a sponge-derived screening hit that can activate the Wnt signaling pathway. This effect of psammaplin A was mediated by its HDAC inhibitory activity rather than through specific modulation of the Wnt pathway (McCulloch et al. 2009). In this assay, the HEK293-based cell line STF3a was engineered to express high levels of Wnt3a and contains a β-catenin responsive luciferase reporter that responds to changes in specific transcriptional activity, which can be related and quantified through luminescence measurements (McCulloch et al. 2009).

HDAC inhibitory activity of natural products has been determined directly using purified enzymes or nuclear cell lysates and fluorogenic peptide substrates, as well as immunoblot analysis and immunohistochemistry using acetylation-specific antibodies. Scintillation counting of ^{3}H-labeled histone substrate has been largely replaced by fluorescence-based techniques (Wegener et al. 2003). High-throughput screening for HDAC inhibitors using in vitro enzymatic assays offers the advantage of direct target identification; however, this may have a drawback compared to phenotypic screening since prodrug HDAC inhibitors may be missed if the assay conditions do not sufficiently activate them to yield the biologically relevant species.

4.5 Challenges and Prospects to the Design of HDAC Inhibitors

4.5.1 Metabolic Stability

One main concern for HDAC inhibitors is their metabolic instability, which consequently affects their in vivo activity (Masuoka et al. 2008; Thaler and Minucci 2011). The trapoxins and related epoxide-bearing cyclic tetrapeptides have been documented to be potent inhibitors of HDAC and growth of cultured cancer cells, but failed to show significant biological activity in animal models. The in vitro IC_{50} value for chlamydocin against P-815 mastocytoma cells (40 h incubation) is 0.36 ng/mL; however, in vivo testing of chlamydocin on mice inoculated with P-815 cells with a dosage of 20–160 mg/kg by daily intraperitoneal injection showed no reduction in tumor and only a 10% optimal increase of life span (Stähelin and Trippmacher 1974). Poor or no significant dose dependencies were observed in in vivo tests with chlamydocin. Rapid metabolic inactivation led to a very low serum level of chlamydocin in rats. A dosage of 50 mg/kg of chlamydocin administered intravenously led to a concentration of 14 μg/L of chlamydocin in rat serum, which, assuming an equal distribution, would correspond to a $t_{1/2}$ of 2.5 min (Stähelin and Trippmacher 1974). Cytostatic activity of chlamydocin is rapidly lost upon incubation with serum and heparinized blood and was used as a measure of its $t_{1/2}$. Incubation of 1 mg/L of chlamydocin gave a $t_{1/2}$ of 8.7 min in freshly prepared rat serum and 0.68–3.6 min in whole blood (Stähelin and Trippmacher 1974). The epoxide moiety for chlamydocin is a metabolic liability. Synthesis of hydroxamic analogs of trapoxin and related cyclic tetrapeptides where the epoxide moiety has been replaced with a hydroxamic acid functionality showed comparable HDAC inhibitory activity and increased $t_{1/2}$ to 50 min (Komatsu et al. 2001).

Trichostatin A undergoes rapid Phase I biotransformation by reduction of the hydroxamic acid to yield trichostatin A amide, *N*-demethylation to give *N*-monomethyl trichostatin A and *N*-monomethyl trichostatin A amide, and oxidative deamination to give trichostatic acid following intraperitoneal injection in female BALB/c mice (Sanderson et al. 2004). It has a $t_{1/2}$ of 6.3 min and 9.6 min for low (0.5 mg/kg) and high (80 mg/kg) doses, respectively. Trichostatin A is rapidly absorbed in the peritoneum and detectable in the plasma within 2 min. Plasma samples collected 20 min postinjection showed that *N*-monomethyl trichostatin A is the only biologically relevant metabolite present and was shown to maintain HDAC inhibitory activity (Sanderson et al. 2004). Other metabolites at this timepoint did not have an intact hydroxamic acid moiety and were inactive for HDAC inhibition. The major metabolite obtained after 60 min is *N*-monomethyl trichostatin A amide; no trichostatin A was detected 24 and 48 h after injection (Sanderson et al. 2004). Further biotransformation pathways for *N*-monomethyl trichostatin A and *N*-monomethyl trichostatin A amide have not been disclosed. In vitro incubation of trichostatin A with microsomes (rat and human) and hepatocytes (rat), however, showed didemethylation of trichostatin A and trichostatin A amide but were not observed in vivo (Elaut et al. 2002).

The problem of rapid metabolic inactivation has been somewhat addressed by natural products themselves by disguising them as cryptic HDAC inhibitors. Aside from romidepsin, spiruchostatins, FR901375, psammaplin A, and largazole, there are several other prodrugs among natural products acting on different targets and that have distinct mechanisms of bioactivation (Kwan and Luesch 2010). The protecting groups can improve cellular permeability, and/or prevent untimely release and sequestration prior to reaching the site of action. In addition to their prodrug moieties, the more elaborate and functionalized structure of cyclic depsipeptide HDAC inhibitors also offers the advantage of protection from biotransformation. Synthesis of analogs of potent HDAC inhibitors utilizing protecting groups and stable functionalities to improve pharmacokinetic and pharmacodynamic profiles as well as oral bioavailability has been a major theme in the development of newer generation HDAC inhibitors. Tumor growth inhibition and more specific biomarkers for in vivo activity such as histone acetylation and p21 expression in tumor cross-sections have been utilized as assay endpoints, which together with information on biotransformation pathway and biologically relevant metabolites in plasma link the in vitro and in vivo activities of HDAC inhibitors.

4.5.2 Target Identification and Isoform Selectivity of HDAC Inhibitors

Trichostatin A was the first secondary metabolite demonstrated to affect histone acetylation and following this, vorinostat and romidepsin were shown to have the same mechanism of action as trichostatin A based on structural similarity and phenotypic effects, respectively (Finnin et al. 1999; Nakajima et al. 1998; Yoshida et al. 1990). Trapoxin B was the first compound shown to interact with HDAC (Taunton et al. 1996a, b). Target identification was achieved through affinity purification with trapoxin B conjugated beads and the then unknown HDAC protein was characterized to show similarity with the yeast transcriptional regulator Rpd3p, later termed as HDAC1 (Taunton et al. 1996b). Even though trichostatin A and trapoxins were unable to reach clinical development, they were mainly utilized as molecular probes to elucidate the structure and function of HDACs. Initial structural information derived from cocrystal structures of hydroxamic acid compounds and HDLP also provided insights into the activity of other classes of HDAC inhibitors. Recent X-ray crystal structures reported for HDAC2 (Bressi et al. 2010), HDAC4 (Bottomley et al. 2008), HDAC7 (Schuetz et al. 2008), and HDAC8 (Somoza et al. 2004; Vannini et al. 2004) showed substantial differences in structure that may be exploited to obtain HDAC isoform selectivity of inhibitors. The selectivity of current natural product HDAC inhibitors is mainly limited to each class rather than individual isoforms. The cyclic depsipeptide HDAC inhibitors are more selective towards class I HDAC isoforms due to their highly functionalized core structure that can interact with divergent amino acids among HDAC classes located >10 Å from the enzyme active site. Isoform selective HDAC inhibition would be difficult to achieve with agents that target homologous regions among canonical HDACs. While class I and II HDACs show significant differences in enzyme structure outside of the catalytic domain, members of class I HDACs are highly homologous. HDAC1 has ~85% sequence identity to HDAC2, while being less related to HDAC3 and HDAC8 with ~50% sequence identity to both (De Ruijter et al. 2003). To date, several synthetic compounds such as tubacin (Haggarty et al. 2003), PCI-34051 (Balasubramanian et al. 2008), and diphenyl methylene hydroxamic acid derivatives (Tessier et al. 2009) have been demonstrated to be HDAC6 or HDAC8 isoform selective inhibitors which target distinct enzyme domains of HDAC isoforms. Structural information from cocrystal structures of existing HDAC inhibitors such as largazole thiol may allow for the rational design of new generation, and isoform-specific HDAC inhibitors. HDAC class selectivity has not yet been demonstrated to improve toxicity profiles of inhibitors and their tumor selectivity, as both pan-selective vorinostat and class I selective romidepsin showed similar side effects among which cardiotoxicity is of utmost concern (Shah et al. 2006). Different members of class I HDAC isoforms are emerging to play different roles in tumor progression, with HDAC1 and HDAC2 being linked to tumor aggressiveness and increased cell proliferation, with the latter also related to sensitization of cancer cells towards chemotherapy (Halkidou et al. 2004; Song et al. 2005). From mouse knockouts of HDAC isoforms, it is evident that these enzymes do not have redundant function. Selective isoform inhibition may improve the safety profile of the HDAC-based anticancer agents while maintaining potent action. Deletion of HDAC1 caused an upregulation of p21 and p27 expression via a p53-dependent mechanism, with concomitant increase in histone acetylation (Witt et al. 2009). HDAC3-null mice showed a delay in cell cycle progression, while deletion of HDAC2 and HDAC5 were lethal due to cardiac defects and enlargement of the heart, respectively. HDAC6 deletion, however, did not affect the level of histone acetylation, but increased α-tubulin acetylation, a known non-histone protein target of HDAC6, indicating that there are expectedly pleiotropic effects of HDACs unrelated to their ability to modify histones (Witt et al. 2009).

4.5.3 Non-Histone Cancer-Relevant Protein Targets of (De)Acetylases

Two recent reports on mass spectrometry-based analysis of human cell lines and liver tissues showed an extensive network of acetylated proteins involved in DNA repair, cytoskeletal function, cell cycle, chromatin remodeling, and metabolism, proving that HATs, HDACs, and SIRTs have a wider array of protein substrates than just histones alone. A total of ~3,600 lysine acetylation sites in ~1,750 proteins were identified from acetylome profiling of Jurkat cells, while ~1,700 acetylation sites in ~1,100 proteins were identified in human liver tissues (Choudhary et al. 2009; Zhao et al. 2010). Improvements in detection, characterization, and absolute quantification methods of acetylated proteins have greatly facilitated the mapping of the acetylome. Transcription factors such as p53, NF-κB, and STATs, the cytoskeletal protein α-tubulin, molecular chaperone heat shock protein 90 (Hsp90), and DNA repair protein Ku70 are among the well studied non-histone proteins posttranslationally modified by acetylation, which may contribute in part to altered cellular response to stress and oncogenic transformation (Glozak et al. 2005). The tumor suppressor protein p53 is acetylated at several Lys residues (Lys305/320/370/372/373/381/382/386) located on the C-terminal regulatory domain, with the HATs p300/CBP and PCAF facilitating this p53 posttranslational modification (Barlev et al. 2001; Wang et al. 2003; Tang et al. 2008; Yamaguchi et al. 2009; Joubel et al. 2009). This modification works in synergy with phosphorylation at the N-terminal domain in modulating the interactions of p53 with its negative regulators Mdm2 and Mdmx (Tang et al. 2008). Mdm2 is a polyubiquitin E3 ligase that catalyzes the degradation of p53 via both ubiquitin-dependent and ubiquitin-independent pathways (Tang et al. 2008). Unmodified Lys can be ubiquitinated, tagging p53 for subsequent degradation. The role of C-terminal acetylation of p53 in DNA binding, however, has not yet been completely established. Two p53 acetylation sites in the DNA binding domain (Lys120/164) may suggest additional roles for this modification that is independent of Mdm2 (Tang et al. 2008). Although deacetylation of individual Lys residues of p53 was not sufficient to abolish p53-dependent expression of p21, complete deacetylation of p53 abrogated p21 expression, thus demonstrating the importance of p53 acetylation in the expression of this cell cycle regulatory protein (Tang et al. 2008). Romidepsin treatment induced acetylation at Lys373/382 together with p21 expression, while BAX activation, ROS production and apoptosis were observed in response to vorinostat treatment, concomitant with p53 acetylation at Lys320/373/382 (Yamaguchi et al. 2009; Zhao et al. 2006). These observations demonstrated the importance of HDACs in controlling p53 posttranslational modification, whereby HDAC1 and SIRT1 are mainly responsible for deacetylation of p53 (Glozak et al. 2005). The major mammalian NF-κB consists of a p50/p65 heterodimer, and both subunits can be acetylated by p300/CBP, which affects the cellular localization and DNA binding affinity of NF-κB. In the inactive state, the p50/p65 heterodimer is complexed with its repressor protein IκB and is localized in the cytoplasm (Quivy and Van Lint 2004). Phosphorylation of IκB by the IκB kinase (IKK) triggers IκB degradation and dissociation from the p50/p65 complex, which then signals the translocation of NF-κB into the nucleus, where it binds to NF-κB target genes and regulates transcription (Quivy and Van Lint 2004). High NF-κB nuclear activity is present in a variety of human malignancies, through constitutive NF-κB activation or inactivating mutations in IκB, leading to altered expression of pro-survival and antiapoptotic genes (Quivy and Van Lint 2004). Acetylation of different Lys residues of p65 has variable effects on cellular localization and DNA binding. Lys221/310 acetylation is required for transcriptional activation of NF-κB target genes, while HDAC1 and HDAC3 are silencing this effect (Chen et al. 2005). Acetylation of Lys310 is dynamically regulated by phosphorylation of Ser276/536 of p65, with phosphorylation providing additional favorable interactions between p65 and the HAT p300/CBP (Quivy and Van Lint 2004; Chen et al. 2005). Acetylation of Lys122/123 residues of p65 by p300/CBP and PCAF has an opposite effect on p65 nuclear localization; acetylation of these residues decreases the affinity of NF-κB for its target DNA sequence and promotes NF-κB–IκB complex formation, thereby rendering the complex inactive and facilitating cytoplasmic export (Kiernan et al. 2003). p300-mediated acetylation of p50 at Lys431/440/441 increases target DNA binding and transcription of the pro-inflammatory genes *iNOS* and *COX-2* (Deng et al. 2003; Deng and Wu 2003). The STAT family of proteins, which mainly function as signal transducers and activators of transcription, are characterized by their cytoplasmic localization in the monomeric form and upon dimerization translocate into the nucleus. Phosphorylation of Ser/Tyr residues at the COOH-terminal region by an upstream kinase induces homodimerization of STAT proteins. STAT3 and STAT5 are generally considered tumor promoters, associated with several malignancies, and also control the expression of cell cycle regulators, antiapoptotic genes, and proangiogenic factors. Mammalian STAT3 dimerizes upon acetylation of Lys685 and, although this posttranslational modification does not completely account for STAT3 activation, this event is postulated to favor a cooperative interaction and to augment dimer formation (Yuan et al. 2005).

Acetylation of cytoplasmic proteins α-tubulin and HSP90 can indirectly affect protein levels rather than gene expression. Acetylation of α-tubulin is inversely related to the mobility of microtubules and, while generally observed in stable microtubules, it does not affect the dynamics of stable microtubule formation. Despite being one of the first non-histone protein characterized to be dynamically regulated by acetylation, the HAT family responsible for modification of Lys40 of α-tubulin has not yet been characterized, while the mainly cytoplasmic HDAC6 and SIRT2 are the primary deacetylases acting on Ac-Lys40 of α-tubulin (Hubbert et al. 2002; Zhang et al. 2003b; Ledizet and Piperno 1987; Piperno et al. 1987; Maruta et al. 1986). HDAC6 and the microtubule network are two critical components of the aggresome pathway, implicated in the transport of misfolded proteins not degraded by the proteasome and preventing accumulation of toxic protein aggregates (Kawaguchi et al. 2003). Alleviation of cellular stress is also accomplished by another HDAC6-targeted protein, HSP90, which stabilizes the protein conformation and protects both normal and oncogenic client proteins from degradation (Aoyagi and Archer 2005; Scroggins et al. 2007). Hyperacetylation of HSP90 disrupts interactions with the client protein and co-chaperones. The inducible isoform (HSP90α) was demonstrated to be a substrate of p300 and modified at Lys69/100/292/327/478/54/558 (Yang et al. 2008). In all, altering acetylation of non-histone proteins presents an additional avenue of targeting cancer independent of chromatin dynamics, by instead modulating key mediators of malignant transformation.

4.5.4 Resistance to HDAC Inhibitors

Resistance to chemotherapeutic agents, either inherent or acquired, has been a primary concern in cancer therapy. Acquired resistance upon prolonged exposure to a single chemotherapeutic agent has been documented for anticancer drugs through either mutation or overexpression of the drug target or drug efflux pumps (Ambudkar et al. 2003). Amino acid mutation of BCR-ABL, the kinase target of imatinib, at the gatekeeper residue (T315I) is sufficient to cause a steric clash between the enzyme and imatinib, and also prevents formation of essential hydrogen bonds between the enzyme and imatinib, leading to ineffective kinase inhibition (Gorre et al. 2001). P-glycoprotein (Pgp), encoded by the multidrug resistance gene 1 (*MDR1*), is responsible for the transport of xenobiotics into the biliary tract and proximal tubules of the kidney with amphipathic and cationic drugs as its major substrates. Vinca alkaloids, anthracyclines, taxanes, and epipodophyllotoxins are all substrates for Pgp which render cancer cells cross-resistant (Ambudkar et al. 2003). FK228 is a substrate for Pgp and also increases the protein and mRNA expression of Pgp, which hence leads to significant resistance to FK228 (Xiao et al. 2005). Although clinical resistance to FK228 has not yet been documented, the response of patients to FK228 has been variable. FK228-resistant cell lines do not show cross-resistance to trichostatin A and SAHA, thus validating that FK228 resistance is independent of the HDAC enzyme and also suggests a non-unified mechanism of drug resistance to HDAC inhibitors (Xiao et al. 2005; Fedier et al. 2007; Dedesa et al. 2009). Resistance to the hydroxamic acid-based inhibitors SAHA, trichostatin A, and panobinostat is independent of P-glycoprotein expression (Fedier et al. 2007). Gene expression profiling of CTCL cell lines with varying susceptibilities to SAHA showed that STAT1, STAT3, and STAT5 expression and their phosphorylation states were biomarkers for responsiveness to treatment (Fantin et al. 2008). Skin biopsies from CTCL patients with advanced mycosis fungoides and Sezarky syndrome enrolled in Phase 2B clinical trials with SAHA were probed for STAT1, STAT3, and STAT5 expression by immunohistochemical staining and correlated to patient response (Fantin et al. 2008). STAT1 expression and nuclear localization of STAT5 were inversely related to treatment response, while low levels of nuclear phosphorylated STAT3 showed good correlation with improved patient prognosis (Fantin et al. 2008). In vitro combination treatment of SAHA with a JAK inhibitor showed a synergistic effect with a combination index of 0.28, preventing the activation of STAT proteins by inhibiting their upstream activators (Fantin et al. 2008). In vitro sensitivity to panobinostat treatment of CTCL cell lines, on the other hand, is related to increased NF-κB and BCL2 expression (Shao et al. 2010). Frameshift mutation in HDAC2 was observed in ~33% of colorectal and ~50% of endometrial carcinoma cell lines that are characterized by microsatellite instability. This mutation leads to deletion of two A and consequently loss in protein expression and enzymatic activity of HDAC2 (Ropero et al. 2006). This mutation was observed in ~20% of primary colon, endometrial, and gastric primary tumors but not in normal tissues (Ropero et al. 2006). Cell lines harboring this mutation were resistant to trichostatin A treatment but not to sodium butyrate or valproate. Trichostatin A failed to induce cell cycle arrest or apoptosis in cell culture and could not reduce tumor size in mouse xenografts bearing HDAC2 deficient cells (Ropero et al. 2006). In a follow up study, it was shown that gene expression of several tumor and cell cycle promoters as well as angiogenic factors can be indirectly or directly regulated by HDAC2 through binding to the 5′-end regulatory regions, causing upregulation in HDAC2 deficient cell lines (Ropero et al. 2008). A related study showed that the hydroxamate-based HDAC inhibitors trichostatin A and SAHA were able to induce histone hyperacetylation but not apoptosis in HDAC2-deficient cell lines (Hanigan et al. 2008). Induction of apoptosis was related to the upregulation of *APAF1* in cell lines with functional HDAC2. APAF1 binds cytochrome c and ATP/dATP to form the apoptosome, a key mediator in the caspase 9 cascade. *APAF1* is directly regulated by HDAC2 through binding to the promoter region (Hanigan et al. 2008). The loss of functional HDAC2 in mutant cell lines was proposed to lead to elevated levels of *APAF1* but is compensated by upregulation of other survival pathways, which leads to observed resistance to hydroxamate-based HDAC inhibitors (Hanigan et al. 2008). Biomarkers of sensitivity and clinical response to HDAC inhibitor treatment in addition to providing clues for possible mechanism of drug resistance can also aid in the design of effective therapeutic regimens to maximize the benefits of HDAC inhibitors in cancer therapy.

4.5.5 *Redefining HDAC Inhibitors Through Combination Therapy*

With the successful demonstration of HDAC inhibitors as monotherapies for hematological malignancies, combination of these agents together with a wide variety of cancer chemotherapeutic agents were introduced in both preclinical and clinical trials to improve existing therapies and possibly curb clinical resistance. Combination therapy with hydroxamate-based HDAC inhibitors and cytosine-based DNMT inhibitors blocks two key epigenetic events associated with long-term gene silencing. In the presence of methylated promoter regions, trichostatin A was unable to upregulate the expression of key cell cycle controllers, *p21* and *p15*, while 5-azadeoxycytidine gave limited reactivation of these genes (Cameron et al. 1999). Combination of trichostatin A and 5-azadeoxycytidine showed significant upregulation of *p21* and *p15* (Cameron et al. 1999). The induction of pro-apoptotic genes and modulation of pro-survival genes by HDAC inhibitors also sensitizes cancer cells to traditional chemotherapeutic agents such as microtubule disrupting agents and topoisomerase inhibitors. In combination with traditional chemotherapeutic agents, HDAC inhibitors lower the apoptotic threshold and render transformed cells susceptible to programmed cell death causing an additive or synergistic effect

(Richon et al. 2009; Kano et al. 2006). Gene expression changes induced by HDAC inhibitors led to the hypothesis that death receptor ligands such as TRAIL could potentiate the effect of HDAC inhibitors and increase progression of the extrinsic apoptotic pathway (Frew et al. 2008).

Protein degradation is another major theme in combination therapy with HDAC inhibitors, exploiting non-histone protein targets of these agents such as α-tubulin and HSP90 (Bolden et al. 2006). Hyperacetylation of α-tubulin and HSP90 prevents aggresome processing and client protein association, respectively, with downstream cellular effects that include increased cellular stress and apoptosis due to elevated levels of misfolded proteins (Kawaguchi et al. 2003; Aoyagi and Archer 2005; Scroggins et al. 2007). Oncogenic proteins ERBB2 and BCR-ABL can evade degradation by interacting with HSP90, however, the protective effect of HSP90 is attenuated by hyperacetylation. HDAC inhibitors together with HSP90-targeting agents (17-allyl-amino-demethoxy geldanamycin) or the corresponding inhibitors of client proteins (trastuzumab or imatinib) have been used in combination regimes (Richon et al. 2009; Kano et al. 2006; Rahmani et al. 2003; George et al. 2005). Accumulation of ubiquitinated and misfolded proteins is attenuated by proteasome-mediated protein degradation and is facilitated by autophagy of perinuclear aggresomes in the absence of a functional proteasome. Central to the aggresome pathway is α-tubulin, which facilitates transport of aggresomes to lysosomes in a dynein-dependent manner (Kawaguchi et al. 2003; Rodriguez-Gonzalez et al. 2008). HDAC6 has been recognized as a key component of the aggresome pathway, where it binds misfolded proteins and dynein motors and associates with the microtubule network to facilitate protein transport (Kawaguchi et al. 2003; Rodriguez-Gonzalez et al. 2008). Disruption of aggresome formation by inhibiting HDAC6, and without a functional proteasome, leads to increased ER stress and apoptosis (Kawaguchi et al. 2003; Rodriguez-Gonzalez et al. 2008). Combination regimens of the proteasome inhibitor bortezomib together with HDAC inhibitors romidepsin, vorinostat, panabinostat, and belinostat are in clinical trials for multiple myeloma therapy (Nawrocki et al. 2006).

Most combination regimens were designed based on observed effects of HDAC inhibitors on downregulation of pro-survival genes, upregulation of pro-apoptotic genes, and pleiotropic effects on non-histone proteins. While initial combination regimens have proved to be successful, systematic and unbiased identification can be facilitated by employing small interfering RNA (siRNA) screens (Whitehurst et al. 2007; Bartz et al. 2006). Loss of gene function in mammalian tumor cells can be related to increased cytotoxicity with otherwise sublethal concentration of the inhibitor. Hence, chemical agents that phenocopy the cellular effect of siRNA knockdowns would potentially act in synergy or additive with the HDAC inhibitor. Such chemical-genetic interaction screening has been demonstrated for several antitumor agents, and screening hits were found to be unique depending on the inhibitor's mechanism of action (Bartz et al. 2006).

4.6 New Natural Modulators of Epigenetic Enzymes

4.6.1 Histone Acetyltransferase Modulators

Plant-derived polyisoprenylated benzophenones make up the majority of documented HAT-modulating natural products. This class of compounds has been reported to have cytotoxic, antioxidant, and antimicrobial properties. Garcinol, isogarcinol, and guttiferones A and E (Fig. 4.12) inhibit HAT activity of p300/CBP, while nemorosone activates p300 HAT, despite the closely related structures among these congeners (Balasubramanyam et al. 2004; Dal Piaz et al. 2010). These natural products are characterized by bicyclic nonane rings that are highly oxygenated and prenylated. Inhibition occurs in both competitive and noncompetitive fashion, with the benzyl substituent occupying the acetyl-CoA binding site, while the prenyl moieties bind to an allosteric site of HAT (Arif et al. 2009). Treatment of HeLa cells with 100 μM of garcinol upregulated the expression of ~1,600 genes while ~600 genes were downregulated; some of the presumably relevant hits included caspase 4, BCL2, and E3 ubiquitin ligase (Balasubramanyam et al. 2004).

4.6.2 SIRT Modulators

In contrast to the well-established role of canonical HDACs in tumor progression, variable effects are observed for different SIRT isoforms. In addition, SIRTs are upregulated and promote cell survival and proliferation during caloric restriction, hence, selective activators of SIRTs that can phenocopy these effects are warranted therapeutics in age-related diseases and life span extension. Known activators of SIRT1 are mostly polyphenolic plant natural products (Fig. 4.12) such as resveratrol, fisetin, and butein that were identified from screening of libraries for SIRT1 modulators (Howitz et al. 2003). These compounds caused a 6–13-fold increase in SIRT1 activity at 100 μM, and SAR studies indicated the importance of a *trans*-stilbene backbone with two hydroxy groups at *meta* position to each other on ring A and a 3′,4′ or 4′ hydroxylation pattern on ring B for SIRT activity (Howitz et al. 2003). These findings on the activation of SIRT1 by resveratrol have been challenged by several independent studies, which suggest the indirect role of resveratrol in SIRT1 activation via resveratrol's downstream cellular effects, instead of direct SIRT1–resveratrol interaction. Incubation of SIRT1 with a nonfluorogenic peptide substrate and resveratrol failed to increase enzyme activity, compared to the observed activation using fluorogenic peptide substrates (Borra et al. 2005; Beher et al. 2009; Pacholec et al. 2010). In contrast to the conflicting kinetic assay results for resveratrol, several in vivo assays showed lifespan extension of budding yeast, *Caenorhabditis elegans*, and *Drosophila* using high resveratrol concentrations (Howitz et al. 2003; Wood et al. 2004; Baur et al. 2006). However, at pharmacologically relevant concentrations of resveratrol that can be achieved through supplementation, ~4% lifespan extension was observed in *C. elegans* (Zarse et al. 2010).

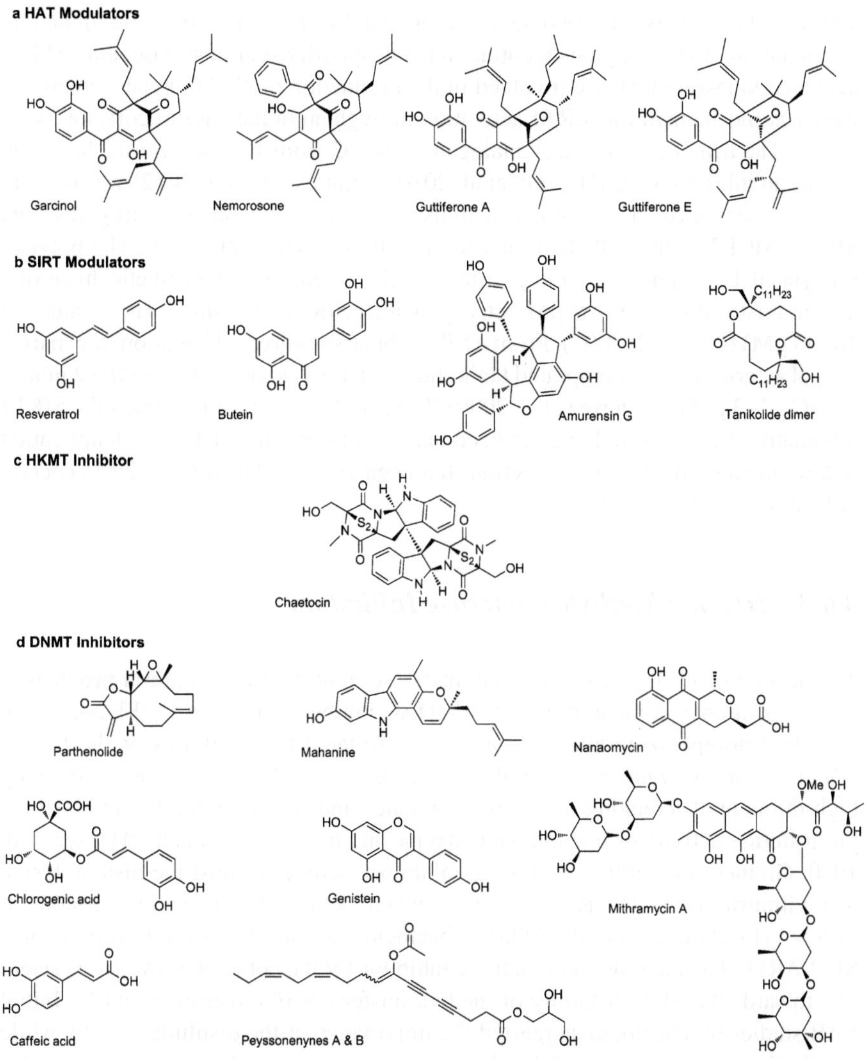

Fig. 4.12 Selected natural modulators of epigenetic enzymes other than canonical HDACs. (**a**) Isoprenylated benzophenones constitute the majority of HAT modulators, with the prenyl and benzyl moieties making critical enzyme interactions. (**b**) Lifespan extension derived from selective SIRT1 activation has been one of the main aims in the discovery of small molecules which can phenocopy this effect, such as polyphenols. (**c**) The fungus-derived chaetocin has been the sole natural product described to selectively inhibit G9a histone lysine methyltransferase. (**d**) Irreversible DNMT inhibitors are oftentimes characterized by an α,β-unsaturated carbonyl moiety that can form a covalent bond with the catalytic Cys residue of DNMT. Polyphenols can also act as non-competitive DNMT inhibitor by depleting the pool of available SAM cofactor molecules that is required for the reaction, and instead yielding SAH which creates a feedback loop in enzyme inhibition

The cellular effects of polyphenolics on SIRT1 activation may be linked to upregulation of key enzymes nicotinamide phosphoribosyl transferase and (AMP)-activated kinases which are involved in the production of NAD^+, hence, creating a feedback mechanism for SIRT1 activation through an in increase in cofactor availability. The trimeric resveratrol analog amuresin G from *Vitis amurensis*, however, acts as an inhibitor of SIRT1 (Oh et al. 2010). Amuresin G (Fig. 4.12) was demonstrated to enhance the cytotoxic activity of doxorubicin on multidrug resistant MCF7 (MCF7/ADR) cells both in vitro and in vivo (Oh et al. 2010). The polyisoprenylated benzophenones (+)-guttiferone G, hyperofin and synthetic hyperofin analog aristorofin have low μM activity against SIRT1 and SIRT2 (IC_{50} values of 10–30 μM) (Gey et al. 2007). From a SIRT2 bioassay-directed isolation, the marine cyanobacteria-derived compound tanikolide dimer was identified as a SIRT2 inhibitor (IC_{50} 2–3 μM) (Gutierrez et al. 2009). The cellular effects of non-phenolic SIRT1 modulators from Nature have not been extensively documented, and identification of key enzyme–modulator interactions has been precluded by difficulties in cocrystallization of these enzymes.

4.6.3 Histone Methyltransferase Inhibitors

The development of histone methyltransferase inhibitors from natural products is still in its infancy compared to that of HDAC inhibitors. Chaetocin (Fig. 4.12), an epidithioketopiperazine alkaloid, previously isolated from cultures of the fungus *Chaetomium minutum*, was identified as a histone methyltransferase inhibitor by high-throughput screening of 3,000 compounds and demonstrated to inhibit methyltransferase activity of recombinant enzyme and in *Drosophila* cells (Hauser et al. 1970; Greiner et al. 2005). It showed inhibitory activity against the histone lysine methyltransferase SU(VAR)3–9 and G9a, which both methylate the Lys9 residue of histone H3 (Greiner et al. 2005). Chaetocin has an IC_{50} of 0.6 μM against SU(VAR)3–9 and acts as a competitive inhibitor for the cofactor SAM and is selective towards the SUV39 family of methyltransferases (Greiner et al. 2005). Initial SAR studies on chaetocin suggested the importance of the disulfide moiety, while the absolute configurations of the chiral centers are apparently not determinants for methyltransferase inhibition (Iwasa et al. 2010). Chaetocin has been previously documented to be an antifungal and antimyeloma agent (Isham et al. 2007). In an earlier study, it was demonstrated to induce apoptosis via a caspase 8-mediated pathway in vitro and showed in vivo activity in a myeloma mouse xenograft (Isham et al. 2007; Teng et al. 2010). The antimyeloma activity of chaetocin was linked to oxidative stress; chaetocin was demonstrated to compete with thioredoxin as a substrate for the oxidative stress remediation enzyme thioredoxin reductase 1 (Tibodeau et al. 2009).

4.6.4 DNA Methyltransferase Inhibitors

Natural products that inhibit DNMTs (Fig. 4.12) can be classified into: (1) covalent inhibitors that form an adduct with the enzyme (parthenolide, nanaomycin) (Liu et al. 2009; Kuck et al. 2010); (2) competitive inhibitor (mithramycin) (Lina et al. 2007); and (3) noncompetitive indirect inhibitors (caffeic acid, chlorogeneic acid, quercetin, fisetin, cathecin, epicathecin) (Lee et al. 2005; Lee and Zhu 2006). In addition, psammaplins A and G, bisaprasin (Pina et al. 2003), mahanine (Jagadeesh et al. 2007), and peyssonenynes A and B (McPhail et al. 2004), have been documented to inhibit DNMT1 using direct enzyme assays; however, their mechanism of enzyme inhibition has not been fully elucidated and requires further study. Aside from the noncompetitive DNMT inhibitors, the majority of these compounds appear to have cytotoxic effects. Parthenolide was demonstrated to reduce tumor size in MV4–11 engrafted nude mice, and tumor sections showed downregulation of DNMT1 (Liu et al. 2009). The covalent inhibitors parthenolide and nanaomycin both possess an α,β-unsaturated carbonyl moiety that can undergo a 1,4-conjugated addition with the catalytic Cys651 of DNMT1 and DNMT3B, respectively, based on molecular docking experiments (Liu et al. 2009; Kuck et al. 2010). Synthesis of a parthenolide analog lacking this moiety was shown to be inactive (Liu et al. 2009). Nanaomycin, unlike parthenolide, does not inhibit DNMT1 but rather DNMT3B, presumably due to additional hydrogen bonding interactions with Glu697, Arg731, and Arg733 (Kuck et al. 2010). The DNMT-inhibitory activity of polyphenols is related to their ability to undergo extensive O-methylation that is mediated by catechol-*O*-methyltransferase and requires the same cofactor, SAM, as DNMTs (Lee et al. 2005). This reaction then depletes the pool of available SAM and converts SAM to *S*-adenosylhomocystine (SAH), a documented potent inhibitor of DNMT, inducing a feedback mechanism for attenuating DNA methylation (Lee et al. 2005; Lee and Zhu 2006).

4.7 Conclusion

HDAC inhibitors from Nature show diversity in both structure and mechanism of enzyme inhibition. While there have been challenges in their development, these compounds yielded new structural templates and potent anticancer agents. As in other therapeutic arenas and/or for other biological targets, natural products science is also at the forefront of discovery of new modulators of DNA and histone methyltransferases and HATs. It has been established that secondary metabolites can modulate the activity of several of these epigenetic enzymes which are currently being validated as drug targets.

Acknowledgments Natural products research in the authors' laboratory is supported by the National Institutes of Health, NIGMS grant P41GM806210 and NCI grants R01CA138544 and R21CA133681, and the Bankhead-Coley Cancer Research Program, grant 1BG07.

References

Abe M, Harpel JG, Metz CN, Nunes I, Loskutoff DJ, Rikfin DB (1994) An assay for transforming growth factor-beta using cells transfected with plasminogen activator inhibitor-1 promoter-luciferase construct. Anal Biochem 216:276–284

Altman SA, Randers L, Rao G (1999) Comparison of trypan blue dye exclusion and fluorometric assays for mammalian cell viability determinations. Biotechnol Prog 9:871–874

Ambudkar SV, Kimchi-Sarfaty C, Sauna ZE, Gottesman MM (2003) P-glycoprotein: from genomics to mechanism. Oncogene 22:7468–7485

Anand R, Marmorstein R (2007) Structure and mechanism of lysine-specific demethylase enzymes. J Biol Chem 282:35425–35429

Aoyagi S, Archer TK (2005) Modulating molecular chaperone Hsp90 functions through reversible acetylation. Trends Cell Biol 15:565–567

Arif M, Pradhan SK, Thanuja GK et al (2009) Mechanism of p300 specific histone acetyltransferase inhibition by small molecules. J Med Chem 52:267–277

Balasubramanian S, Ramos J, Luo W, Sirisawad M, Verner E, Buggy JJ (2008) A novel histone deacetylase 8 (HDAC8)-specific inhibitor PCI-34051 induces apoptosis in T-cell lymphomas. Leukemia 22:1026–1034

Balasubramanyam K, Altaf M, Varier RA et al (2004) Polyisoprenylated benzophenone, garcinol, a natural histone acetyltransferase inhibitor, represses chromatin transcription and alters global gene expression. J Biol Chem 279:33716–33726

Balcerczyk A, Pirola L (2010) Therapeutic potential of activators and inhibitors of sirtuins. Biofactors 36:383–393

Barlev NA, Liu L, Chehab NH et al (2001) Acetylation of p53 activates transcription through recruitment of coactivators/histone acetyltransferases. Mol Cell 8:1243–1254

Bartz SR, Zhang Z, Burchard J et al (2006) Small interfering RNA screens reveal enhanced cisplatin cytotoxicity in tumor cells having both BRCA network and TP53 disruptions. Mol Cell Biol 26:9377–9386

Baur JA, Pearson KJ, Price NL et al (2006) Resveratrol improves health and survival of mice on a high-calorie diet. Nature 444:337–342

Beher D, Wu J, Cumine S et al (2009) Resveratrol is not a direct activator of SIRT1 enzyme activity. Chem Biol Drug Des 74:619–624

Berndsen CE, Albaugh BN, Tan S, Denu JM (2007) Catalytic mechanism of a MYST family histone acetyltransferase. Biochemistry 46:623–629

Bolden J, Peart MJ, Johnstone RW (2006) Anticancer activities of histone deacetylase inhibitors. Nat Rev Drug Discov 5:769–784

Borra MT, Smith BC, Denu JM (2005) Mechanism of human SIRT1 activation by resveratrol. J Biol Chem 280:17187–17195

Bottomley MJ, Lo Surdo P, Di Giovine P et al (2008) Structural and functional analysis of the human HDAC4 catalytic domain reveals a regulatory structural zinc-binding domain. J Biol Chem 283:26694–26704

Bowers AA, Greshock TJ, West N et al (2009) Synthesis and conformation-activity relationships of the peptide isosteres of FK228 and largazole. J Am Chem Soc 131:2900–2905

Breslow R, Jursic B, Yan ZF et al (1991) Potent cytodifferentiating agents related to hexamethylenebisacetamide. Proc Natl Acad Sci U S A 88:5542–5546

Bressi JC, Jennings AJ, Skene R et al (2010) Exploration of the HDAC2 foot pocket: synthesis and SAR of substituted N-(2-aminophenyl)benzamides. Bioorg Med Chem Lett 20:3142–3145

Cameron EE, Bachman KE, Myöhänen S, Herman JG, Baylin SB (1999) Synergy of demethylation and histone deacetylase inhibition in the re-expression of genes silenced in cancer. Nat Genet 21:103–107

Chen Y, Gambs C, Abe Y, Wentworth PJ, Janda KD (2003) Total synthesis of the depsipeptide FR-901375. J Org Chem 68:8902–8905

Chen L, Williams SA, Mu Y et al (2005) NF-κB RelA phosphorylation regulates RelA acetylation. Mol Cell Biol 25:7966–7975

Cheng X, Blumenthal RM (2008) Mammalian DNA methyltransferases: a structural perspective. Structure 16:341–350

Choudhary C, Kumar C, Gnad F et al (2009) Lysine acetylation targets protein complexes and co-regulates major cellular functions. Science 325:834–843

Clark GJ, Cox A, Graham SM, Der CJ (1995) Biological assays for Ras transformation. In: Balch WE, Der CJ, Hall A (eds) Methods in enzymology. Academic, New York

Closse A, Huguenin R (1974) Isolierung und Strukturaufklärung von Chlamydocin. Helv Chim Acta 57:533–544

Cole KE, Dowling DP, Boone MA, Phillips AJ, Christianson DW (2011) Structural basis of the antiproliferative activity of largazole, a depsipeptide inhibitor of the histone deacetylases. J Am Chem Soc 133:12474–12477

Culhane JC, Cole PA (2007) LSD1 and the chemistry of histone demethylation. Curr Opin Chem Biol 11:561–568

Dal Piaz F, Tosco A, Eletto D et al (2010) The identification of a novel natural activator of p300 histone acetyltransferase provides new insights into the modulation mechanism of this enzyme. ChemBioChem 11:818–827

De Ruijter AJM, Van Gennip AH, Caron HN, Kemp S, Van Kuilenburg ABP (2003) Histone deacetylases (HDACs): characterization of the classical HDAC family. Biochem J 370:737–749

Dedesa KJ, Dedesa I, Imescha P, von Buerenb AO, Finka D, Fediera A (2009) Acquired vorinostat resistance shows partial cross-resistance to 'second-generation' HDAC inhibitors and correlates with loss of histone acetylation and apoptosis but not with altered HDAC and HAT activities. Anticancer Drugs 20:321–333

Deng W, Wu KK (2003) Regulation of inducible nitric oxide synthase expression by p300 and p50 acetylation. J Immunol 171:6581–6588

Deng W, Zhu Y, Wu KK (2003) Up-regulation of p300 binding and p50 acetylation in tumor necrosis factor-a-induced cyclooxygenase-2 promoter activation. J Biol Chem 278:4770–4777

Di Lorenzo A, Bedford MT (2011) Histone arginine methylation. FEBS Lett 585:2024–2031

Di Micco S, Terracciano S, Bruno I et al (2008) Molecular modeling studies toward the structural optimization of new cyclopeptide-based HDAC inhibitors modeled on the natural product FR235222. Bioorg Med Chem 16:8635–8642

Dittenhafer-Reed KE, Feldman JL, Denu JM (2011) Catalysis and mechanistic insights into sirtuin activation. ChemBioChem 12:281–289

Elaut G, Torok G, Vinken M et al (2002) Major phase I biotransformation pathways of trichostatin A in rat hepatocytes and in rat and human liver microsomes. Drug Metab Dispos 30:1320–1328

Fantin VR, Loboda A, Paweletz CP et al (2008) Constitutive activation of signal transducers and activators of transcription predicts Vorinostat resistance in cutaneous T-cell lymphoma. Cancer Res 68:3785–3794

Fedier A, Dedes KJ, Imesch P, Von Bueren AO, Fink D (2007) The histone deacetylase inhibitors suberoylanilide hydroxamic (Vorinostat) and valproic acid induce irreversible and MDR1-independent resistance in human colon cancer cells. Int J Oncol 31:633–641

Ficner R (2009) Novel structural insights into Class I and II histone deacetylases. Curr Top Med Chem 9:235–240

Finnin MS, Donigian JR, Cohen A et al (1999) Structures of a histone deacetylase homologue bound to the TSA and SAHA inhibitors. Nature 401:188–193

Frew AJ, Lindemann RK, Martin BP et al (2008) Combination therapy of established cancer using a histone deacetylase inhibitor and a TRAIL receptor agonist. Proc Natl Acad Sci U S A 105:11317–11322

Fry DG, Hurlin PJ, Maher VM, McCormick JJ (1988) Transformation of diploid human fibroblasts by transfection with the v-*sis*, PDGF2/c-*sis*, or T24 H-*ras* genes. Mutat Res 199:341–351

Furumai R, Matsuyama A, Kobashi N et al (2002) FK228 (Depsipeptide) as a natural prodrug that inhibits Class I histone deacetylases. Cancer Res 62:4916–4921

George P, Bali P, Annavarapu S et al (2005) Combination of the histone deacetylase inhibitor LBH589 and the hsp90 inhibitor 17-AAG is highly active against human CML-BC cells and AML cells with activating mutation of FLT-3. Blood 105:1768–1776

Gey C, Kyrylenko S, Hennig L et al (2007) Phloroglucinol derivatives guttiferone G, aristoforin, and hyperforin: inhibitors of human sirtuins SIRT1 and SIRT2. Angew Chem Int Ed 46:5219–5222

Glaser KB, Staver MJ, Waring JF, Stender J, Ulrich RG, Davidsen SK (2003) Gene expression profiling of multiple histone deacetylase (HDAC) inhibitors: defining a common gene set produced by HDAC inhibition in T24 and MDA carcinoma cell lines. Mol Cancer Ther 2:151–163

Glozak MA, Sengupta N, Zhang X, Seto E (2005) Acetylation and deacetylation of non-histone proteins. Gene 363:15–23

Gorre ME, Mohammed M, Ellwood K et al (2001) Clinical resistance to STI-571 cancer therapy caused by BCR-ABL gene mutation or amplification. Science 293:876–880

Gregoretti IV, Lee Y, Goodson HV (2004) Molecular evolution of the histone deacetylase family: functional implications of phylogenetic analysis. J Mol Biol 338:17–31

Greiner D, Bonaldi T, Eskeland R, Roemer E, Imhof A (2005) Identification of a specific inhibitor of the histone methyltransferase SU(VAR)3-9. Nat Chem Biol 1:143–145

Gu W, Cueto M, Jensen PR, Fenical W, Silverman RB (2007) Microsporins A and B: new histone deacetylase inhibitors from the marine-derived fungus *Microsporum* cf. *gypseum* and the solid-phase synthesis of microsporin A. Tetrahedron 63:6535–6541

Gutierrez M, Andrianasolo EH, Shin WK et al (2009) Structural and synthetic investigations of tanikolide dimer, a SIRT2 selective inhibitor, and tanikolide seco-acid from the Madagascar marine cyanobacterium *Lyngbya majuscula*. J Org Chem 74:5267–5275

Haggarty SJ, Koeller KM, Wong JC, Grozinger CM, Schreiber SL (2003) Domain-selective small-molecule inhibitor of histone deacetylase 6 (HDAC6)-mediated tubulin deacetylation. Proc Natl Acad Sci U S A 100:4389–4394

Halkidou K, Gaughan L, Cook S, Leung HY, Neal DE, Robson CN (2004) Upregulation and nuclear recruitment of HDAC1 in hormone refractory prostate cancer. Prostate 59:177–189

Hanigan CL, Van Engeland M, De Bruine AP et al (2008) An inactivating mutation in HDAC2 leads to dysregulation of apoptosis mediated by *APAF1*. Gastroenterology 135:1654–1664

Hauser D, Weber HP, Sigg HP (1970) Isolation and structure elucidation of chaetocin. Helv Chim Acta 53:1061–1073

Hirota A, Suzuki A, Suzuki H, Tamura S (1973) Isolation and biological activity of Cyl-2, a metabolite of Cylindrocladium scoparium. Agric Biol Chem 37:643–647

Hodawadekar SC, Marmorstein R (2007) Chemistry of acetyl transfer by histone modifying enzymes: structure, mechanism and implications for effector design. Oncogene 26:5528–5540

Hong J, Luesch H (2012) Largazole: From discovery to broad-spectrum therapy. Nat Prod Rep 29: 449–456

Hoshino I, Hisahiro M, Akutsu Y et al (2007) Gene expression profiling induced by histone deacetylase inhibitor, FK228, in human esophageal squamous cancer cells. Oncol Rep 18:585–592

Howitz KT, Bitterman KJ, Cohen HY et al (2003) Small molecule activators of sirtuins extend *Saccharomyces cerevisiae* lifespan. Nature 425:191–196

Hubbert C, Guardiola A, Shao R et al (2002) HDAC6 is a microtubule-associated deacetylase. Nature 417:455–458

Isham CR, Tibodeau JD, Jin W, Xu R, Timm MM, Bible KC (2007) Chaetocin: a promising new antimyeloma agent with in vitro and in vivo activity mediated via imposition of oxidative stress. Blood 109:2579–2588

Itazaki H, Nagashima K, Sugita K et al (1990) Isolation and structural elucidation of new cyclotetrapeptides, trapoxins A and B, having detransformation activities as antitumor agents. J Antibiot 43:1524–1532

Iwasa E, Hamashima Y, Fujishiro S et al (2010) Total synthesis of (+)-chaetocin and its analogues: their histone methyltransferase G9a inhibitory activity. J Am Chem Soc 132:4078–4079

Jagadeesh S, Sinha S, Pal BC, Bhattacharya S, Banerjee PP (2007) Mahanine reverses an epigenetically silenced tumor suppressor gene RASSF1A in human prostate cancer cells. Biochem Biophys Res Commun 362:212–217

Jones PA, Baylin SB (2002) The fundamental role of epigenetic events in cancer. Nat Rev Genet 3:415–428

Jones PA, Baylin SB (2007) The epigenomics of cancer. Cell 128:683–692

Joubel A, Chalkley RJ, Medzihradszky KF, Hondermarck H, Burlingame AL (2009) Identification of new p53 acetylation sites in COS-1 cells. Mol Cell Proteomics 8:1167–1173

Jurkowska RZ, Jurkowski TP, Jeltsch A (2011) Structure and function of mammalian DNA methyltransferases. ChemBioChem 12:206–222

Kano Y, Akutsu M, Tsunoda S et al (2006) Cytotoxic effects of histone deacetylase inhibitor FK228 (depsipeptide, formally named FR901228) in combination with conventional anti-leukemia/lymphoma agents against human leukemia/lymphoma cell lines. Invest New Drugs 25:31–40

Kawaguchi Y, Kovacs JJ, McLaurin A, Vance JM, Ito A, Yao T (2003) The deacetylase HDAC6 regulates aggresome formation and cell viability in response to misfolded protein stress. Cell 115:727–738

Kiernan R, Bres V, Ng RWM et al (2003) Post-activation turn-off of NF-κB-dependent transcription is regulated by acetylation of p65. J Biol Chem 278:2758–2766

Kijima M, Yoshida M, Sugita K, Horinouchi S, Beppu T (1993) Trapoxin, an antitumor cyclic tetrapeptide, is an irreversible inhibitor of mammalian histone deacetylase. J Biol Chem 268:22429–22435

Kim DH, Shin J, Kwon HJ (2007) Psammaplin A is a natural prodrug that inhibits class I histone deacetylase. Exp Mol Med 39:47–55

Komatsu Y, Tomizaki K, Tsukamoto M et al (2001) Cyclic hydroxamic-acid-containing peptide 31, a potent synthetic histone deacetylase inhibitor with antitumor activity. Cancer Res 61:4459–4466

Kornberg RD, Lorch Y (1999) Twenty-five years of the nucleosome, fundamental particle of the eukaryote chromosome. Cell 98:285–294

Kouzarides T (2007) Chromatin modifications and their function. Cell 128:693–705

Kuck D, Caulfield T, Lyko F, Medina-Franco JL (2010) Nanaomycin A selectively inhibits DNMT3B and reactivates silenced tumor suppressor genes in human cancer cells. Mol Cancer Ther 9:3015–3023

Kwan JC, Luesch H (2010) Weapons in disguise—activating mechanisms and protecting group chemistry in Nature. Chem Eur J 16:13020–13029

Kwon T, Chang JH, Kwak E et al (2003) Mechanism of histone lysine methyl transfer revealed by the structure of SET7/9-AdoMet. EMBO J 22:292–303

Ledizet M, Piperno G (1987) Identification of an acetylation site of Chlamydomonas α-tubulin. Proc Natl Acad Sci U S A 84:5720–5724

Lee WJ, Zhu BT (2006) Inhibition of DNA methylation by caffeic acid and chlorogenic acid, two common catechol-containing coffee polyphenols. Carcinogenesis 279:269–277

Lee WJ, Shim J, Zhu BT (2005) Mechanisms for the inhibition of DNA methyltransferases by tea catechins and bioflavonoids. Mol Pharmacol 68:1018–1030

Lee J, Smith E, Shilatifard A (2010) The language of histone crosstalk. Cell 142:682–685

Li B, Carey M, Workman JL (2007) The role of chromatin during transcription. Cell 128:707–719

Liang G, Lin JCY, Wei V et al (2004) Distinct localization of histone H3 acetylation and H3-K4 methylation to the transcription start sites in the human genome. Proc Natl Acad Sci U S A 101:7357–7362

Lin Y, Fletcher CM, Zhou J, Allis CD, Wagner G (1999) Solution structure of the catalytic domain of GCN5 histone acetyltransferase bound to coenzyme A. Nature 400:86–89

Lina R, Hsu C, Wang Y (2007) Mithramycin A inhibits DNA methyltransferase and metastasis potential of lung cancer cells. Anticancer Drugs 18:1157–1164

Liu X, Wang L, Zhao K et al (2008) The structural basis of protein acetylation by the p300/CBP transcriptional coactivator. Nature 451:846–850
Liu Z, Liu S, Xie Z et al (2009) Modulation of DNA methylation by a sesquiterpene lactone parthenolide. J Pharmacol Exp Ther 329:505–514
Liu Y, Salvador LA, Byeon S et al (2010) Anticolon cancer activity of largazole, a marine-derived tunable histone deacetylase inhibitor. J Pharmacol Exp Ther 335:351–361
Luger K, Mader AW, Richmond RK, Sargent DF, Richmond TJ (1997) Crystal structure of the nucleosome core particle at 2.8 Å resolution. Nature 389:251–260
Marks PA, Breslow R (2007) Dimethyl sulfoxide to vorinostat: development of this histone deacetylase inhibitor as an anticancer drug. Nat Biotechnol 25:84–90
Maruta H, Greet K, Rosenbaum JL (1986) The acetylation of alpha-tubulin and its relationship to the assembly and disassembly of microtubules. J Cell Biol 103:571–579
Masuoka Y, Nagai A, Shin-ya K et al (2001) Spiruchostatins A and B, novel gene expression-enhancing substances produced by *Pseudomonas* sp. Tetrahedron Lett 42:41–44
Masuoka Y, Shindoh N, Inamura N (2008) Histone deacetylase inhibitors from microorganisms: the Astellas experience. In: Amstutz R, Petersen F (eds) Progress in drug research. Birkhauser Verlag, Basel
Maulucci N, Chini MG, Di Micco S et al (2007) Molecular insights into azumamide E histone deacetylases inhibitory activity. J Am Chem Soc 129:3007–3012
McCulloch MWB, Coombs GS, Banerjee N et al (2009) Psammaplin A as a general activator of cell-based signaling assays via HDAC inhibition and studies on some bromotyrosine derivatives. Bioorg Med Chem 17:2189–2198
McPhail KL, France D, Cornell-Kennon S, Gerwick WH (2004) Peyssonenynes A and B, novel enediyne oxylipins with DNA methyltransferase inhibitory activity from the red marine alga *Peyssonnelia caulifera*. J Nat Prod 67:1010–1013
Min J, Feng Q, Li Z, Zhang Y, Xu R (2003) Structure of the catalytic domain of human DOT1L, a non-SET domain nucleosomal histone methyltransferase. Cell 112:711–723
Mori H, Urano Y, Abe F et al (2003a) FR235222, a fungal metabolite, is a novel immunosuppressant that inhibits mammalian histone deacetylase (HDAC). I. Taxonomy, fermentation, isolation and biological activities. J Antibiot 56:72–79
Mori H, Abe F, Furukawa S et al (2003b) FR235222, a fungal metabolite, is a novel immunosuppressant that inhibits mammalian histone deacetylase (HDAC). II. Biological activities in animal models. J Antibiot 56:80–86
Mori H, Urano Y, Kinoshita T, Yoshimura S, Takase S, Hino M (2003c) FR235222, a fungal metabolite, is a novel immunosuppressant that inhibits mammalian histone deacetylase. III. Structure determination. J Antibiot 56:181–185
Mosammaparast N, Shi Y (2010) Reversal of histone methylation: biochemical and molecular mechanisms of histone demethylases. Annu Rev Biochem 79:155–179
Mosmann T (1983) Rapid colorimetric assay for cellular growth and survival: application to proliferation and cytotoxicity assays. J Immunol Methods 65:55–63
Nakagawa M, Oda Y, Eguchi T et al (2007) Expression profile of class I histone deacetylases in human cancer tissues. Oncol Rep 18:769–774
Nakajima H, Kim YB, Terano H, Yoshida M, Horinouchi S (1998) FR901228, a potent antitumor antibiotic, is a novel histone deacetylase inhibitor. Exp Cell Res 241:126–133
Nakao Y, Yoshida S, Matsunaga S et al (2006) Azumamides A–E: histone deacetylase inhibitory cyclic tetrapeptides from the marine sponge *Mycale izuensis*. Angew Chem Int Ed 45:7553–7557
Nawrocki ST, Carew JS, Pino MS et al (2006) Aggresome disruption: a novel strategy to enhance Bortezomib-induced apoptosis in pancreatic cancer cells. Cancer Res 66:3773–3781
Newman DJ, Cragg GM (2012) Natural products as sources of new drugs over the 30 years from 1981 to 2010. J. Nat Prod 75:311–335
Nie L, Ueki M, Kakeya H, Osada H (2001) A facile and effective screening method for $p21^{WAF1}$ promoter activators from microbial metabolites. J Antibiot 54:783–788
Noh JH, Song JH, Eun JW et al (2009) Systemic cell-cycle suppression by apicidin, a histone deacetylase inhibitor, in MDA-MB-435 cells. Int J Mol Med 24:205–226

O'Brien J, Wilson I, Orton T, Pognan F (2000) Investigation of the Alamar Blue (resazurin) fluorescent dye for the assessment of mammalian cell cytotoxicity. Eur J Biochem 267:5421–5426

Oh WK, Cho KB, Hien TT et al (2010) Amurensin G, a potent natural SIRT1 inhibitor, rescues doxorubicin responsiveness via down-regulation of multidrug resistance. Mol Pharmacol 78:855–864

Oliver SS, Denu JM (2011) Dynamic interplay between histone H3 modifications and protein interpreters: emerging evidence for a "histone language". ChemBioChem 12:299–307

Pacholec M, Bleasdale JE, Chrunyk B et al (2010) SRT1720, SRT2183, SRT1460, and resveratrol are not direct activators of SIRT1. J Biol Chem 285:8340–8345

Papazisis KT, Geromichalos GD, Dimitriadis KA, Kortsaris AH (1997) Optimization of the sulforhodamine B colorimetric assay. J Immunol Methods 208:151–158

Pina IC, Gautschi JT, Wang G et al (2003) Psammaplins from the sponge *Pseudoceratina purpurea*: inhibition of both histone deacetylase and DNA methyltransferase. J Org Chem 68:3866–3873

Piperno G, LeDizet M, Chang X (1987) Microtubules containing acetylated α-tubulin in mammalian cells in culture. J Cell Biol 104:289–302

Quivy V, Van Lint C (2004) Regulation at multiple levels of NF-κB-mediated transactivation by protein acetylation. Biochem Pharmacol 68:1221–1229

Rahmani M, Yu C, Dai Y et al (2003) Coadministration of the heat shock protein 90 antagonist 17-allylamino-17-demethoxygeldanamycin with suberoylanilide hydroxamic acid or sodium butyrate synergistically induces apoptosis in human leukemia cells. Cancer Res 63:8420–8427

Reuben RC, Wife RL, Breslow R, Rifkind RA, Marks PA (1976) A new group of potent inducers of differentiation in murine erythroleukemia cells. Proc Natl Acad Sci U S A 73:862–866

Richon VM, Webb Y, Merger R et al (1996) Second generation hybrid polar compounds are potent inducers of transformed cell differentiation. Proc Natl Acad Sci U S A 93:5705–5708

Richon VM, Garcia-Vargas J, Hardwick JS (2009) Development of vorinostat: current applications and future perspectives for cancer therapy. Cancer Lett 280:201–210

Rodriguez-Gonzalez A, Lin T, Ikeda AK, Simms-Waldrip T, Fu C, Sakamoto KM (2008) Role of the aggresome pathway in cancer: targeting histone deacetylase 6-dependent protein degradation. Cancer Res 68:2557–2560

Ropero S, Fraga MF, Ballestar E et al (2006) A truncating mutation of HDAC2 in human cancers confers resistance to histone deacetylase inhibition. Nat Genet 38:566–569

Ropero S, Ballestar E, Alaminos M, Arango D, Schwartz S Jr, Esteller M (2008) Transforming pathways unleashed by a HDAC2 mutation in human cancer. Oncogene 27:4008–4012

Sanders BD, Jackson B, Marmorstein R (2010) Structural basis for sirtuin function: what we know and what we don't. Biochim Biophys Acta 1804:1604–1616

Sanderson L, Taylor GW, Aboagye EO et al (2004) Plasma pharmacokinetics and metabolism of the histone deacetylase inhibitor trichostatin A after intraperitoneal administration to mice. Drug Metab Dispos 32:1132–1138

Schuetz A, Min J, Allali-Hassani A et al (2008) Human HDAC7 harbors a class IIa histone deacetylase-specific zinc binding motif and cryptic deacetylase activity. J Biol Chem 283:11355–11363

Scroggins BT, Robzyk K, Wang D et al (2007) An acetylation site in the middle domain of Hsp90 regulates chaperone function. Mol Cell 25:151–159

Shah MH, Binkley P, Chan K et al (2006) Cardiotoxicity of histone deacetylase inhibitor depsipeptide in patients with metastatic neuroendocrine tumors. Clin Cancer Res 12:3997–4003

Shao W, Growney JD, Feng Y et al (2010) Activity of deacetylase inhibitor panobinostat (LBH589) in cutaneous T-cell lymphoma models: defining molecular mechanisms of resistance. Int J Cancer 127:2199–2208

Sharma S, Kelly TK, Jones PA (2010) Epigenetics in cancer. Carcinogenesis 31:27–36

Shigematsu N, Ueda H, Takase S, Tanaka H (1994) FR901228, a novel antitumor bicyclic depsipeptide produced by *Chromobacterium violaceum* No. 968. II. Structure determination. J Antibiot 47:311–314

Shimada J, Kwon HJ, Sawamura M, Schreiber S (1995) Synthesis and cellular characterization of the detransformation agent, (–)-depudecin. Chem Biol 2:517–525
Singh SB, Zink DL, Polishook JD et al (1996) Apicidins: novel cyclic tetrapeptides as coccidiostats and antimalarial agents from *Fusarium pallidoroseum*. Tetrahedron Lett 37:8077–8080
Singh SB, Zink DL, Liesch JM et al (2001) Structure, histone deacetylase, and antiprotozoal activities of apicidins B and C, congeners of apicidin with proline and valine substitutions. Org Lett 3:2815–2818
Sippl W, Jung M (2009) DNA methyltransferase inhibitors. Methods Princ Med Chem 42:163–183
Smith BC, Denu JM (2009) Chemical mechanisms of histone lysine and arginine modifications. Biochim Biophys Acta 1789:45–57
Somoza JR, Skene RJ, Katz BA et al (2004) Structural snapshots of human HDAC8 provide insights into the Class I histone deacetylases. Structure 12:1325–1334
Song J, Noh JH, Lee JH et al (2005) Increased expression of histone deacetylase 2 is found in human gastric cancer. APMIS 113:264–268
Spannhoff A, Hauser A, Heinke R, Sippl W, Jung M (2009) The emerging therapeutic potential of histone methyltransferase and demethylase inhibitors. ChemMedChem 42:1568–1582
Stähelin H, Trippmacher A (1974) Cytostatic activity of chlamydocin, a rapidly inactivated cyclic tetrapeptide. Eur J Cancer 103:801–808
Strober W (2001) Trypan blue exclusion test of cell viability. Curr Protoc Immunol Appendix 3:A.3B.1–A.3B.2
Tanaka M, Levy J, Terada M, Breslow R, Rifkind RA, Marks PA (1975) Induction of erythroid differentiation in murine virus infected erythroleukemia cells by highly polar compounds. Proc Natl Acad Sci U S A 72:1003–1006
Tang Y, Zhao W, Chen Y, Zhao Y, Gu W (2008) Acetylation is indispensable for p53 activation. Cell 133:612–626
Tanner KG, Trievel RC, Kuoi M et al (1999) Catalytic mechanism and function of invariant glutamic acid 173 from the histone acetyltransferase GCN5 transcriptional coactivator. J Biol Chem 274:18157–18160
Taori K, Paul VJ, Luesch H (2008) Structure and activity of largazole, a potent antiproliferative agent from the Floridian marine cyanobacterium *Symploca* sp. J Am Chem Soc 130:1806–1807
Taunton J, Collins JL, Schreiber SL (1996a) Synthesis of natural and modified trapoxins, useful reagents for exploring histone deacetylase function. J Am Chem Soc 118:10412–10422
Taunton J, Hassig CA, Schreiber SL (1996b) A mammalian histone deacetylase related to the yeast transcriptional regulator Rpd3p. Science 272:408–411
Teng Y, Iuchi K, Iwasa E et al (2010) Unnatural enantiomer of chaetocin shows strong apoptosis-inducing activity through caspase-8/caspase-3 activation. Bioorg Med Chem Lett 20:5085–5088
Tessier P, Smil DV, Wahhab A et al (2009) Diphenylmethylene hydroxamic acids as selective class IIa histone deacetylase inhibitors. Bioorg Med Chem Lett 19:5684–5688
Thaler F, Minucci S (2011) Next generation histone deacetylase inhibitors: the answer to the search for optimized epigenetic therapies? Expert Opin Drug Discov 6:393–404
Tibodeau JD, Benson LM, Isham CR, Owen WG, Bible KC (2009) The anticancer agent chaetocin is a competitive substrate and inhibitor of thioredoxin reductase. Antioxid Redox Signal 11:1097–1106
Ueda H, Nakajima H, Hori Y et al (1994a) FR901228, a novel antitumor bicyclic depsipeptide produced by *Chromobacterium violaceum* No. 968. I. Taxonomy, fermentation, isolation, physico-chemical and biological properties, and antitumor activity. J Antibiot 47:301–310
Ueda H, Manda T, Matsumoto S et al (1994b) FR901228, a novel antitumor bicyclic depsipeptide produced by *Chromobacterium violaceum* No. 968 III. Antitumor activities on experimental tumors in mice. J Antibiot 47:315–324
Uekia M, Teruya T, Nie L, Usamib R, Yoshida M, Osada H (2001) New trichostatin derivative, trichostatin RK, from *Streptomyces* sp. RK98-A74. J Antibiot 54:1093–1095

Umehara K, Nakahara K, Kiyoto S et al (1983) Studies on WF-3161, a new antitumor antibiotic. J Antibiot 36:478–483

Vannini A, Volpari C, Filocamo G et al (2004) Crystal structure of a eukaryotic zinc-dependent histone deacetylase, human HDAC8, complexed with a hydroxamic acid inhibitor. Proc Natl Acad Sci U S A 101:15064–15069

Voigt W (2005) Sulforhodamine B assay and chemosensitivity. Methods Mol Med 110:39–48

Wang Y, Tsay Y, Chin-Ming Tan B, Lo W, Lee S (2003) Identification and characterization of a novel p300-mediated p53 acetylation site, lysine 305. J Biol Chem 278:25568–25576

Wegener D, Wirsching F, Riester D, Schwienhorst A (2003) A fluorogenic histone deacetylase assay well suited for high-throughput activity screening. Chem Biol 10:61–68

Whitehurst AW, Bodemann BO, Cardenas J et al (2007) Synthetic lethal screen identification of chemosensitizer loci in cancer cells. Nature 446:815–819

Wilson AJ, Byun DS, Popova N et al (2006) Histone deacetylase 3 (HDAC3) and other class I HDACs regulate colon cell maturation and p21 expression and are deregulated in human colon cancer. J Biol Chem 281:13548–13558

Witt O, Deubzer HE, Oehme I (2009) HDAC family: what are the cancer relevant targets? Cancer Lett 277:8–21

Wood JG, Rogina B, Lavu S et al (2004) Sirtuin activators mimic caloric restriction and delay ageing in metazoans. Nature 430:686–689

Xiao B, Jing C, Wilson JR et al (2003) Structure and catalytic mechanism of the human histone methyltransferase SET7/9. Nature 421:652–656

Xiao JJ, Huang Y, Dai Z et al (2005) Chemoresistance to depsipeptide FK228 [(*E*)-(1*S*,4*S*,10*S*,21*R*)-7-[(*Z*)-Ethylidene]-4,21-diisopropyl-2-oxa-12,13-dithia-5,8,20,23-tetraazabicyclo[8,7,6]-tricos-16-ene-3,6,9,22-pentanone] is mediated by reversible MDR1 induction in human cancer cell lines. J Pharmacol Exp Ther 314:467–475

Yamaguchi H, Woods NT, Piluso LG et al (2009) p53 acetylation is crucial for its transcription-independent proapoptotic functions. J Biol Chem 284:11171–11183

Yang M, Culhane JC, Szewczuk LM et al (2007) Structural basis of histone demethylation by LSD1 revealed by suicide inactivation. Nat Struct Mol Biol 14:535–539

Yang Y, Rao R, Shen J et al (2008) Role of acetylation and extracellular location of heat shock protein 90A in tumor cell invasion. Cancer Res 68:4833–4842

Ying Y, Taori K, Kim H, Hong J, Luesch H (2008) Total synthesis and molecular target of largazole, a histone deacetylase inhibitor. J Am Chem Soc 130:8455–8459

Yoshida M, Kijima M, Akita M, Beppu T (1990) Potent and specific inhibition of mammalian histone deacetylase both *in vivo* and *in vitro* by trichostatin A. J Biol Chem 265:17174–17179

Young CW, Fanucchi MP, Walsh TD et al (1988) Phase I trial and clinical pharmacological evaluation of hexamethylene bisacetamide administration by ten-day continuous intravenous infusion at twenty-eight-day intervals. Cancer Res 48:7304–7309

Yuan Z, Guan Y, Chatterjee D, Chin YE (2005) Stat3 dimerization regulated by reversible acetylation of a single lysine residue. Science 307:269–273

Zarse K, Schmeisser S, Birringer M, Falk E, Schmoll D, Ristow M (2010) Differential effects of resveratrol and SRT1720 on lifespan of adult *Caenorhabditis elegans*. Horm Metab Res 42:837–839

Zhang X, Cheng X (2003) Structure of the predominant protein arginine methyltransferase PRMT1 and analysis of its binding to substrate peptides. Structure 11:509–520

Zhang X, Yang Z, Khan SI et al (2003a) Structural basis for the product specificity of histone lysine methyltransferases. Mol Cell 12:177–185

Zhang Y, Li N, Caron C et al (2003b) HDAC-6 interacts with and deacetylates tubulin and microtubules *in vivo*. EMBO J 22:1168–1179

Zhao Y, Lu S, Wu L et al (2006) Acetylation of p53 at lysine 373/382 by the histone deacetylase inhibitor depsipeptide induces expression of $p21^{Waf1/Cip1}$. Mol Cell Biol 26:2782–2790

Zhao S, Xu W, Wenqing J et al (2010) Regulation of cellular metabolism by protein lysine acetylation. Science 327:1000–1004

Chapter 5
Natural Product Cytotoxins as Payloads for Antibody Drug Conjugates

Frank E. Koehn

Abstract Natural products from plants and microorganisms are unsurpassed as a source of effective anticancer drugs. However it remains a major challenge to discover and develop natural product chemotherapeutics because of their dose-limiting toxicity. Antibody drug conjugates (ADC) offer a modality by which the cell-killing potency of a cytotoxin is coupled with the selectivity of a monoclonal antibody (mAb) for delivery to the tumor. An ADC consists of a mAb which binds a tumor-associated antigen, and to which one or more molecules of a potent cytotoxin payload are covalently bound. All clinically validated ADC payloads to date are directly derived from natural products. Natural product cytotoxins are well suited for the role of ADC payload for they often posses the potency, mechanism of action, chemical tractability, and structural properties required for effectiveness. This chapter will detail the functional role and requirements for effective ADC payloads. It then details current ADC payloads along with opportunities for new natural product payload candidates.

5.1 Natural Product Anticancer Drugs and Antibody Drug Conjugates

Natural products from plants and microorganisms are unsurpassed as a source of effective anticancer drugs. Indeed, over the past 30 years just under 50% of all cancer drugs are either natural products themselves, or are directly derived from them (Newman and Cragg 2012). Natural products structures represent a privileged class

F.E. Koehn (✉)
Natural Products, Oncology Worldwide Medicinal Chemistry, Pfizer World-Wide Research and Development, 445 Eastern Point Road, Groton, CT 06340, USA
e-mail: Frank.koehn@pfizer.com

F.E. Koehn (ed.), *Natural Products and Cancer Drug Discovery*, Cancer Drug Discovery and Development, DOI 10.1007/978-1-4614-4654-5_5,

of biologically selected chemical scaffolds which are biased to bind protein targets, and which can be optimized further for selectivity and drug-like properties through modification of structure. The past decade has provided numerous natural product anticancer agents which are either the naturally occurring compound itself, or a direct structural analog. These include the native natural products trabectedin (Cuevas and Francesch 2009), and romidepsin (Bertino and Otterson 2011), the semisynthetic rapamycin derivatives temsirolimus, everolimus, and zotarolimus (Reddy et al. 2006), the taxane analog carbazitaxel (Villanueva et al. 2011), the epothilone analog ixabepilone (Borzilleri and Vite 2009) and the fully synthetic eribulin (Yu et al. 2011). Natural products continue to be the well represented in cancer clinical trials (Butler 2010).

Despite this record of success, it remains a major challenge to effectively discover and develop natural product chemotherapeutics which have the desired levels of safety and efficacy, particularly for the treatment of solid tumors. One of the main reasons for this is the dose-limiting toxicity associated with many of these potential chemotherapeutics. This toxicity is due in great part to the narrow therapeutic index afforded by their nonselective mechanism of action. As a result, many effective cancer cell-killing agents cannot be dosed at concentrations high enough to eradicate the tumor (Chen et al. 2007). Function-blocking monoclonal antibodies (mAb), on the other hand, are highly specific, but have shown only minimal antitumor efficacy against solid carcinomas, predominantly because of their limited potency. Thus, these biologics are often used in combination therapy with small molecule cytotoxins, examples being the combination of cetuximab and irinotecan, or bevacizumab combined with 5-flourouracil for colorectal cancer (Pasetto et al. 2006).

Antibody drug conjugates (ADC) offer a modality by which the cell-killing potency of a “small molecule” cytotoxin is literally coupled to the exquisite selectively of a mAb as a means of delivery to the tumor (Carter and Senter 2008). An ADC consists of a mAb which binds a tumor-associated antigen (target), to which one or more molecules of a potent cytotoxin (payload) are covalently bound (conjugated) by means of a chemical linker (Fig. 5.1). The specificity of the mAb for the tumor cell target serves to deliver the cytotoxin to the tumor cell and spares peripheral tissues from exposure to the cytotoxin. Upon target binding, the ADC-antigen complex is internalized and the payload is released either by cleavage of the linker or by catabolism of the antibody. Once released, the payload or payload-linker antibody fragment is free to engage the intracellular drug-target and kill the cell. The overall design of the ADC approach is straightforward and elegant in principle. However, the chemical, biochemical, and cellular processes necessary for the mode of action are each highly complex, and are challenging to interrogate via assay. Thus, prospective design and optimization of ADC’s is only now beginning to emerge from what was initially a highly empirical approach for the first generation of ADC’s. Nonetheless, the recent approval of brentuximab vedotin (SGN-35) (Gualberto 2012) and the late stage clinical success of transtuzumab emtansine (T-DM1) (LoRusso et al. 2011) and inotuzumab ozogamicin (CMC-544) (Kantarjian

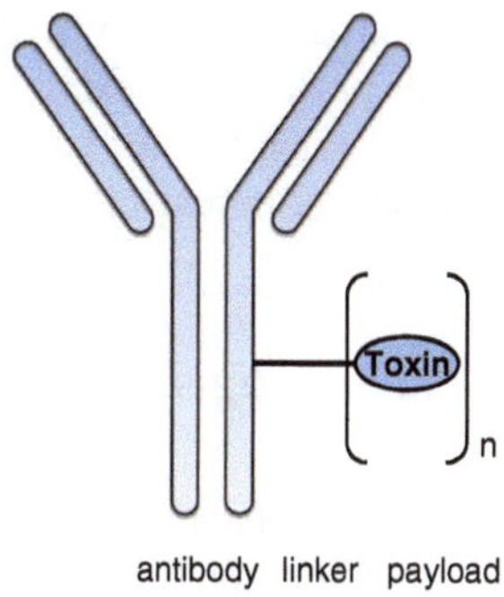

Fig. 5.1 Antibody drug conjugate (ADC)

et al. 2012) demonstrate the potential for ADC's to be a major modality for cancer therapy in the years ahead. Indeed, there are more than ten ADC programs undergoing clinical evaluation against solid tumors at the time of this writing (Sapra et al. 2011). We will see that natural product cytotoxins, by virtue of their diverse, potent mechanisms of action, and privileged chemical scaffolds, are a promising source of payloads for the next generation of ADC's. Rather than an extensive review of the status of selected ADC's in clinical development, this chapter instead will provide insights, based on recent published results with current ADC's, as to the current role and future potential of natural product cytotoxins as payloads for ADC's. For detailed reviews of current ADC's against various cancer types, the reader is referred to recent monographs (Polson et al. 2011; Sapra et al. 2011; Senter 2010).

5.2 Antibody Drug Conjugates: Processes and Components

The three main components of an ADC: antibody, linker, and payload, are each critical to the efficacy and safety of the ADC, and their roles are highly interdependent within the overall ADC mode of action. These processes involved with ADC function, and the role or dependency of the various ADC components within them, are described here.

5.2.1 Systemic Circulation

Naked antibodies possess very long half-lives in blood circulation, and the ADC must mimic this in order to achieve systemic exposure and tissue distribution sufficient to engage the tumor (Boswell et al. 2011). However, payload loss (deconjugation) can often degrade the pharmacokinetic properties of the ADC, particularly at high loading. Once in circulation it is essential that the bound payload molecule itself be stable to serum degradation, and that the payload-linker be stable to deconjugation in order to avoid payload loss by hydrolysis or by serum esterases, otherwise systemic exposure to free payload will cause toxicity. Should premature shedding of the payload occur, it is desirable that the payload itself be subject to high systemic clearance and have low cellular permeability, so as to minimize exposure to nontumor tissue.

5.2.2 Antigen Binding and Intracellular Payload Release

The choice of antibody target is critical to the efficacy and safety of an ADC (Teicher 2009). Numerous antigens have been the subject of past and current ADC trials and important target-related principals are emerging. One of these is the desirability of so-called "clean targets," i.e., antigens that are highly expressed on tumor cells but not in nontumor tissues. In practice it has proven to be difficult to identify these targets, as most cell surface antigens expressed on solid tumors are also present in normal tissues (Sapra et al. 2011). It is also important that the target have a high copy number, because in principle the number of ADC surface binding sites places a ceiling on the number of payload molecules delivered to the internalization process. Once the ADC binds, the surface antigen-ADC complex undergoes rapid internalization and efficient trafficking to the lysosome, where it then undergoes catabolism (Sutherland et al. 2006). For cleavable linkers, the payload is released when the linker is cleaved by lysosomal enzymes or by the chemical environment of the lysosome. For non-cleavable linkers, the ADC is catabolized by proteases, resulting in a released specie that consists of the payload-linker bound to the remaining conjugated amino acid residue (Doronina et al. 2006).

5.2.3 Linker Design

Over the past 15 years significant new chemistry has been incorporated into linker design, the result being linkers which show improved serum stability, pharmacokinetics, and intracellular release kinetics (Ducry and Stump 2010). While not the central topic of this chapter, the linker is a critical component of an ADC, and some background description of linkers is appropriate as a basis for discussing payloads. Examples of the main cleavable and non-cleavable linker systems used in clinically validated ADC's are shown in Fig. 5.2. Cleavable linkers release free drug or a simple derivative, usually by hydrolytic or proteolytic means. They generally provide more potent ADCs and can release permeable payloads. The chemically cleavable acid-labile 4-(4′-acetylphenoxy) butanoic acid (Ac-But) linker system in inotuzumab ozogamicin (CMC-544, Pfizer) is used to conjugate gamma calicheamicin dimethylhydrazide (DMH) to a humanized anti-CD22 IgG4 antibody (DiJoseph et al. 2004). In it the 4-(4-acetylphenoxy)butanoic acid is conjugated to surface-exposed lysine side chains of the antibody by an amide linkage. The other end forms an acyl hydrazone linkage with a 3-dimethylthiopropionate that is linked via a disulfide to the calicheamicin. In the acidic pH of the lysosome, the hydrazone is

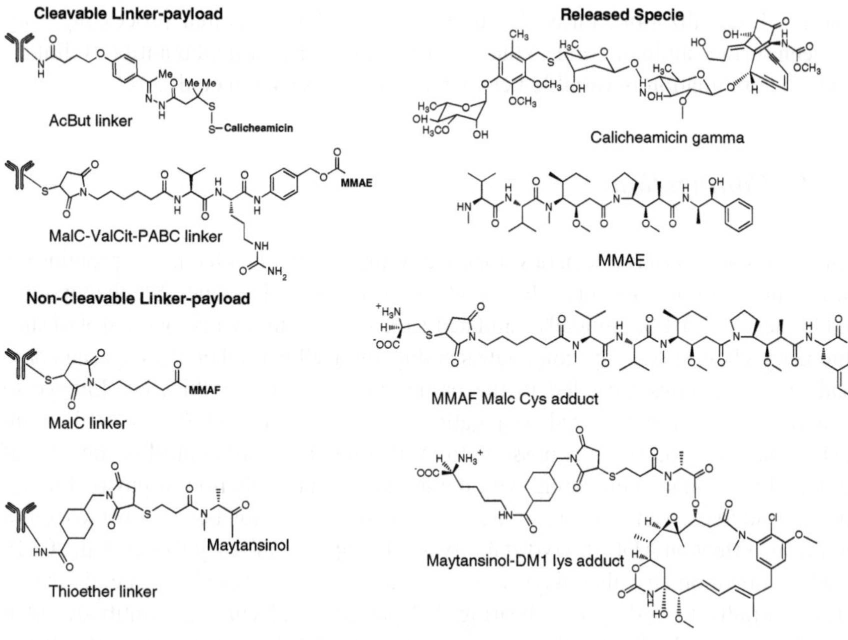

Fig. 5.2 ADC Linker-Payloads—cleavable, non-cleavable, and released species

cleaved to liberate the calicheamicin prodrug, which is then activated by reductive cleavage of the disulfide. The second cleavable linker shown in Fig. 5.2 is the proteolytically cleavable *self-imolative* MalC-ValCit-PABC linker used in the recently approved Adcetris™ ADC (Gualberto 2012). In this design, the linker is conjugated at one end to one or more cysteine thiol groups on the antibody by means of a maleimide precursor. Once in the lysosome, the amide bond between the citrulline residue and the *p*-aminobenzylcarbamate is cleaved by cathepsins and the remaining portion then undergoes a 1,6 elimination to liberate *para*-iminoquinone, CO_2, and the free amine of the payload.

Non-cleavable linkers rely on the full catabolism of the antibody in order to form the active released specie. Two examples are shown in Fig. 5.2. The first is the maleimidocaproyl (MalC) linker used in the SGN-75 of Seattle Genetics, an anti-CD70 antibody bearing the auristatin MMAF (Ryan et al. 2010). In this ADC design, once in the lysosome the antibody is catabolized while the linker remains intact. The final released specie in ADC's of this type has been shown to be the MMAF-caproylmaleido-cysteine adduct (Doronina et al. 2006). The second example is the thioether-based linker used in the transtuzumab-DM1 ADC of Immunogen (Erickson et al. 2006). In this case the released specie is the intact maytansinoid payload and linker attached to lysine (Fig. 5.2). In both non-cleavable examples the production of payload-containing released specie is dependent upon intracellular trafficking and lysosomal processing, and the cell-killing efficacy is dependent on the amino acid-linker payload adduct maintaining sufficient potency for intracellular target binding. Internalization, intracellular trafficking, and lysosomal processing vary between surface antigens, and so the effectiveness of any particular antibody-linker-payload combination is variable between cancer types (Sapra et al. 2011).

5.2.4 Conjugation

Numerous methods have been developed to conjugate small molecules to proteins via the functional groups present on the side chains of amino acid residues (Stephanopoulos and Francis 2011). For clinically validated ADCs the requirements for control of stoichiometry, chain specificity, conjugate stability, intracellular linker cleavage, and payload release kinetics have led to the predominance of either thioether linkage to cysteine or lysine amide-based conjugation methods (Ducry and Stump 2010). With each of these approaches it is possible to exert some degree of control of the sites of payload-linker attachment along with the number and distribution of payload molecules bound to each antibody (loading). The site of conjugation been shown to exert a significant effect on ADC in vivo stability and therapeutic efficacy (Shen et al. 2012). Loading stoichiometry also plays an important role in the overall therapeutic index, where partially loaded species bearing 2–4 payload molecules per antibody often show better overall efficacy than more highly loaded ADC's with 6–8 payloads per antibody (McDonagh et al. 2006). The drug–antibody conjugation can be performed with so-called "random" loading, where a population of cysteine or lysine residues are modified, resulting in a population of ADC molecules bearing a distribution of loading. Alternatively, conjugation can be performed in a "site selective" fashion by means of site-specific antibody mutants (Voynov et al. 2010). In this case, the resulting ADC has a tightly defined residue of attachment and stoichiometry of loading.

5.2.5 Payload Design

ADC payload requirements cover a number of molecular characteristics and properties. Given the relatively small amount of payload that can be delivered to a tumor by an ADC it is probably a safe assumption that high payload potency is a requisite. This is exemplified by those highly potent payloads—calicheamicin, auristatin, and maytansinoids that show clinical efficacy. The payload should possess a cell-killing mechanism that provides efficacy against the targeted cancer but also has reduced toxicity toward normal cells. The payload must also be chemically tractable, meaning that it must bear one or more functional sites to which a linker can be attached for conjugation to the antibody, and which either can be regenerated after the linker is cleaved, or which is not essential for potency. The linker-payload must be stable in serum for over the circulation time of the ADC, which can be weeks. It is desirable that the free payload have limited cellular permeability on its own, so as to minimize its action on peripheral tissues if it is liberated in serum prior to reaching the tumor. Finally, for efficacy against certain tumors it must be a poor substrate for efflux.

Natural Product	Natural Producer	Payload	Potency (IC_{50})	Mechanism of Action	Status of ADC
ansamitosin P-3 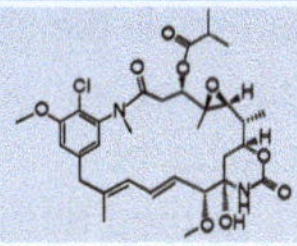	*Actinosynnema pretiosum*	Maytansine DM-1, DM-4	10^{-10}- 10^{-11} 10^{-11}- 10^{-12}	Antimitotic-tubulin depolymerizer	Phase III
dolastatin 10 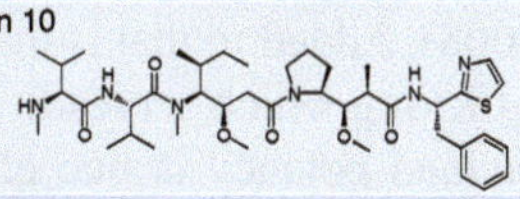	Marine cyano-bacterium	Auristatins- MMAE MMAF	10^{-11}- 10^{-9} 10^{-8}- 10^{-7}	Antimitotic-tubulin depolymerizer	Market
calicheamicin 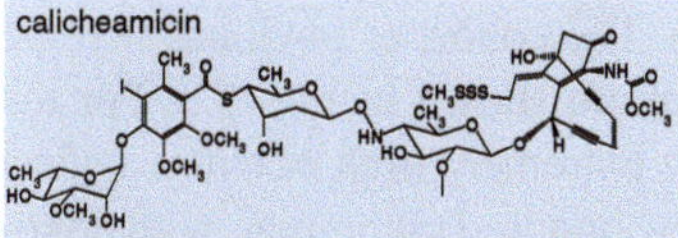	*Micromonospora sp.*	calicheamicin	10^{-10}- 10^{-9}	DNA cleavage	Phase III
CC-1065/duocarmycin 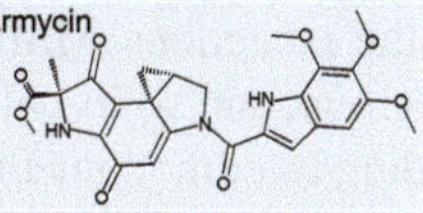	*Streptomyces sp.*	MGBA	$< 10^{-9}$	DNA alkylation	Phase I
camptothecin 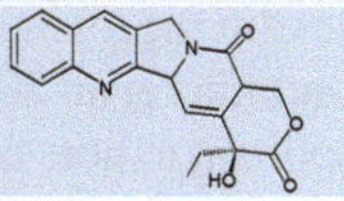	*Camptotheca acuminata*	SN-38	10^{-8}	Topoisomerase I inhibition	Phase I
doxorubicin 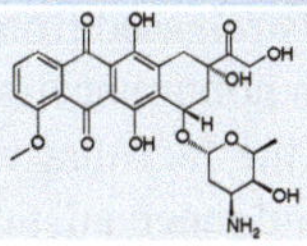	*Streptomyces peucetius*	doxorubicin	10^{-7}	DNA intercalation/ Topoisomerase II inhibition	Phase I/II

Fig. 5.3 Natural product ADC payloads

5.2.5.1 Potency and Mechanism of Action

A defining characteristic of many natural products is their highly potent cytotoxicity and unique cell-killing mechanism of action. As a result, they are especially well-positioned as ADC payload candidates. Perhaps this should come as no surprise, because many natural products are utilized by their producing hosts as defense mechanisms. It is the scope of biochemical mechanisms coupled with the remarkable structural diversity that make natural products a truly rich source for ADC payloads (Mayer and Gustafson 2008). This biological rationale is fortified by a recent network analysis of the biochemical targets of known natural products, which shows a high bias towards targets related to cell death (Dancik et al. 2010). The potent cytotoxicity of natural products has led to many clinical trials of natural product cytotoxins as small molecule drugs over the last several decades (Butler 2008). A number of these compounds, shown in Fig. 5.3, have found a role as ADC payloads. Three classes—calicheamicin, dolastatins, and maytansines have led to payloads on approved or late clinical stage ADC's.

The ADC model is conceptually both elegant and simple in design. However, there are fundamental limitations to the number of cytotoxin payload molecules that can be delivered into a target tumor cell. One of these limitations is the absolute number of antigen molecules available on the tumor cell surface for antibody binding. Many tumor antigens are frequently down-regulated by cancer cells as a means of evading the immune system, resulting for some targets in a limited number of target molecules for the ADC to bind (Begley and Ribas 2008). Added to that is the effect of slow internalization or inefficient intracellular trafficking which serves to reduce the intracellular target exposure of the payload. Payload potency is one effective means by which these limitations can be overcome. The cohort of payloads that are at the most advanced stage of clinical development are shown in Fig. 5.3. For the most part the more effective payloads to date show extremely high potency against cell lines in their non-conjugated state, typically a few hundred picomolar, with doxorubicin and monomethylauristatin F (MMAF) being the exceptions. MMAF has limited cell permeability due to its free carboxylic acid group, and so its cell potency is roughly 10–200 times less than that of the corresponding ester monomethylauristatin E (MMAE). However, mAb-Val-Cit-MMAF conjugates are approximately ten times more potent than corresponding MMAE conjugates (Doronina et al. 2004), reflecting the high intracellular target potency of MMAF. It is important to note that the cellular cytotoxicity is a liability for payloads which are prematurely released from the ADC, as systemic exposure to free payload will lead to toxicity in nontumor tissues.

One of the particular strengths of natural products is the fact that within each mode of action there are often several distinct structural and mechanistic classes of molecules. One such group of natural products are the tubulin-interactive antimitotic agents, a group which has produced several anticancer chemotherapeutics and two clinically validated ADC payloads (Kingston 2009). The tubulin antimitotics group can be divided into two main classes (1) agents which inhibit the polymerization of tubulin such as the *Vinca* alkaloids, combretastatins, halichondrin B, dolastatins 10 and 15, rhizoxin, maytansine (ansamitocin), and hemiasterlin, and (2) agents which are tubulin stabilizers (polymerization promoters) such as the epothilones, taxanes, discodermolide, and dictyostatin (Fig. 5.4). Diversifying even further—within each functional class there are multiple binding modes. For example, although both the *Vinca* alkaloids and colchicines are tubulin depolymerizers, they bind the tubulin heterodimer at distinctly different sites. Several members of this group have become marketed cancer drugs and two—dolastatin 10 and maytansine have been developed into ADC payloads.

The antimitotic natural products in Fig. 5.4 boast a broad diversity of structural classes—indole alkaloids, peptides, polyketides, phenolics, and diterpenes. Structural complexity ranges from the relatively simple colchicines and combretastatin to the complex polyketide macrocycles such as dictyostatin and maytansine. The structural diversity and multifunctionalization of natural products structures is important with regard to ADC payload development. A payload candidate must bear a chemically accessible amine, hydroxyl or thiol group in order to enable linker attachment in a manner which does not interfere with the molecular mechanism of action. These complex structures can often offer SAR-independent chemical handles of this type.

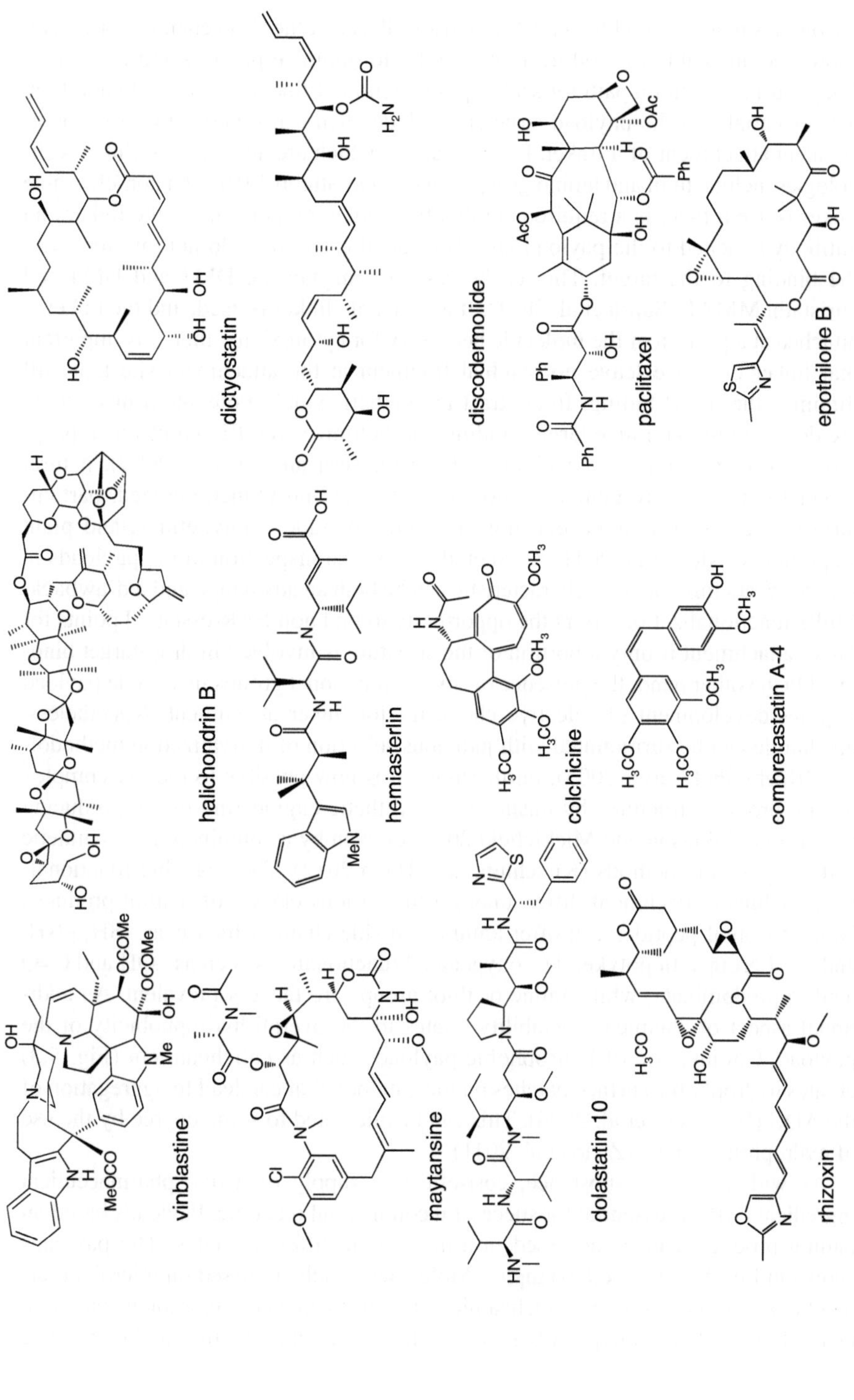

Fig. 5.4 Natural product tubulin-interactive antimitotic agents

5.2.5.2 Chemical Tractability

Cytotoxins that are suitable as ADC payloads, like any other lead chemical structure, should be amenable to modification in order to optimize potency and properties. They must also offer a path for scale-up production of final candidates. In addition, it is essential that the payload structure include chemical functionality to support covalent attachment of a linker. Functionalities which are amenable to linker technologies include thiol and amino groups (Ducry and Stump 2010). If a non-cleavable linker is used, there is a requirement that the remaining portions of the linker and antibody attached to the payload after lysosomal catabolism do not interfere with the binding to the target. This is the case for maytansine DM1 and DM4 and auristatin MMAF (Sapra et al. 2011). If a cleavable linker is used, and the linker is attached at a portion of the molecule necessary for cytotoxicity, then it is important that linker cleavage leave no residual fragment at the attachment site that will diminish the target biding. In contrast to synthetic small molecule drugs, which are designed to maximize target binding and optimize ADME and physical properties on a low molecular weight scaffold (Keseru and Makara 2009), natural product structures are generally more complex and polyfunctionalized, particularly classes such as microbial non-ribosomal peptides, polyketides, and plant isoprenoids (Alex et al. 2011; Koch et al. 2005). An inspection of the payloads in Fig. 5.3 bears this out. This characteristic can be both an advantage and a drawback. Multi-functionalization offers the opportunity to find non-SAR-essential points for linker attachment if only a portion of the structure is involved in drug-target binding. On the other hand, the presence of several functional groups in a single payload requires development of selective chemistry for linker attachment. Nonetheless, this hurdle can be surmounted with judicious selection of derivatization methodology (Koehn and Carter 2005). Furthermore, it is now possible to modify complex natural product structures by means of biosynthetic engineering of the producing organism (Wilkinson and Micklefield 2007) or even by a combination of synthetic and biosynthetic methods (Kirschning and Hahn 2012). The available functionalities for linker attachment differ amongst the various classes of natural products. Non-ribosomal peptides can offer amino acid side chains which bear –SH, –NH, and –OH groups. In polyketides, oxygenated functionalities such as –OH and C=O tend to predominate, while amine or thiol groups are far less prevalent. An additional aspect of chemical tractability relates to the overall hydrophobicity of the payload. Conjugation of hydrophobic payloads such as calicheamicin (Fig. 5.3) creates hydrophobic surface patches on the antibody that can lead to aggregation of the ADC (Hollander et al. 2008). This can be alleviated to some degree by the use of hydrophilic linkers (Zhao et al. 2011).

As with any drug substance, cost-effective supply of active pharmaceutical ingredient (API) is essential for successful commercial outcome. Payloads based on natural products can be accessed in a number of different routes. The payloads shown in Fig. 5.3 are good examples. Molecules which are based on microbial fermentation products, such as calicheamicin, ansamitocin, and doxorubicin can often be produced in kilogram quantities as a result of strain improvement and large-scale

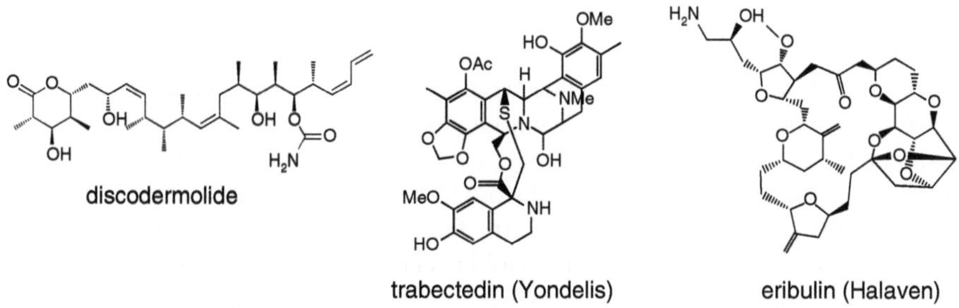

Fig. 5.5 Natural product drugs made by chemical synthesis

cultivation. This can be an arduous task, often requiring significant investment in time, but one that often leads ultimately to success. Two prominent examples in this area can be cited amongst the current clinical stable of natural product ADC payloads. The first is calicheamicin. The original wild type producing strain of *Micromonospora echinospora* ssp. *calichensis* produced calicheamicin in only trace quantities, requiring more than 100,000 L of fermentation to produce sufficient material for structural and biological characterization and lead advancement (Lee et al. 1991). Later work focused on bioprocess development in order to produce sufficient quantities to support advancement as an ADC payload. Recently biosynthetic engineering methodologies have been very effective in improving production of the enediyne family of polyketides (Chen et al. 2011).

The second example of an ADC natural product payload enabled by microbial fermentation and biosynthetic engineering is the maytansoid ansamitocin—the payload found in transtuzumab-T-DM1. Following the original isolation of maytansine from plant species of the genus *Maytenus*, various members of this class of antimitotics were isolated from other plant genera, mosses, and from the Actinomycete, *Actinosynnema pretiosum*. Maytansine became the focus of clinical investigation in the 1980s, culminating in a role as an ADC payload (Kirschning et al. 2008). To advance the compound clinically it was necessary to develop a high producing variant of *A. pretiosum* and a high yielding bioprocess as well (Kuo et al. 2005). Recently, the development of new, potent ansamitocin analogs has been achieved through biosynthetic engineering of the microbial producer (Taft et al. 2008).

Finally, natural product payload lead optimization and development can also be advanced by total chemical synthesis. While the synthesis of complex natural products can be quite daunting, there are numerous cases where lead optimization and advancement to clinical supplies have relied on either partial synthesis from a precursor or total synthesis from simple cost-effective starting materials. Some important benchmark examples include the marine sponge-derived antimitotic polyketide discodermolide (Paterson and Florence 2009), the marine alkaloid trabectedin (Cuevas and Francesch 2009), and another sponge-derived polyketide eribulin (Yu et al. 2011), all of which are shown in Fig. 5.5. Finally, the auristatins MMAE and MMAF (Figs. 5.2 and 5.3) represent a class of totally synthetic ADC payloads which are based a family of marine natural product peptides. The role of chemical synthesis in the advancement of the auristatins is described in the section below.

5.2.5.3 Cellular Permeability

The permeability of a payload plays a significant role in the overall safety of an ADC as well as its efficacy against solid tumors. From the aspect of safety, one can rationalize that a highly non-permeable payload is desirable, so that the effect of any early systemic release of payload would be minimized. On the other hand, non-permeable payloads have shown limited effectiveness against solid tumors for lack of a "bystander effect" (Kovtun and Goldmacher 2007). This relates to the ability of a payload, once it has been internalized, released from the ADC and has killed the initial antigen-expressing cell, to then diffuse to adjacent cells and kill them. The ability to kill neighboring cells of a tumor by a non-internalizing ADC mechanism is perhaps fundamental in an ADC's effectiveness, because the level of antigen expression in solid tumors can be quite heterogenous (Christiansen and Rajasekaran 2004). In studies measuring the bystander effect in mixed cell cultures consisting of both antigen positive and antigen negative cells using ADC's bearing maytansinoid- or CC1065-based payloads, it was shown that bystander cell killing was dependent on the presence of antigen positive cells, and also on the linker type (Kovtun et al. 2006). Conjugates linked via a reducible disulfide bond were capable of exerting the bystander effect whereas equally potent conjugates linked via a nonreducible thioether bond were not. Presumably, the disulfide-linked payload yielded a permeable released specie while the thioether did not. In the maytansinoid-based ADC's it was shown that the permeability of the released specie plays a major role in the bystander effect (Erickson et al. 2006). Erickson and coworkers investigated the intracellular processing of huC242-maytansinoid conjugates containing either a disulfide linker (huC242-SPDB-DM4) or a thioether linker (huC242-SMCC-DM1), and found that lysosomal processing of either ADC was necessary for cell-killing activity, but also that the disulfide-linked DM4 ADC was able to kill adjacent cells, while the thioether-linked DM1 ADC was not. It was shown that the sole maytansinoid thioether-linked product was the highly polar lysine adduct, while the disulfide-linked ADC gave the lipophilic, and presumably permeable, *S*-methyl-DM4. Natural products scaffolds often have the ability to maintain permeability properties, even though they have larger, more complex structures. This is a function of their ability to use conformational mobility to reduce polar surface area in order to maintain favorable LogP values (Ganesan 2008).

5.2.5.4 Payload Efflux

Efflux of anticancer drugs mediated by multidrug resistance-associated proteins is an underlying resistance mechanism for many tumors. This holds true for payloads used in ADCs as well, for most ADC payloads currently under clinical investigation are substrates for the multidrug transporter MDR1 in their small molecule chemical form. The calicheamicin-based drug conjugate CMC-544 shows limited effectiveness against B-cell lymphocytic leukemia and lymphoma cell lines which over-express P-glycoprotein efflux pumps (Takeshita et al. 2009). There is good evidence that the ADC format can be leveraged to mitigate MDR, for Kovtun and coworkers showed that even though the DM1 maytansinoid payload is subject to MDR efflux, this can be circumvented by the use of hydrophilic ADC linkers (Kovtun et al. 2010). One of the keys to this approach lies in the structure activity relationships of the payload itself, for the DM1 structural tolerance makes it possible to append the desired linkers to reduce MDR without severely degrading the ability to bind tubulin. Furthermore, now that the clinical effectiveness of tubulin depolymerization agents has been demonstrated, it is reasonable to expect that other antimitotics which are not subject to MDR efflux will be the subject of new investigations. There are also several classes of natural product inhibitors of P-glycoprotein and other ABC transporters (Klepsch et al. 2010). However, these agents suffer from significant systemic toxicity due to the widespread transporter involvement in normal cellular function as well as drug uptake, disposition, and metabolism. It is not inconceivable however that efflux pump inhibitors could also be selectively delivered as ADC payloads as a means of mitigating their toxicity, or it may be possible to design hybrid payload molecules which incorporate both cytotoxicity and MDR resistance (Klepsch et al. 2010).

5.3 Current ADC Natural Product Payloads

5.3.1 Calicheamicin

The calicheamicin family of polyketide enediyne antitumor antibiotics was discovered by the Lederle group of the American Cyanamid Company, who first isolated the compounds from a fermentation broth of *M. echinospora calichensis* on the basis of their striking antibiotic activity and cytotoxicity (Lee et al. 1989; Maiese et al. 1989). The structure of calicheamicin (Fig. 5.3) was solved after a herculean effort to produce sufficient material (a few milligrams) for extensive spectroscopic and chemical analysis (Lee et al. 1992). Calicheamicin causes double strand scission of DNA by binding with high specificity to the minor groove followed by generation of a 1,4-dehydrobenzene diradical ion which causes double hydrogen ion abstraction from the DNA backbone (Zein et al. 1989), an event that is triggered by the nucleophilic attack of intracellular thiol on the trisulfide moiety (Chatterjee et al. 1996) (Fig. 5.6). The initial binding is highly specific for the (T-C-C-T)Â·(A-G-G-A) regions, and is mediated by the sugar residues and the unique conformational preferences of the ten-member polyketide ring (Ellestad 2011). The lethality of double strand DNA lesions are what give calicheamicin its exquisite cellular potency (Elmroth et al. 2003).

Calicheamicin shows picomolar cytotoxicity against a broad array of tumor cell lines, but the nonselective mechanism of action makes the compound unsuitable as a therapeutic agent. It was recognized however, that this inherent potency could be harnessed by targeted delivery with an antibody, and early conjugates were made

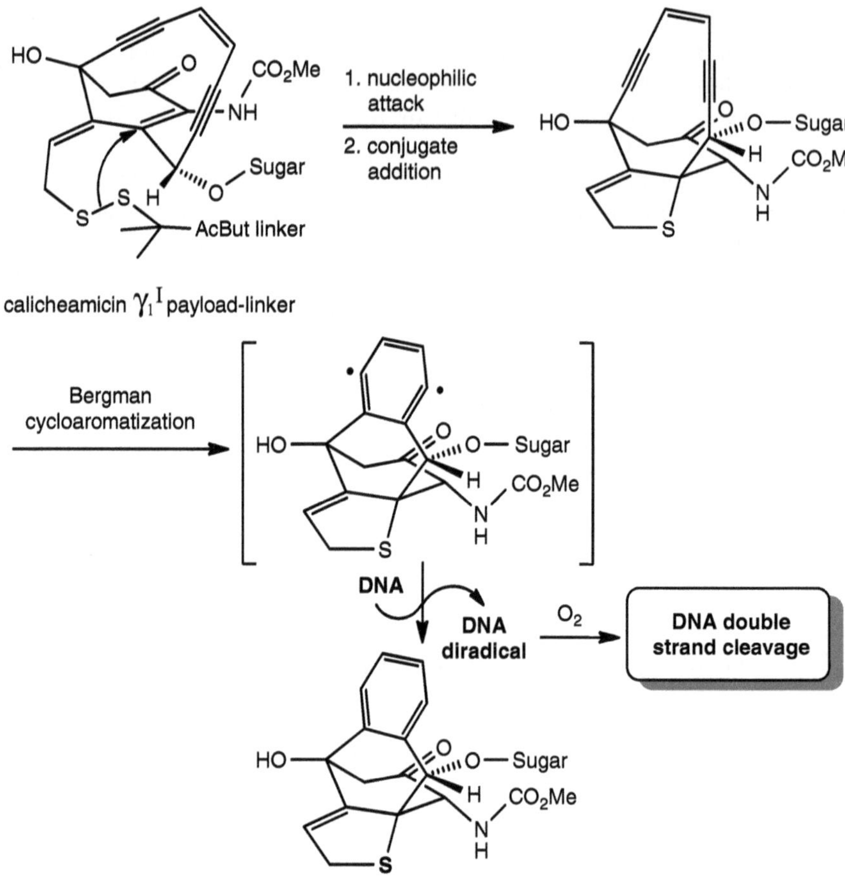

Fig. 5.6 Mechanism of calicheamicin

by linking to an internalizing anti-polyepithelial mucin antibody (Hinman et al. 1993). The calicheamicin conjugates retained the immunoreactivity of the unmodified antibody and were specifically cytotoxic toward antigen bearing tumor cells in vitro and in vivo. It was also found that improvement in the linker stability led to an increase in therapeutic index. Subsequent studies with additional antibodies led to the conjugation of *N*-acetyl-calicheamicin gamma to a humanized anti-CD33 mAb using a chemically cleavable Ac-But linker (Stasi 2008). This ADC, Mylotarg™ was approved in 2000 by the FDA for the treatment of CD33-positive refractory or relapsed acute myeloid leukemia. The same calicheamicin payload-linker combination is currently used on CMC-544 (Inotuzumab ozogamicin), a humanized anti-CD22 monoclonal ADC currently undergoing Phase II/III clinical trial (Thomas 2012).

5.3.2 *Maytansine*

The antimitotic ansa macrolide maytansine (Fig. 5.4) was first isolated by Kupchan and coworkers from the plant *Maytenus ovatus*, who recognized early its potent antileukemic activity and unique chemical structure (Kupchan et al. 1972; Remillard et al. 1975). Later studies demonstrated that the potent cytotoxicity and in vivo antitumor activity were the result of potent binding to tubulin at the *Vinca* alkaloid site (Mandelbaum-Shavit et al. 1976). Through the 1970s the powerful antitumor activity coupled with the relative scarcity of the compound inspired a great deal of synthetic chemistry aimed at preparing quantities for clinical trial (Anon 1983). These efforts were aided considerably by the discovery that a closely related group of antitumor ansa macrocyclic polyketides, the ansamitocins, was produced by the soil microorganism *Nocardia* (Higashide et al. 1977). Later studies showed that ansa macrolides were also produced by a wider variety of plants and bacteria (Nettleton et al. 1981; Tanida et al. 1980). The enhanced availability of maytansine facilitated several safety and efficacy clinical trials against numerous cancers in the mid-1970s (Douros et al. 1979), and additional clinical trials continued well into the 1980s. Overall, the outcomes of these trials were disappointing, with the major complications being dose-limiting toxicity and lack of efficacy (Blum et al. 1978; Borden et al. 1982; Creech et al. 1982; Franklin et al. 1980; Rosenthal et al. 1980). In order to mitigate the systemic maytansinoid toxicity, two groups initially explored the suitability of the class as payloads for ADCs. The Okamoto group constructed a bispecific mAb that bound ansamitocin P-3 and also the human transferrin receptor. When tested in A431 tumor xenograft models in nude mice, the bound ansamitocin delivered by the bispecific mAb was more effective in suppressing tumor growth than was ansamitocin P-3 alone. Eventual tumor eradication was achieved, as well as reduced toxicity (Okamoto et al. 1992). During the same period Chari and coworkers at Immunogen reported the synthesis of high potency maytansine analogs and their conjugation to antibodies via disulfide linkages to release fully active maytansine inside cells (Chari et al. 1992). An initial Phase I clinical trial of an immunoconjugate consisting of maytansine coupled to an anti-CD44v6 antibody was not successful and led to the termination of CD44v6 as an ADC target (Riechelmann et al. 2008). However, a following trial with maytansine bound to a humanized antibody against the CanAg antigen demonstrated improved safety and tolerability (Rodon et al. 2008). Subsequent Phase II trials of Her2- and CD19-based ADC's bearing maytansine DM1 have shown robust clinical efficacy in breast and B-cell malignancies respectively (Blanc et al. 2011; Burris et al. 2011), and numerous additional efficacy and Phase III studies are underway.

5.3.3 Auristatins

The antimitotic auristatin ADC payloads have their origins in the dolastatin group of marine natural products, an extensive family of non-ribosomal peptides originally discovered by the George Pettit group as antitumor constituents of the Indian Ocean sea hare *Dolabella auricularia* (Pettit et al. 1981). Approximately 20 dolastatins have been isolated to date, possessing a broad scope of structural diversity, mechanism of action, pharmacological activity, and often stunning potency (Pettit 1997). The dolastatins present a number of intriguing aspects in terms of their biosynthesis and chemistry. It has been shown for several members of the class that the producing organism is actually a cyanobacterial endo-symbiont of *D. auricularia* (Namikoshi and Rinehart 1996; Harrigan et al. 2000). While complex in structure, the peptide nature of several members of the family nonetheless makes them synthetically accessible, an aspect which has facilitated studies of their structure–activity relationships as well as enabling clinical investigation (Pettit et al. 1995). The initial promise of the dolastatins inspired several chemical syntheses, which furnished API for clinical trial as well as new analogs with improved safety and properties (Singh et al. 2008). Prominent members of the family include dolastatin 10, due to its powerful inhibition of microtubule assembly, tubulin-dependent GTP hydrolysis, and by the fact that it binds the same site on tubulin as the *Vinca* alkaloids (Bai et al. 1990) and dolastatin 15, a seven residue peptide which induces apoptosis and bcl-2 phosphorylation (Ali et al. 1998). Both of these agents were investigated in initial clinical trials, either as the native structure in the case of dolastatin 10, or in the form of a water-soluble analog LU-103793 in the case of dolastatin 15 (de Arruda et al. 1995). Initial clinical safety evaluation of these agents showed an acceptable profile, with some dose-limiting toxicities (Pitot et al. 1999; Villalona-Calero et al. 1998). Subsequent efficacy studies proved disappointing—LU 103793 lacked efficacy against non-small cell lung cancer (Marks et al. 2003) and metastatic breast cancer (Kerbrat et al. 2003), and showed only modest efficacy at best against malignant melanoma (Smyth et al. 2001). Similarly, dolastatin 10 showed no objective response or significant clinical efficacy against advanced colorectal cancer (Saad et al. 2002), metastatic soft tissue sarcoma (von Mehren et al. 2004), platinum sensitive ovarian carcinoma (Hoffman et al. 2003), prostate (Vaishampayan et al. 2000), or non-small-cell lung cancer (Krug et al. 2000). Additional trials of dolastatin 10 analogs such as TZT-1027 (Solbidotin) as single agents continue, but clinical efficacy remains elusive (Patel et al. 2006; Riely et al. 2007).

A new class of dolastatin 10 synthetic analogs, the auristatins, were developed with a simplified replacement of selected rare amino acid residues, making them easier to synthesize while maintaining potency (Pettit et al. 1995). Among this cohort of analogs were two compounds, auristatin E and monomethylauristatin E, which showed comparable potency to dolastatin 10 itself, with improved drug properties (Pettit et al. 1998) (Fig. 5.2). Later, scientists at Seattle Genetics were able to show that these auristatins could be effective ADC payloads in vitro and in vivo when conjugated to cBR96 anti-Lewis Y and cAC10 anti-CD30 mAbs (Doronina et al. 2003). MMAE has reached the market as the payload for the anti-CD33 ADC Adcetris™ (Brentuximab vedotin).

An additional auristatin MMAF (Fig. 5.2), a C-terminated phenylalanine auristatin derivative that contains a free carboxyl group, was shown to be effective in vivo when conjugated to anti-CD30 antibodies (Doronina et al. 2006), and has now been conjugated to additional antibodies for future clinical evaluation (Sapra et al. 2011). MMAF alone has marginal activity against hematological tumor cell lines. This is presumably due to its poor permeability, a consequence of the free carboxylate group of the terminal phenylalanine residue. The ADC effectively serves to deliver the payload intracellularly, and results in a potency increase of 2,200-fold as well as reduced toxicity against non-antigen-expressing cells. As a result of these and other developments, several clinical trials are ongoing with ADCs bearing auristatin payloads.

5.3.4 Other Natural Products ADC Payloads

Several other natural product classes are in clinical trial or advanced preclinical development as ADC payloads (Fig. 5.3). These include payloads based on the DNA alkylating agents duocarmycin CC1065 (Tietze et al. 2009), the ansamycin alkylator doxorubicin (Stein et al. 2009) and the intercalating topoisomerase I inhibitor SN-38 (derived from the plant metabolite camptothecin (Cardillo et al. 2011). While the total number of investigations of these ADC is impressive, the payloads employed still represent only a fraction of the possible natural product cytotoxins which could be considered. No doubt, this narrow focus is a result in large part of difficulties in accessing those rare agents which could significantly expand the payload space.

5.4 Conclusions

For a small molecule to serve as an effective payload for ADCs, it must possess a unique combination of potency, mechanism, chemical tractability, structure, and physical properties. Natural product cytotoxins by virtue of their biological activity, mode of action, privileged chemical structures and properties, fit these requirements quite well and have proven to be a rich source of ADC payloads and payload candidates. To date, all clinically validated ADC payloads are directly derived from natural products, as are those payloads in early clinical and late stage preclinical development. A vast number of highly potent, mechanistically novel natural products remain to be investigated as ADC payloads. Given the track record to date, and the relatively early stage of ADC development, we can anticipate that additional natural product cytotoxins will find utility in this role.

Acknowledgements The author is grateful to Pfizer Oncology colleagues for many helpful discussions in the preparation of this manuscript.

References

Alex A, Millan DS, Perez M, Wakenhut F, Whitlock GA (2011) Intramolecular hydrogen bonding to improve membrane permeability and absorption in beyond rule of five chemical space. MedChemComm 2:669–674

Ali MA, Rosati R, Pettit GR, Kalemkerian GP (1998) Dolastatin 15 induces apoptosis and BCL-2 phosphorylation in small cell lung cancer cell lines. Anticancer Res 18:1021–1026

Anon (1983) Maytansine. Synform 1:278–282

Bai R, Pettit GR, Hamel E (1990) Binding of dolastatin 10 to tubulin at a distinct site for peptide antimitotic agents near the exchangeable nucleotide and vinca alkaloid sites. J Biol Chem 265:17141–17149

Begley J, Ribas A (2008) Targeted therapies to improve tumor immunotherapy. Clin Cancer Res 14:4385–4391

Bertino EM, Otterson GA (2011) Romidepsin: a novel histone deacetylase inhibitor for cancer. Expert Opin Investig Drugs 20:1151–1158

Blanc V, Bousseau A, Caron A, Carrez C, Lutz RJ, Lambert JM (2011) SAR3419: an anti-CD19-maytansinoid immunoconjugate for the treatment of B-cell malignancies. Clin Cancer Res 17:6448–6458

Blum RH, Wittenberg BK, Canellos GP, Mayer RJ, Skarin AT, Henderson IC, Parker LM, Frei E III (1978) A therapeutic trial of maytansine. Cancer Clin Trials 1:113–117

Borden EC, Ash A, Enterline HT, Rosenbaum C, Laucius JF, Paul AR, Falkson G, Lerner H (1982) Phase II evaluation of dibromodulcitol, ICRF-159, and maytansine for sarcomas. Am J Clin Oncol 5:417–420

Borzilleri RM, Vite GD (2009) Case history: discovery of ixabepilone (IXEMPRA), a first-in-class epothilone analog for treatment of metastatic breast cancer. Annu Rep Med Chem 44:301–322

Boswell CA, Mundo EE, Zhang C, Bumbaca D, Valle NR, Kozak KR, Fourie A, Chuh J, Koppada N, Saad O, Gill H, Shen B-Q, Rubinfeld B, Tibbitts J, Kaur S, Theil F-P, Fielder PJ, Khawli LA, Lin K (2011) Impact of drug conjugation on pharmacokinetics and tissue distribution of anti-STEAP1 antibody-drug conjugates in rats. Bioconjug Chem 22:1994–2004

Burris HA III, Rugo HS, Vukelja SJ, Vogel CL, Borson RA, Limentani S, Tan-Chiu E, Krop IE, Michaelson RA, Girish S, Amler L, Zheng M, Chu Y-W, Klencke B, O'Shaughnessy JA (2011) Phase II study of the antibody drug conjugate trastuzumab-DM1 for the treatment of human epidermal growth factor receptor 2 (HER2)-positive breast cancer after prior HER2-directed therapy. J Clin Oncol 29:398–405

Butler MS (2008) Natural products to drugs: natural product-derived compounds in clinical trials. Nat Prod Rep 25:475–516

Butler MS (2010) A snapshot of natural product-derived compounds in late stage clinical development at the end of 2008. Royal Society of Chemistry, London, pp 321–354

Cardillo TM, Govindan SV, Sharkey RM, Trisal P, Goldenberg DM (2011) Humanized anti-trop-2 IgG-SN-38 conjugate for effective treatment of diverse epithelial cancers: preclinical studies in human cancer xenograft models and monkeys. Clin Cancer Res 17:3157–3169

Carter PJ, Senter PD (2008) Antibody-drug conjugates for cancer therapy. Cancer J 14:154–169

Chari RV, Martell BA, Gross JL, Cook SB, Shah SA, Blattler WA, McKenzie SJ, Goldmacher VS (1992) Immunoconjugates containing novel maytansinoids: promising anticancer drugs. Cancer Res 52:127–131

Chatterjee M, Smith PJ, Townsend CA (1996) The role of the aminosugar and helix binding in the thiol-induced activation of calicheamicin for DNA cleavage. J Am Chem Soc 118:1938–1948

Chen Y, Jungsuwadee P, Vore M, Butterfield DA, St Clair DK (2007) Collateral damage in cancer chemotherapy: oxidative stress in nontargeted tissues. Mol Interv 7:147–156

Chen Y, Yin M, Horsman GP, Shen B (2011) Improvement of the enediyne antitumor antibiotic C-1027 production by manipulating its biosynthetic pathway regulation in Streptomyces globisporus. J Nat Prod 74:420–424

Christiansen J, Rajasekaran AK (2004) Biological impediments to monoclonal antibody-based cancer immunotherapy. Mol Cancer Ther 3:1493–1501

Creech RH, Stanley K, Vogl SE, Ettinger DS, Bonomi PD, Salazar O (1982) Phase II study of cisplatin, maytansine, and chlorozotocin in small cell lung carcinoma (EST 2578). Cancer Treat Rep 66:1417–1419

Cuevas C, Francesch A (2009) Development of Yondelis (trabectedin, ET-743). A semisynthetic process solves the supply problem. Nat Prod Rep 26:322–337

Dancik V, Seiler KP, Young DW, Schreiber SL, Clemons PA (2010) Distinct biological network properties between the targets of natural products and disease genes. J Am Chem Soc 132:9259–9261

de Arruda M, Cocchiaro CA, Nelson CM, Grinnell CM, Janssen B, Haupt A, Barlozzari T (1995) LU103793 (NSC D-669356): a synthetic peptide that interacts with microtubules and inhibits mitosis. Cancer Res 55:3085–3092

DiJoseph JF, Armellino DC, Boghaert ER, Khandke K, Dougher MM, Sridharan L, Kunz A, Hamann PR, Gorovits B, Udata C, Moran JK, Popplewell AG, Stephens S, Frost P, Damle NK (2004) Antibody-targeted chemotherapy with CMC-544: A CD22-targeted immunoconjugate of calicheamicin for the treatment of B-lymphoid malignancies. Blood 103:1807–1814

Doronina SO, Toki BE, Torgov MY, Mendelsohn BA, Cerveny CG, Chace DF, DeBlanc RL, Gearing RP, Bovee TD, Siegall CB, Francisco JA, Wahl AF, Meyer DL, Senter PD (2003) Development of potent monoclonal antibody auristatin conjugates for cancer therapy. Nat Biotechnol 21:778–784

Doronina S, Mendelsohn B, Toki B, Alley S, Meyer D, Hamblett K, Francisco J, Cerveny C, Wahl A, Senter P (2004) Immunoconjugates comprised of drugs with impaired cellular permeability: a new approach to targeted therapy. American Chemical Society, Washington, DC, p MEDI-010

Doronina SO, Mendelsohn BA, Bovee TD, Cerveny CG, Alley SC, Meyer DL, Oflazoglu E, Toki BE, Sanderson RJ, Zabinski RF, Wahl AF, Senter PD (2006) Enhanced activity of monomethylauristatin F through monoclonal antibody delivery: effects of linker technology on efficacy and toxicity. Bioconjug Chem 17:114–124

Douros J, Suffness M, Chiuten D, Adamson R (1979) Maytansine. Pergamon, New York, pp 59–73

Ducry L, Stump B (2010) Antibody-drug conjugates: linking cytotoxic payloads to monoclonal antibodies. Bioconjug Chem 21:5–13

Ellestad GA (2011) Structural and conformational features relevant to the anti-tumor activity of calicheamicin Î³1I. Chirality 23:660–671

Elmroth K, Nygren J, Martensson S, Ismail IH, Hammarsten O (2003) Cleavage of cellular DNA by calicheamicin Î³1. DNA Repair 2:363–374

Erickson HK, Park PU, Widdison WC, Kovtun YV, Garrett LM, Hoffman K, Lutz RJ, Goldmacher VS, Blaettler WA (2006) Antibody-maytansinoid conjugates are activated in targeted cancer cells by lysosomal degradation and linker-dependent intracellular processing. Cancer Res 66:4426–4433

Franklin R, Samson MK, Fraile RJ, Abu-Zahra H, O'Bryan R, Baker LH (1980) A phase I-II study of maytansine utilizing a weekly schedule. Cancer 46:1104–1108

Ganesan A (2008) The impact of natural products upon modern drug discovery. Curr Opin Chem Biol 12:306–317

Gualberto A (2012) Brentuximab Vedotin (SGN-35), an antibody-drug conjugate for the treatment of CD30-positive malignancies. Expert Opin Investig Drugs 21:205–216

Harrigan GG, Luesch H, Moore RE, Paul VJ (2000) The anticancer dolastatins as cyanobacterial metabolites. Spec Publ R Soc Chem 257:126–139

Higashide E, Asai M, Ootsu K, Tanida S, Kozai Y, Hasegawa T, Kishi T, Sugino Y, Yoneda M (1977) Ansamitocin, a group of novel maytansinoid antibiotics with antitumour properties from Nocardia. Nature 270:721–722

Hinman LM, Hamann PR, Wallace R, Menendez AT, Durr FE, Upeslacis J (1993) Preparation and characterization of monoclonal antibody conjugates of the calicheamicins: a novel and potent family of antitumor antibiotics. Cancer Res 53:3336–3342

Hoffman MA, Blessing JA, Lentz SS (2003) A phase II trial of dolastatin-10 in recurrent platinum-sensitive ovarian carcinoma: a Gynecologic Oncology Group study. Gynecol Oncol 89:95–98

Hollander I, Kunz A, Hamann PR (2008) Selection of reaction additives used in the preparation of monomeric antibody-calicheamicin conjugates. Bioconjug Chem 19:358–361

Kantarjian H, Thomas D, Jorgensen J, Jabbour E, Kebriaei P, Rytting M, York S, Ravandi F, Kwari M, Faderl S, Rios MB, Cortes J, Fayad L, Tarnai R, Wang SA, Champlin R, Advani A, O'Brien S (2012) Inotuzumab ozogamicin, an anti-CD22-calecheamicin conjugate, for refractory and relapsed acute lymphocytic leukaemia: a phase 2 study. Lancet Oncol 13:403–411

Kerbrat P, Dieras V, Pavlidis N, Ravaud A, Wanders J, Fumoleau P (2003) Phase II study of LU 103793 (dolastatin analogue) in patients with metastatic breast cancer. Eur J Cancer 39:317–320

Keseru GM, Makara GM (2009) The influence of lead discovery strategies on the properties of drug candidates. Nat Rev Drug Discov 8:203–212

Kingston DGI (2009) Tubulin-interactive natural products as anticancer agents. J Nat Prod 72:507–515

Kirschning A, Hahn F (2012) Merging chemical synthesis and biosynthesis: a new chapter in the total synthesis of natural products and natural product libraries. Angew Chem Int Ed 51:4012–4022

Kirschning A, Harmrolfs K, Knobloch T (2008) The chemistry and biology of the maytansinoid antitumor agents. C R Chim 11:1523–1543

Klepsch F, Jabeen I, Chiba P, Ecker GF (2010) Pharmacoinformatic approaches to design natural product type ligands of ABC-transporters. Curr Pharm Des 16:1742–1752

Koch MA, Schuffenhauer A, Scheck M, Wetzel S, Casaulta M, Odermatt A, Ertl P, Waldmann H (2005) Charting biologically relevant chemical space: a structural classification of natural products (SCONP). Proc Natl Acad Sci U S A 102:17272–17277

Koehn FE, Carter GT (2005) The evolving role of natural products in drug discovery. Nat Rev Drug Discov 4:206–220

Kovtun YV, Goldmacher VS (2007) Cell killing by antibody-drug conjugates. Cancer Lett 255:232–240

Kovtun YV, Audette CA, Ye Y, Xie H, Ruberti MF, Phinney SJ, Leece BA, Chittenden T, Blaettler WA, Goldmacher VS (2006) Antibody-drug conjugates designed to eradicate tumors with homogeneous and heterogeneous expression of the target antigen. Cancer Res 66:3214–3221

Kovtun YV, Audette CA, Mayo MF, Jones GE, Doherty H, Maloney EK, Erickson HK, Sun X, Wilhelm S, Ab O, Lai KC, Widdison WC, Kellogg B, Johnson H, Pinkas J, Lutz RJ, Singh R, Goldmacher VS, Chari RVJ (2010) Antibody-maytansinoid conjugates designed to bypass multidrug resistance. Cancer Res 70:2528–2537

Krug LM, Miller VA, Kalemkerian GP, Kraut MJ, Ng KK, Heelan RT, Pizzo BA, Perez W, McClean N, Kris MG (2000) Phase II study of dolastatin-10 in patients with advanced non-small-cell lung cancer. Ann Oncol 11:227–228

Kuo C, Byng GS, Widdison WC (2005) Methods for the production of ansamitocins, U.S. Pat. Appl., US 20050170475 A1 29pp

Kupchan SM, Komoda Y, Court WA, Thomas GJ, Smith RM, Karim A, Gilmore CJ, Haltiwanger RC, Bryan RF (1972) Tumor inhibitors: LXXIII. Maytansine, a novel antileukemic ansa macrolide from Maytenus ovatus. J Am Chem Soc 94:1354–1356

Lee MD, Manning JK, Williams DR, Kuck NA, Testa RT, Borders DB (1989) Calichemicins, a novel family of antitumor antibiotics: 3. Isolation, purification and characterization of calichemicins Î²1Br, Î³1Br, Î±2I, Î±3I, Î²1I, Î³1I, and Î´1I. J Antibiot 42:1070–1087

Lee MD, Ellestad GA, Borders DB (1991) Calicheamicins: discovery, structure, chemistry, and interaction with DNA. Acc Chem Res 24:235–243

Lee MD, Dunne TS, Chang CC, Siegel MM, Morton GO, Ellestad GA, McGahren WJ, Borders DB (1992) Calicheamicins, a novel family of antitumor antibiotics: 4. Structure elucidation of calicheamicins Î²1Br, Î³1Br, Î±2I, Î±3I, Î²1I, Î³1I, and Î´1I. J Am Chem Soc 114:985–997

LoRusso PM, Weiss D, Guardino E, Girish S, Sliwkowski MX (2011) Trastuzumab emtansine: a unique antibody-drug conjugate in development for human epidermal growth factor receptor 2-positive cancer. Clin Cancer Res 17:6437–6447

Maiese WM, Lechevalier MP, Lechevalier HA, Korshalla J, Kuck N, Fantini A, Wildey MJ, Thomas J, Greenstein M (1989) Calicheamicins, a novel family of antitumor antibiotics: taxonomy, fermentation and biological properties. J Antibiot 42:558–563

Mandelbaum-Shavit F, Wolpert-DeFilippes MK, Johns DG (1976) Binding of maytansine to rat brain tubulin. Biochem Biophys Res Commun 72:47–54

Marks RS, Graham DL, Sloan JA, Hillman S, Fishkoff S, Krook JE, Okuno SH, Mailliard JA, Fitch TR, Addo F (2003) A phase II study of the dolastatin 15 analogue LU 103793 in the treatment of advanced non-small-cell lung cancer. Am J Clin Oncol 26:336–337

Mayer AMS, Gustafson KR (2008) Marine pharmacology in 2005-2006: antitumour and cytotoxic compounds. Eur J Cancer 44:2357–2387

McDonagh CF, Turcott E, Westendorf L, Webster JB, Alley SC, Kim K, Andreyka J, Stone I, Hamblett KJ, Francisco JA, Carter P (2006) Engineered antibody-drug conjugates with defined sites and stoichiometries of drug attachment. Protein Eng Des Sel 19:299–307

Namikoshi M, Rinehart KL (1996) Bioactive compounds produced by cyanobacteria. J Ind Microbiol Biotechnol 17:373–384

Nettleton DF Jr, Balitz DM, Brown M, Moseley JE, Myllymaki RW (1981) Large scale isolation of maytansine and other maytansinoids from the seeds of Maytenus rothiana using preparative LC. J Nat Prod 44:340–347

Newman DJ, Cragg GM (2012) Natural products as sources of new drugs over the 30 years from 1981 to 2010. J Nat Prod 75:311–335

Okamoto K, Harada K, Ikeyama S, Iwasa S (1992) Therapeutic effect of ansamitocin targeted to tumor by a bispecific monoclonal antibody. Jpn J Cancer Res 83:761–768

Pasetto LM, Bortolami A, Falci C, Sinigaglia G, Monfardini S (2006) Recent progress in target therapy in colorectal cancer. Anticancer Res 26:3973–3981

Patel S, Keohan ML, Saif MW, Rushing D, Baez L, Feit K, DeJager R, Anderson S (2006) Phase II study of intravenous TZT-1027 in patients with advanced or metastatic soft-tissue sarcomas with prior exposure to anthracycline-based chemotherapy. Cancer 107:2881–2887

Paterson I, Florence GJ (2009) The chemical synthesis of discodermolide. Top Curr Chem 286:73–119

Pettit GR (1997) The dolastatins. Prog Chem Org Nat Prod 70:1–79

Pettit GR, Kamano Y, Fujii Y, Herald CL, Inoue M, Brown P, Gust D, Kitahara K, Schmidt JM (1981) Antineoplastic agents. Part 72. Marine animal biosynthetic constituents for cancer chemotherapy. J Nat Prod 44:482–485

Pettit GR, Srirangam JK, Barkoczy J, Williams MD, Durkin KPM, Boyd MR, Bai R, Hamel E, Schmidt JM, Chapius J-C (1995) Antineoplastic agents. 337. Synthesis of dolastatin 10 structural modifications. Anticancer Drug Des 10:529–544

Pettit GR, Srirangam JK, Barkoczy J, Williams MD, Boyd MR, Hamel E, Pettit RK, Hogan F, Bai R, Chapuis J-C, McAllister SC, Schmidt JM (1998) Antineoplastic agents 365. Dolastatin 10 SAR probes. Anticancer Drug Des 13:243–277

Pitot HC, McElroy EA Jr, Reid JM, Windebank AJ, Sloan JA, Erlichman C, Bagniewski PG, Walker DL, Rubin J, Goldberg RM, Adjei AA, Ames MM (1999) Phase I trial of dolastatin-10 (NSC 376128) in patients with advanced solid tumors. Clin Cancer Res 5:525–531

Polson AG, Ho WY, Ramakrishnan V (2011) Investigational antibody-drug conjugates for hematological malignancies. Expert Opin Investig Drugs 20:75–85

Reddy GK, Mughal TI, Rini BI (2006) Current data with mammalian target of rapamycin inhibitors in advanced-stage renal cell carcinoma. Clin Genitourin Cancer 5:110–113

Remillard S, Rebhun LI, Howie GA, Kupchan SM (1975) Antimitotic activity of the potent tumor inhibitor maytansine. Science 189:1002–1005

Riechelmann H, Sauter A, Golze W, Hanft G, Schroen C, Hoermann K, Erhardt T, Gronau S (2008) Phase I trial with the CD44v6-targeting immunoconjugate bivatuzumab mertansine in head and neck squamous cell carcinoma. Oral Oncol 44:823–829

Riely GJ, Gadgeel S, Rothman I, Saidman B, Sabbath K, Feit K, Kris MG, Rizvi NA (2007) A phase 2 study of TZT-1027, administered weekly to patients with advanced non-small cell lung cancer following treatment with platinum-based chemotherapy. Lung Cancer 55:181–185

Rodon J, Garrison M, Hammond LA, de Bono J, Smith L, Forero L, Hao D, Takimoto C, Lambert JM, Pandite L, Howard M, Xie H, Tolcher AW (2008) Cantuzumab mertansine in a three-times a week schedule: a phase I and pharmacokinetic study. Cancer Chemother Pharmacol 62:911–919

Rosenthal S, Harris DT, Horton J, Glick JH (1980) Phase II study of maytansine in patients with advanced lymphomas: an Eastern Cooperative Oncology Group pilot study. Cancer Treat Rep 64:1115–1117

Ryan MC, Kostner H, Gordon KA, Duniho S, Sutherland MK, Yu C, Kim KM, Nesterova A, Anderson M, McEarchern JA, Law CL, Smith LM (2010) Targeting pancreatic and ovarian carcinomas using the auristatin-based anti-CD70 antibody-drug conjugate SGN-75. Br J Cancer 103:676–684

Saad ED, Kraut EH, Hoff PM, Moore DF Jr, Jones D, Pazdur R, Abbruzzese JL (2002) Phase II study of dolastatin-10 as first-line treatment for advanced colorectal cancer. Am J Clin Oncol 25:451–453

Sapra P, Hooper AT, O'Donnell CJ, Gerber H-P (2011) Investigational antibody drug conjugates for solid tumors. Expert Opin Investig Drugs 20:1131–1149

Senter PD (2010) Recent advancements in the use of antibody drug conjugates for cancer therapy. Biotechnol Pharm Aspects 11:309–322

Shen B-Q, Xu K, Liu L, Raab H, Bhakta S, Kenrick M, Parsons-Reponte KL, Tien J, Yu S-F, Mai E, Li D, Tibbitts J, Baudys J, Saad OM, Scales SJ, McDonald PJ, Hass PE, Eigenbrot C, Nguyen T, Solis WA, Fuji RN, Flagella KM, Patel D, Spencer SD, Khawli LA, Ebens A, Wong WL, Vandlen R, Kaur S, Sliwkowski MX, Scheller RH, Polakis P, Junutula JR (2012) Conjugation site modulates the in vivo stability and therapeutic activity of antibody-drug conjugates. Nat Biotechnol 30:184–189

Singh R, Sharma M, Joshi P, Rawat DS (2008) Clinical status of anti-cancer agents derived from marine sources. Anticancer Agents Med Chem 8:603–617

Smyth J, Boneterre ME, Schellens J, Calvert H, Greim G, Wanders J, Hanauske A (2001) Activity of the dolastatin analogue, LU103793, in malignant melanoma. Ann Oncol 12:509–511

Stasi R (2008) Gemtuzumab ozogamicin: an anti-CD33 immunoconjugate for the treatment of acute myeloid leukaemia. Expert Opin Biol Ther 8:527–540

Stein R, Smith MR, Chen S, Zalath M, Goldenberg DM (2009) Combining milatuzumab with bortezomib, doxorubicin, or dexamethasone improves responses in multiple myeloma cell lines. Clin Cancer Res 15:2808–2817

Stephanopoulos N, Francis MB (2011) Choosing an effective protein bioconjugation strategy. Nat Chem Biol 7:876–884

Sutherland MSK, Sanderson RJ, Gordon KA, Andreyka J, Cerveny CG, Yu C, Lewis TS, Meyer DL, Zabinski RF, Doronina SO, Senter PD, Law C-L, Wahl AF (2006) Lysosomal trafficking and cysteine protease metabolism confer target-specific cytotoxicity by peptide-linked anti-CD30-auristatin conjugates. J Biol Chem 281:10540–10547

Taft F, Bruenjes M, Floss HG, Czempinski N, Grond S, Sasse F, Kirschning A (2008) Highly active ansamitocin derivatives: mutasynthesis using an AHBA-blocked mutant. Chembiochem 9:1057–1060

Takeshita A, Shinjo K, Yamakage N, Ono T, Hirano I, Matsui H, Shigeno K, Nakamura S, Tobita T, Maekawa M, Ohnishi K, Sugimoto Y, Kiyoi H, Naoe T, Ohno R (2009) CMC-544 (inotuzumab ozogamicin) shows less effect on multidrug resistant cells: analyses in cell lines and cells from patients with B-cell chronic lymphocytic leukaemia and lymphoma. Br J Haematol 146:34–43

Tanida S, Hasegawa T, Hatano K, Higashide E, Yoneda M (1980) Ansamitocins, maytansinoid antitumor antibiotics. Producing organism, fermentation, and antimicrobial activities. J Antibiot 33:192–198

Teicher BA (2009) Antibody-drug conjugate targets. Curr Cancer Drug Targets 9:982–1004

Thomas X (2012) in the treatment of B-cell acute lymphoblastic leukemia. Expert Opin Investig Drugs 21:871–878

Tietze LF, Krewer B, von Hof JM, Frauendorf H, Schuberth I (2009) Determination of the biological activity and structure activity relationships of drugs based on the highly cytotoxic duocarmycins and CC-1065. Toxins 1:134–150
Vaishampayan U, Glode M, Du W, Kraft A, Hudes G, Wright J, Hussain M (2000) Phase II study of dolastatin-10 in patients with hormone-refractory metastatic prostate adenocarcinoma. Clin Cancer Res 6:4205–4208
Villalona-Calero MA, Baker SD, Hammond L, Aylesworth C, Eckhardt SG, Kraynak M, Fram R, Fischkoff S, Velagapudi R, Toppmeyer D, Razvillas B, Jakimowicz K, Von HDD, Rowinsky E (1998) Phase I and pharmacokinetic study of the water-soluble dolastatin 15 analog LU103793 in patients with advanced solid malignancies. J Clin Oncol 16:2770–2779
Villanueva C, Awada A, Campone M, Machiels JP, Besse T, Magherini E, Dubin F, Semiond D, Pivot X (2011) A multicentre dose-escalating study of cabazitaxel (XRP6258) in combination with capecitabine in patients with metastatic breast cancer progressing after anthracycline and taxane treatment: a phase I/II study. Eur J Cancer 47:1037–1045
von Mehren M, Balcerzak SP, Kraft AS, Edmonson JH, Okuno SH, Davey M, McLaughlin S, Beard MT, Rogatko A (2004) Phase II trial of dolastatin-10, a novel anti-tubulin agent, in metastatic soft tissue sarcomas. Sarcoma 8:107–111
Voynov V, Chennamsetty N, Kayser V, Wallny H-J, Helk B, Trout BL (2010) Design and application of antibody cysteine variants. Bioconjug Chem 21:385–392
Wilkinson B, Micklefield J (2007) Mining and engineering natural-product biosynthetic pathways. Nat Chem Biol 3:379–386
Yu MJ, Zheng W, Seletsky BM, Littlefield BA, Kishi Y (2011) Case history: discovery of eribulin (HALAVEN), a halichondrin B analogue that prolongs overall survival in patients with metastatic breast cancer. Annu Rep Med Chem 46:227–241
Zein N, Poncin M, Nilakantan R, Ellestad GA (1989) Calicheamicin Î³1I and DNA: molecular recognition process responsible for site-specificity. Science 244:697–699
Zhao RY, Wilhelm SD, Audette C, Jones G, Leece BA, Lazar AC, Goldmacher VS, Singh R, Kovtun Y, Widdison WC, Lambert JM, Chari RVJ (2011) Synthesis and evaluation of hydrophilic linkers for antibody-maytansinoid conjugates. J Med Chem 54:3606–3623

Part II
Advances in Natural Product Cancer Drug Discovery

Chapter 6
Natural Product Scaffolds in Cancer Therapy

Christian P. Mang and Lars Ole Haustedt

Abstract Natural products are a unique pool of structurally amazing and so far unprecedented novel molecular scaffolds. The sources for these compounds are as diverse as their respective structures and this diversity is reflected by the variety of their mechanisms of action. Even though these compounds selected by evolution were not designed to fight cancer in the first place, they are indispensable in modern cancer therapy. This review gives insight into sources, structural features, and structure activity relationships of a selection of natural products frequently used in cancer therapy. This analysis will be carried out on taxol, eribulin, geldanamycin, camptothecin, epothilone, podophyllotoxin, and the anthracycline scaffolds.

6.1 Introduction

> Ex nihilo nihil fit (Descartes, Principia philosophiae, Part I, Article 49)
> Nothing comes out of nothing
> A child of five would understand this. Send someone to fetch a child of five (Groucho Marx).

Natural products are an invaluable source of new structures with unforeseen properties. They are of paramount importance in medicine since they have been used in the form of plant parts, extracts, concoctions etc. probably since the early days of mankind. The first historical records are reported from ancient India and Sumer, Egypt and China and from this time onwards, written records can be found in every culture (Borchardt 2002; Sonnendecker and Kremers 1986). Besides minerals and ashes the main ingredients usually consist of parts from plants or animals. This tradition still continues to the present day, with 65% formulations of traditional medicines

C.P. Mang (✉) • L.O. Haustedt (✉)
AnalytiCon Discovery GmbH, Hermannswerder Haus 17, 14471 Potsdam, Germany
e-mail: c.mang@ac-discovery.com; l.haustedt@ac-discovery.com

F.E. Koehn (ed.), *Natural Products and Cancer Drug Discovery*, Cancer Drug Discovery and Development, DOI 10.1007/978-1-4614-4654-5_6,

being derived from plants (Johnson 1999; Moerman 1986; Newman and Cragg 2010). The treatment of cancer with natural products also dates back to the early days even though the definition of cancer is hard to trace in historical records. With the onset of medicinal chemistry, the attempts to find a cure for cancer have been an ongoing quest in this research area. Since then, an incredible rich and vivid scene has emerged, with a sheer unconceivable multitude of approaches, ideas and in some cases solutions. However, to date no convincing cure for cancer has been found.

With the advent of combinatorial chemistry and HTS technology, great optimism arose that the key technologies to successfully tackle the medical challenges still unmet had been identified, first and foremost of course the discovery of new antitumor agents. This optimism faded rapidly, especially since the new process failed to deliver the numbers of leads or candidate structures that were initially anticipated. In retrospect, it was diagnosed that one main problem lay in the structures generated during early parallel chemistry synthesis programs which were created for the highly efficient new screening technologies. Characteristics of these "easy to make" compound libraries were mostly flat, molecular architectures with fused and otherwise linked aromatic rings. Later commentators suggested that the molecules apparently did not possess the relevant biological space.

With the rising popularity of new chemistry and screening methods, natural products became successively less popular. The main reason for this consisted in the difficulties usually associated with working with natural products. With the looming failure of the new technologies, the scene began to change slowly and the demand for good compounds with perspectives to be developed into NCEs rose. The focus slowly shifted back onto natural products. Their potential as an immensely rich starting point for medicinal chemical developments in biological activity again became clear. Statistical investigations also demonstrated quite impressively the potential of natural products for drug discovery. It was shown, that roughly half of all NCEs over a 20 years span had been derived in one way or another from a natural product (Newman 2008).

The main fields of application for drugs derived from natural products are the antibiotic and the cytostatic areas. While antibiotic research seemed to be able to successfully solve most of the problems posed (at least for many years), cancer research—despite all spectacular results—seemed way less successful. However, the vastly demanding scientific challenge combined with the prospect of a hugely profitable market led to a great number of marvelous approaches to solve the problem. A plethora of effective molecules of natural origin, with a vast structural variety was found to be effective against cancer and many of these have found their way into cancer therapy.

One of the major drawbacks of natural products—apart from their often quite complex structures and the resulting difficulties in structural optimization—is the often limited access to sufficient quantities of the respective compound from natural sources. Due to often quite complex molecular architectures, total synthesis in most cases is not an option for the production of sufficient amounts (with a few bright exceptions, e.g. eribulin, discodermolide). Semi-synthetic routes to natural products seem to be highly underappreciated—especially in academic research—compared to the higher art of total synthesis. Considering the tremendous impact on health and economic well being, as many of the following examples will show, this seems to be a serious misconception.

After all a great number of natural products are currently being used in cancer therapy with an impressive number of different modes of action. It would exceed the scope of this review to discuss all the different molecules. Several more comprehensive review articles have been published in the past and are highly recommended for further reading (see e.g. Cragg et al. 2009).

6.2 Taxol

Although the initial publication of the structure elucidation and description of cellular activity approached its 40th anniversary in May 2011 (Wani et al. 1971), the fame and attention surrounding the taxane family of natural products has in no way diminished. While accounts exist in ancient literature describing the toxic properties of the European yew tree (Mann 1994; Caesar), its beneficial features were only first discovered in the 1960s. Whereas the toxic properties are associated with the European yew tree (*Taxus baccata*), the American yew tree (*Taxus brefolia*) does not contain these toxic compounds. The two substances responsible for the toxicity of the European yew tree were just recently identified to be taxine A and taxine B (Wilson et al. 2001). These two compounds seem to be responsible for livestock poisoning. As stated by Itokawa (Itokawa and Lee 2003) if the biomaterial used in the discovery of taxol would have been the European yew, the whole story would have been sadly different due to the toxic properties of these extracts (Scheme 6.1).

The stony path to reaping health benefits for millions of patients worldwide started even earlier in the late 1950s of the last century. The U.S. Department of Agriculture joined forces with the National Cancer Institute to initiate a program which is even impressive in the era of high-throughput screening. From 1961 to 1981, 144,045 plant extracts and 16,000 extracts from animals were screened in multiple cellular assays. This program comprised a plant sample from the pacific yew tree, collected by Arthur S. Barclay in the Gifford Pinchot National Forrest Washington near the foot of Mt. St. Helens in August 1962 (Stephenson 2002).

In Durham, North Carolina, Monroe Wall and his assistant, Mansukh Wani were performing probably the first bioassay guided fractionation process established on

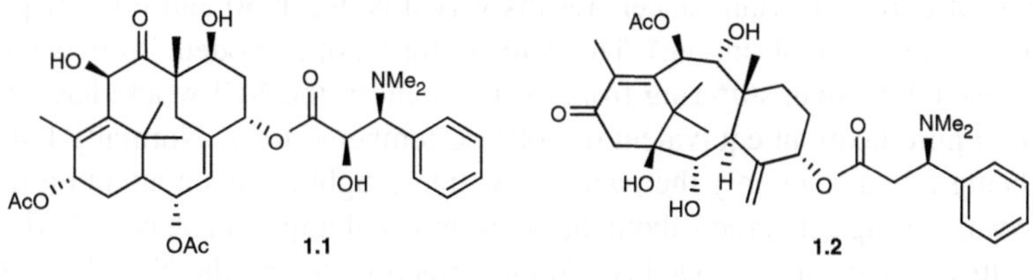

Scheme 6.1 Taxine A and B

Scheme 6.2 Taxol

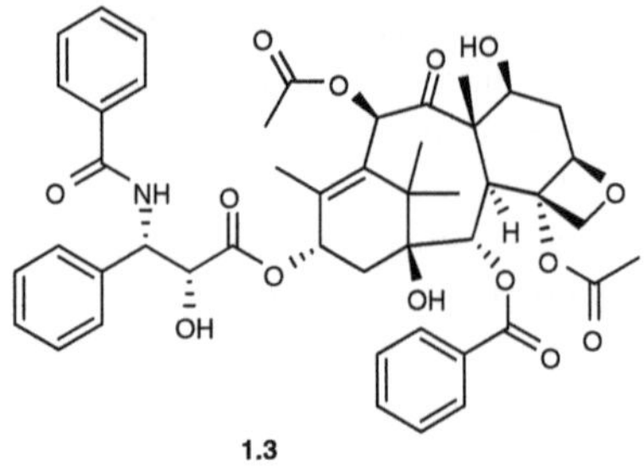

1.3

so called 9 kb cancer tissue cultures (Leighton and Kalla 1958). In December 1963, they isolated a sample and named it K172. K172 showed the broadest spectrum of activity ever noted in their screening on mouse leukemia cell lines.

For the structure elucidation, Wall ordered 45 lb of bark, twigs and needles of the pacific yew (*Taxus brevifolia*). Right from the beginning, the major problem of procurement was surfacing. Wall and Wani first announced the observed in vitro activity in 1967 at the 153rd ACS meeting in Miami Beach, FL. Four years later, after tedious structure elucidation by ^{1}H-NMR and together with Coggon and McPhail from Duke University using X-ray, Wall and Wani published the structure, including the correct absolute stereochemistry of K172 and named it taxol. Now, a second major obstacle was becoming obvious. In spite of being stunning, the molecular arrangement of functional groups around a unique, so far unprecedented, diterpene scaffold is a very demanding challenge in terms of synthetic organic chemistry. In fact, this challenging structure triggered one of the toughest races in the field of total synthesis (Scheme 6.2).

The publication attracted so much attention that the NCI in 1977 had already ordered 7,000 lb of bark from the pacific yew (corresponding to 1,500 trees). At this stage, the issue of procurement increasingly became a pressing matter. On the other hand, it was very clear that taxol was the most promising lead the NCI was pursuing. The interest in taxol was even more fuelled in 1979 when Susan Horwitz published a so far unprecedented mechanism of action (Schiff et al. 1979). In contrast to the vinca alkaloids, taxol was promoting in vitro microtubule assembly resulting in cell cycle arrest in the G_2 and/or M phase.

In 1984 the NCI compiled enough data to enter phase I clinical trials. At this point in time, 10 tons of pacific yew bark were required. Promising results in phase I paved the way for entering clinical phase II trials in 1985. With the publication of phase II data in 1988, the public was becoming aware of the potential taxol inherited in terms of curing ovarian cancer (results varied between 30 and 60% of patients with observed tumor shrinking). The demand for taxol exploded. To being able to cure every US woman suffering from ovarian cancer, the NCI would have needed 240 lb of pure taxol an equivalent of 360,000 lumbered trees. Normally left to rot after being felled, suddenly the pacific yew tree caught so much attention that the public was getting concerned about the environmental impact (The New York Times 1987). Just as demand exploded so did the costs invested by the NCI. By 1988 the NCI had already spent 25 million USD on the taxol program. Due to this reason, in

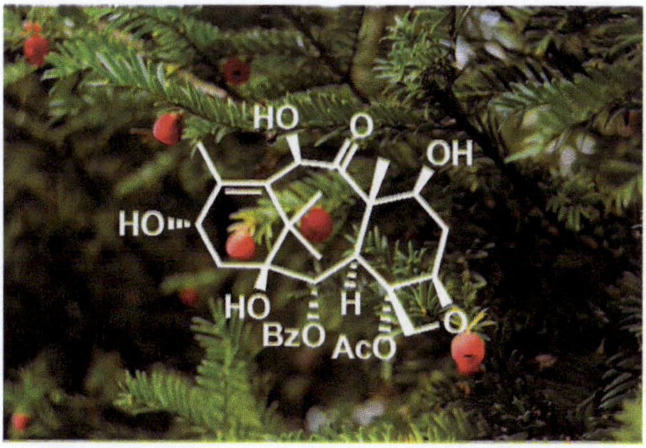

Fig. 6.1 10-Deacetyl-baccatin III and *Taxus baccata* (*background*)

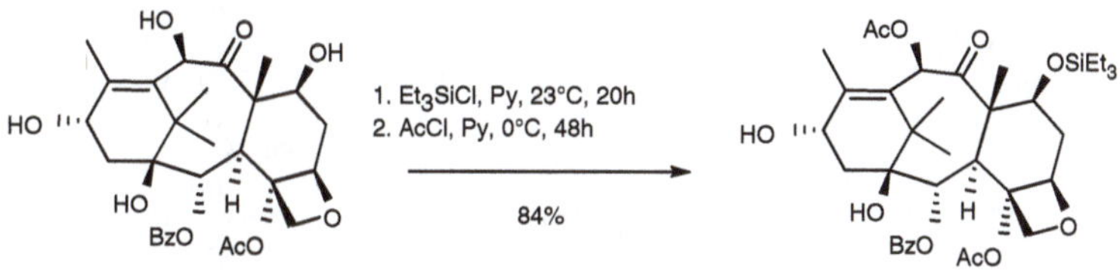

Scheme 6.3 Selective protection of deacetyl-baccatin III by Poitier et al.

August 1989, the NCI advertised a *Cooperative Research and Development Agreement* (CRADA). In December 1989, Bristol-Meyer signed the agreement with the NCI getting full access to the data the NCI had assembled so far. Furthermore, Bristol-Meyer got the exclusive right to commercialize taxol without being able to file a patent of matter.

As a starting material for the synthesis of taxol, 10-deacetyl-baccatin III was identified by Potier et al. (Fig. 6.1). They performed a tubulin assay by screening diverse parts of felled *T. baccata* L. (European species), which had to give way for a new approach on the Gif-sur-Yvette campus (Guénard et al. 1993). The extraction yields from the leaves of the European yew tree could be improved from 0.02 to 0.1%. Although 10-deacetyl-baccatin III only showed weak bioactivity compared to taxol, it was an excellent renewable starting material for the semi-synthesis of taxol (Sènilh et al. 1984) (Scheme 6.3).

As 10-deacetyl-baccatin III is sensitive to acid (A-ring Wagner–Meerwein rearrangement and oxetane opening) and base (C2 benzoyl group hydrolysis, C4 acetyl group hydrolysis, and oxetane opening to tetrahydrofuran system), Potier et al. had to establish an orthogonal set of protecting groups, which allowed the flexible introduction of the side chain at C13.

After the identification of an excellent starting material for a commercially viable synthesis of taxol, the need for an efficient side chain synthesis had become

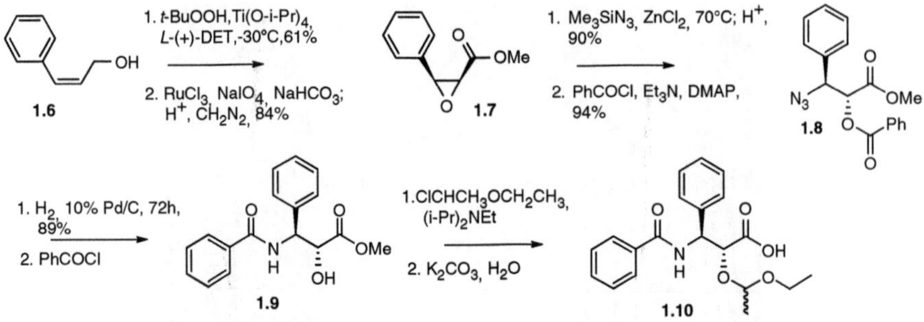

Scheme 6.4 Side chain synthesis by Greene et al.

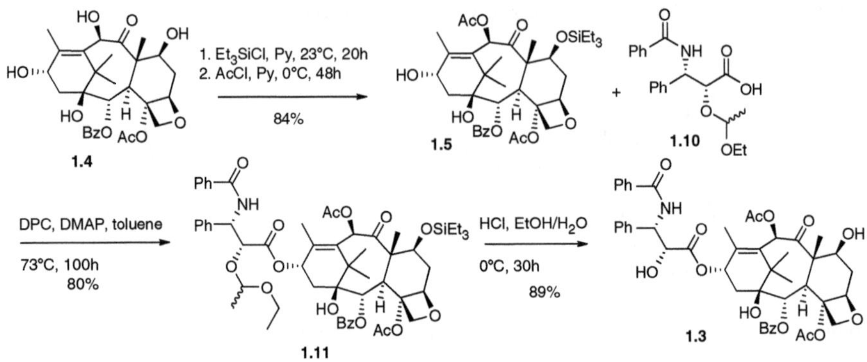

Scheme 6.5 Potier–Greene semi-synthesis of taxol

more and more urgent. In 1986 Greene et al. published a synthetic approach to the natural taxol side chain (Denis et al. 1986) (Scheme 6.4).

From here on, the way was paved for a renewable approach to taxol. In 1988, Pierre Potier and Andrew Green published the first semi-synthetic approach to taxol (Denis et al. 1988) (Scheme 6.5).

This synthetic approach could further be used to synthesize side chain analogs. The *t*-butyl side chain analog could also be synthesized by the efficient methodology published by Denis et al. (1986) (Scheme 6.6).

A patent filed in July 1986 by the CNRS and Rhône-Poulenc (Colin et al. 1989 FR 8610401) comprised the structure later to be named Taxotere, now the most successful active compound from the taxane structural family. Found in tubulin binding assays, taxotere showed slightly better activity on tubulin and also in cellular screenings (Guèritte-Voegelein et al. 1991).

The synthesis of taxotere can be achieved in a similar manner as taxol by using the same universal taxane precursor (Collin et al. 1990 EP 0253738) (Scheme 6.7).

Robert Holton like Pierre Potier also realized the desperate need for an efficient and practical synthetic access to taxol (Holton, 1991 US 5,015,744; Holton, 1992

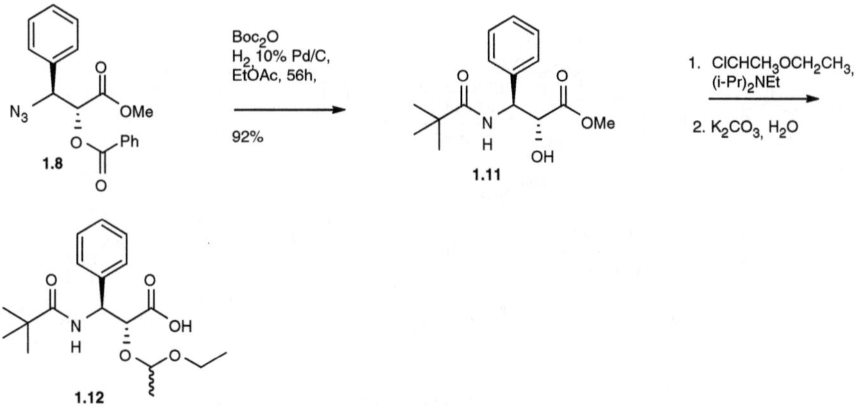

Scheme 6.6 Side chain synthesis of taxotere by Greene et al.

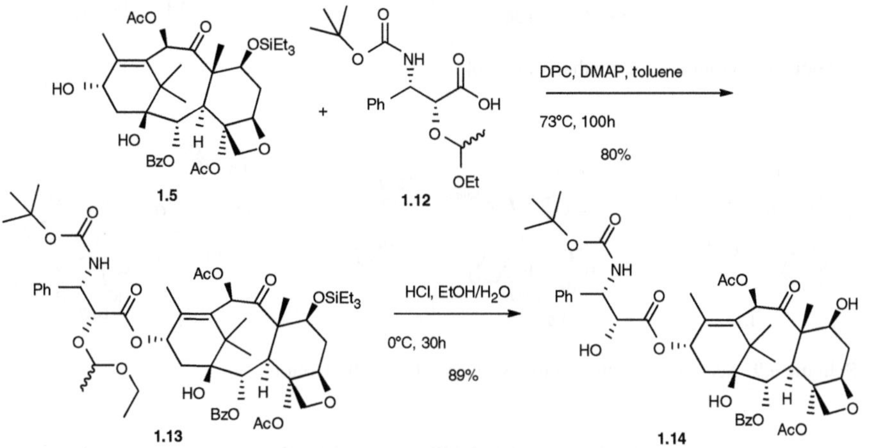

Scheme 6.7 Semi-synthesis of taxotere by Potier et al.

US 5,175,315). Holton did also realize the fundamental role of **1.4** in a commercially successful synthesis. Apart from a total synthesis of both enantiomers published in 1994 (Holton et al. 1994a, b), Holton addressed the need for an efficient semi-synthesis of taxol realizing that a total synthesis might not solve the procurement issue within a reasonable time and cost frame. Key steps in Holton's semi-synthesis are a [2+2]-cycloaddition to establish the two contiguous stereocenters of the side chain and an efficient coupling with a selectively protected baccatin III established by Potier et al. (Scheme 6.8).

The efficient semi-synthetic approach was filed by Holton in April 1992. Holton described a way to directly couple the baccatin III intermediate with the β-lactam

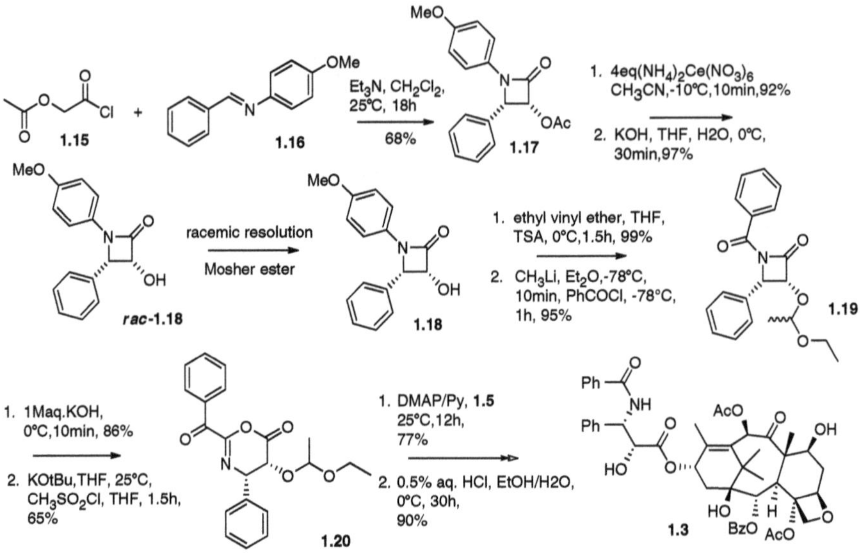

Scheme 6.8 Holton semi-synthesis of taxol

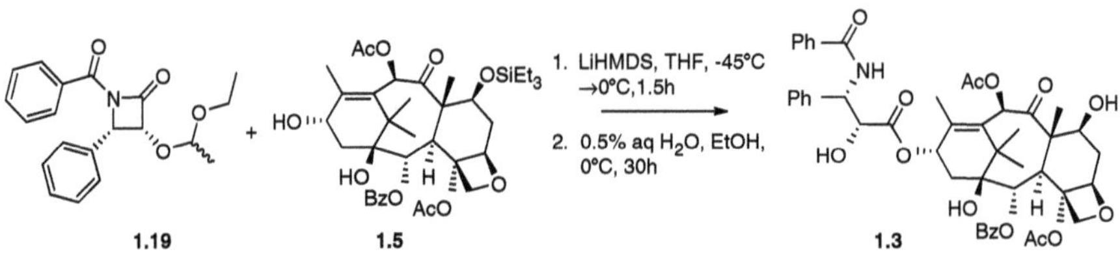

Scheme 6.9 Second generation semi-synthesis of taxol by Holton

using a metal alkoxide (Holton, 1993 US 5,274,124). By deprotonation of the C13 alcohol with lithium hexadisilazide the β-lactam could be selectively opened to form the natural side chain. This not only was the shortest and most efficient semi-synthesis for taxol it does also opened a way for side chain analogs (Scheme 6.9).

Ojima et al. contributed a further improved enantioselective side chain synthesis using an enolate imine cyclocondensation reaction (Ojima et al. 1991) (Schemes 6.10 and 6.11).

The three major breakthroughs of a renewable and easily adjustable intermediate, an unproblematic accessible side chain precursor and an efficient coupling method made the major success of semi-synthetic taxol possible. The process was in licensed by Bristol-Myers and in 1994 the Food and Drug Administration (FDA) approved semi-synthetic taxol for cancer treatment.

It was not only the invention of a very efficient synthetic access (80% in four steps), it was also going beyond an all too common prejudice: It is not worth pursuing a semi-synthetic approach compared to the synthetic Mount Everest of a total synthesis.

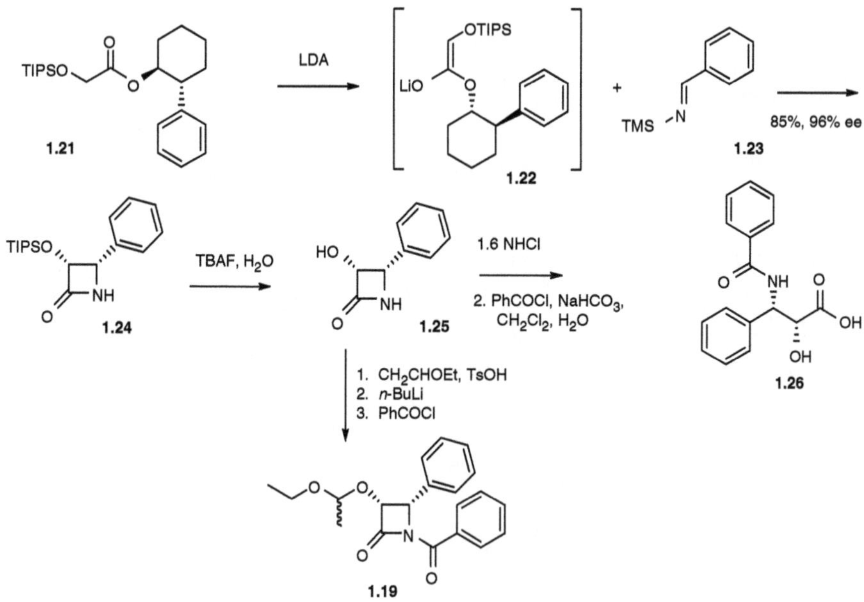

Scheme 6.10 Side chain synthesis by Ojima et al.

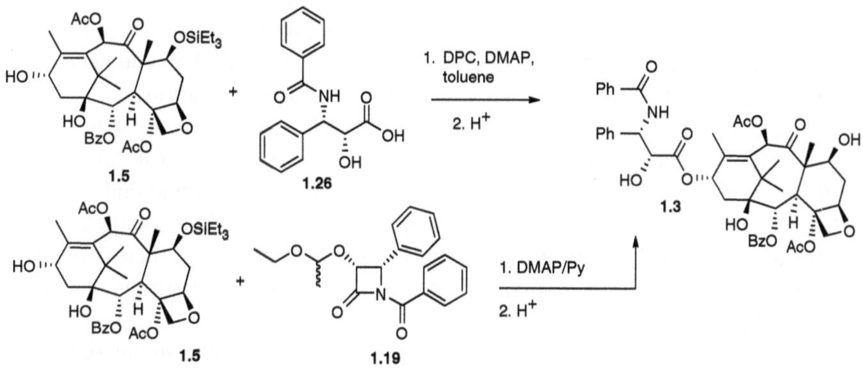

Scheme 6.11 Semi-synthesis of taxol by Ojima et al.

It was Robert Holton's farsighted view to set the priorities right. Although he also climbed the Mount Everest by reaching the top first (Holton et al. 1994a, b), he published his total synthesis after Nicolaou et al. in 1994 (other total synthesis were published by Danishefsky et al., Wender et al. and Mukaiyama et al.) (Holton et al. 1994a, b; Nicolaou et al. 1994; Danishefsky et al. 1996; Wender et al. 1997; Mukaiyama et al. 1999). He pursued first the route having the highest impact for cancer patients worldwide.

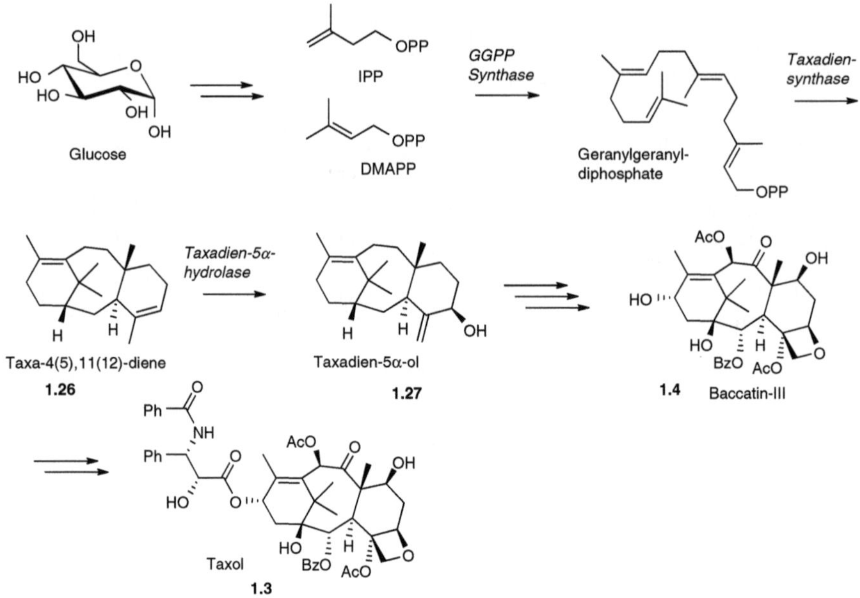

Scheme 6.12 Taxol biosynthesis

In addition to the interest in a semi-synthetic and synthetic access of paclitaxel the biosynthesis of this molecule was equally important (Croteau et al. 2006). The biosynthesis of baccatin-III as an intermediate in the paclitaxel biosynthesis is furthermore of mutual importance for a sustainable semi-synthetic access.

By gaining a profound insight into the paclitaxel biosynthesis a production of baccatin III has been established using cell culture technology (Yukimune et al. 1996). This process was used to guarantee the supply for the commercial synthesis of paclitaxel and taxotere. Even a heterologous expression of the synthetic gene-cluster in bacteria can be used for the production of the semi-synthetic precursor and will eagerly lower the production costs for drugs belonging to the taxane family (Ajikumar et al. 2010) (Scheme 6.12).

Beside from the accessibility of the drug there was a second major drawback connected to taxol. The compound was virtually insoluble in water. This problem was addressed by the NCI using formulation called Cremophore EL made out of Castor oil. Cremophore EL is a widely used polyoxyethylated castor oil. With the solubility being a show stopper at this time, this was the only solution at hand. It later turned out to be a serious problem and a driving force for invention. Associated with the Castor oil formulation is a hypersensitivity reaction in many patients, making a premedication with high doses of corticosteroids and H1, H2 antagonists necessary. Even with this premedication side reactions were observed in 41–44% of the treatments. In 1.5–3% these side effects were life threatening. Besides hypersensitivity, neurotoxicity seems also at least in part be associated with the castor oil formulation (Gelderblom et al. 2001). To address the solubility and formulation issues many efforts have been made ranging from structural modifications to alternate

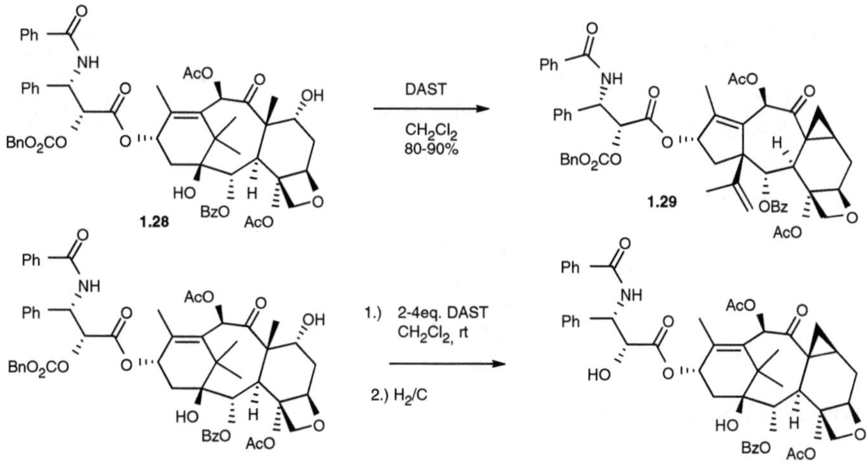

1.28

1.29

1.28

1.30

Scheme 6.13 Synthesis of a C7-C8 cyclopropyl analog

formulation vehicles. Abraxis Bioscience developed a solvent free paclitaxel formulation. Paclitaxel is bound to human albumin in this nanoparticle injectable. In 2005 Abraxane was approved by the FDA for breast cancer treatment. The highly functionalized di-terpenoid nucleus of taxol not only provides a synthetic challenge but is also subject to unusual and often unexpected transformations. One of these transformations found by serendipity even led to a compound now in phase III clinical trials. Unexpected transformation and skeleton rearrangements always have the potential advantage to alter the ligand backbone which may eventually lead to a compound more efficient in multidrug resistant (MDR) tumors. While the overall pharmacophore stays in place subtle alterations of the backbone may lead to compounds being less effectively excluded by efflux pumps or transporters. A group at Bristol-Meyers-Squibb trying to establish a more thorough SAR around the C7 position of taxol found by chance an interesting transformation. Upon treatment of 7-epi-taxol with DAST (Diethyl amino sulfur trifluoride) the C7C8 cyclopropyl forms in conjunction with a ring contraction of the A-ring to form a five-membered ring. By lowering the reaction temperature and the equivalents of DAST the A-ring contraction could be avoided (Chen et al. 1993). After hydrogenation the generated cyclopropyl analog of taxol showed very similar bioactivity compared to the parent compound (Scheme 6.13).

The initially formed 5-7-6 backbone showed remarkably diminished activity on cancer cell lines (KB cell culture) though interestingly enough compounds from the same backbone showed similar activities in tubulin depolymerization assays to taxol itself (Samaranayake et al. 1991).

Combining the taxotere side chain with the cyclopropyl di-terpenoid scaffold Sanofi-Aventis introduced a compound later named Larotaxel into clinical trials. Larotaxel is currently in clinical trials phase III against metastatic breast cancer pretreated with taxanes (Diéras et al. 2008). Especially against P-gp-mediated efflux

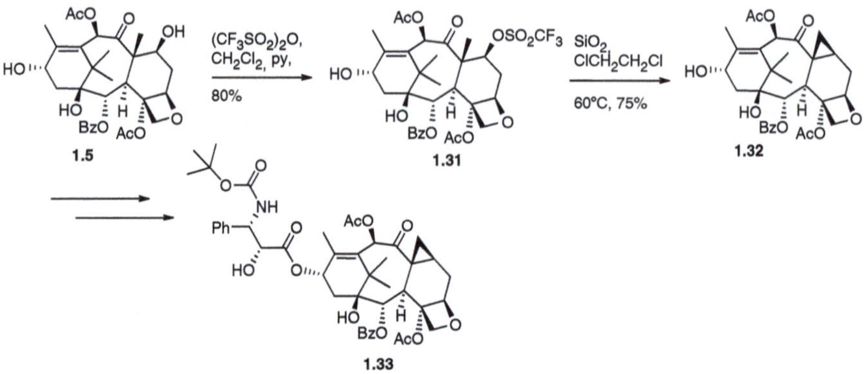

Scheme 6.14 Synthesis of larotaxel

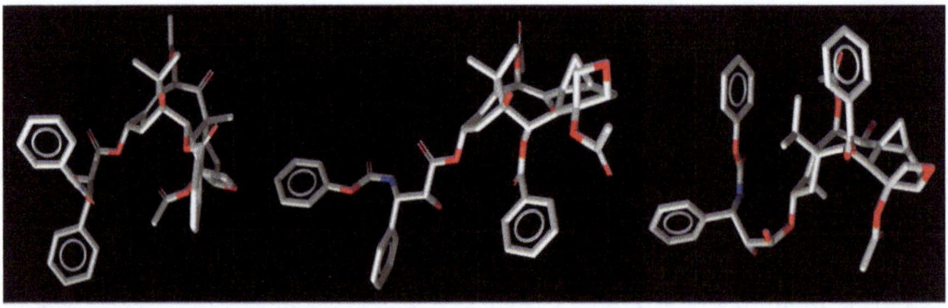

Scheme 6.15 Comparison of taxol, CP analog of taxol, and the A-ring contracted CP analog

larotaxel showed promising results compared to the parent compounds taxol and taxotere (Schemes 6.14 and 6.15).

An C7,C10 bis-methoxy analog of taxotere got recent FDA approval for the treatment of hormone refractory prostate cancer in combination with prednisone. Sanofi-Aventis marketed this cabazitaxel named compound under the trade name Jevtana. Even this very subtle structural change seems to influence the solubility and PK properties compared to the parent compound. The elegant semi-synthesis starts from 10-deacetyl-baccatin III. The 7 and 10 position of the tetrol were selectively alkylated under low temperature conditions (Didier et al. 1999 US 5,962,702; Bissery, 2002 US 6,403,634) (Scheme 6.16).

As depicted in Scheme 6.7 next to paclitaxel, taxotere, carbazitaxel, and abraxane there are at least six further taxanes in clinical trials. Larotaxel is in phase III clinical trials for pancreatic cancer, milataxel in phase II clinical trials for mesothelioma, ortataxel is in phase II clinical trials for taxane-resistant tumors, tesataxel is in phase II clinical trials for gastric cancer, colorectal cancer, and melanoma, BMS-275183 was in phase II clinical trials for none small cell lung cancer (clinical trial was terminated because of highly variable pharmacokinetics), TPI-287 is in phase II clinical trials for prostate cancer and pediatric CNS cancer (Dumontet and Jordan 2010) (Scheme 6.17).

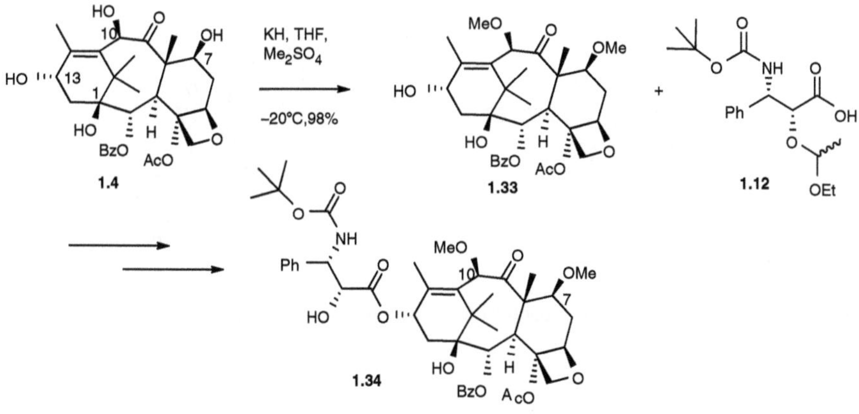

Scheme 6.16 Semi-synthesis of cabazitaxel

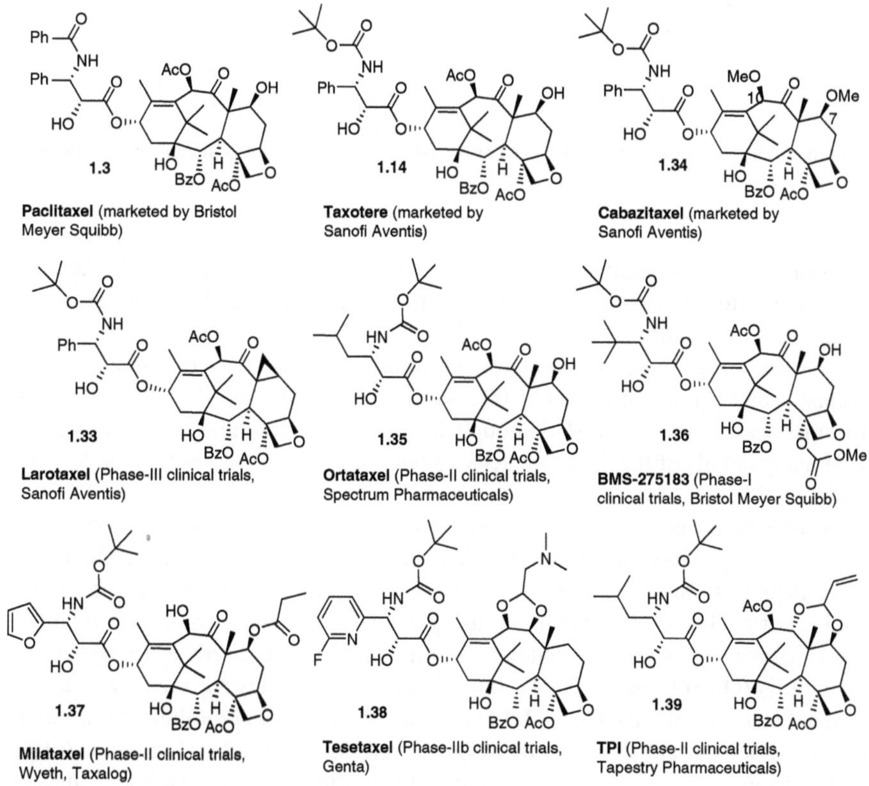

Scheme 6.17 Taxanes on the market or in clinical trials

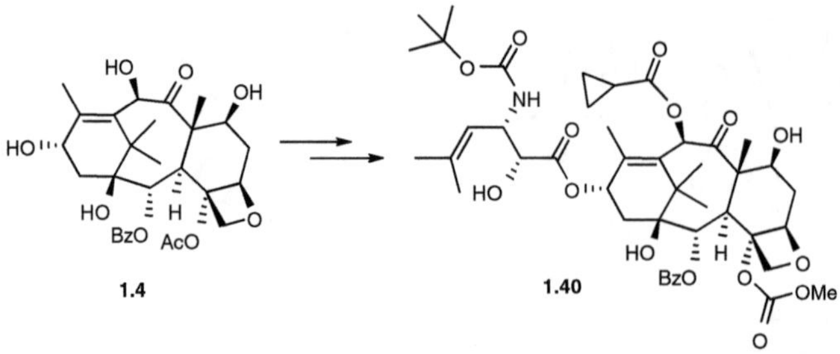

Scheme 6.18 Ojimas synthesis of paclitaxel anolog **1.40**

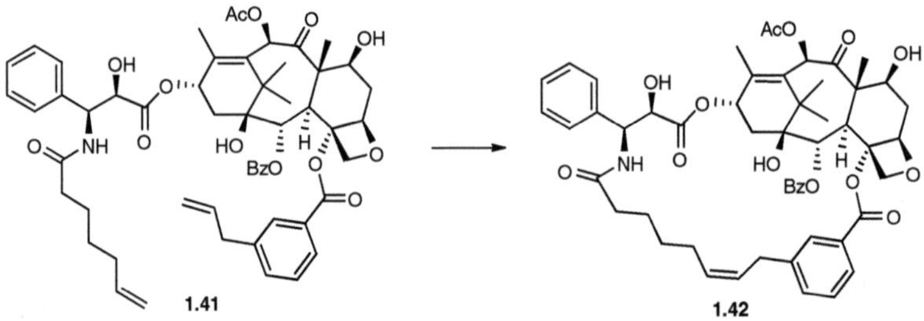

Scheme 6.19 Synthesis of conformational restricted paclitaxel analogs by Guéritte and Dubois et al.

Alongside the compounds already in clinical trials or on the market a tremendous amount of effort was used to develop even more selective analogs which can evolve from the taxane scaffold.

As many other groups and pharmaceutical companies the groups of Houlton, Kingston and Ojima (Ojima et al. 2008) synthesized a plethora of analogous starting from 10-deacetyl-baccatin III. Although there is a lot of structural information available many details of the structure activity relationship are still not well understood (Scheme 6.18).

Snyder et al. postulated a T-shaped bioactive conformation based on crystal structures of paclitaxel on β-tubulin (Synder et al. 2001). To challenge this model the groups of Ojima, Guéritte, and Kingston designed conformationally restricted analogs of paclitaxel. Guéritte et al. using a side chain amine and a substituent on the benzoate side chain to install two terminal olefins utilized in a ring closing metathesis (Querolle et al. 2004). Interestingly both the acyclic precursor and the bridged macrocyclic derivative **1.42** showed comparable microtubule disassembly inhibitory activity and cytotoxicity to cancer cell lines as docetaxel (Scheme 6.19).

Kingston et al. generated conformationally restricted bridged paclitaxel analogs by ring closing metathesis. Compounds **1.44** and **1.45** showed up to 50-fold improvements in terms of their activity against ovarian cancer cells (A2780) (Ganesh et al. 2007). The activity in the tubulin polymerization assay and against pancreatic

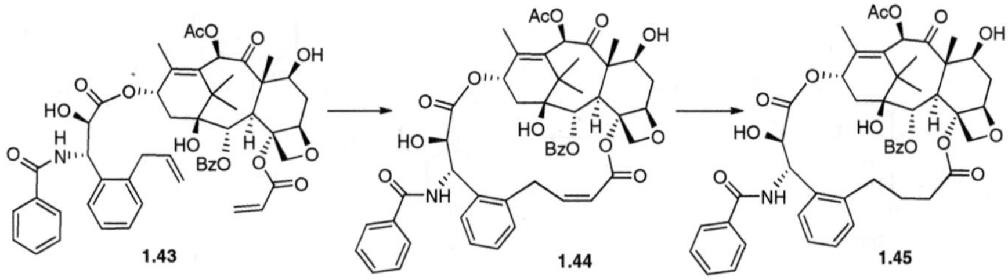

Scheme 6.20 Synthesis of conformational restricted paclitaxel derivatives by Kingston et al.

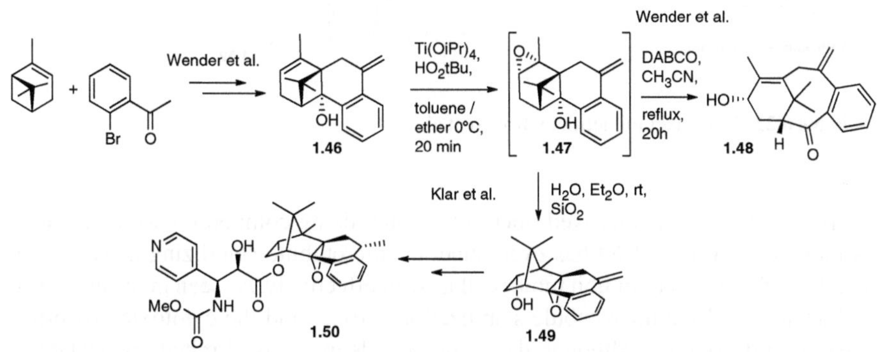

Scheme 6.21 Synthesis of borneol derivatives starting from a synthetic precursor of taxol by Klar et al.

cancer cells are in a comparable range to paclitaxel. Despite the impressive cellular activity these analogs come with the penalty of increased synthetic complexity. None of the less they give important insights into the bioactive conformation of paclitaxel (Scheme 6.20).

Despite the strong antitumor activity of paclitaxel, docetaxel and the other taxane analogs in clinical trials or on the market there are many modes of resistance (e.g. P-gp efflux and tubulin variability) which might be related to the taxane scaffold itself. The structure-related resistance and the still cost intense treatments for active compounds starting from 10-deactyl baccatin III make it worthwhile to search for structurally less complex scaffolds directing a similar arrangement of pharmacophoric groups as the parent compounds.

Klar et al. at Schering redirected a synthetic protocol developed by Wender et al. during the pinene route to taxol (DE 4416 374 1995; DE 195 13 040 1996; Klar et al. 1998; Wender and Mucciaro 1992), generating borneol derivatives. In the attempt to isolate the epoxide intermediate after aqueous work-up Klar et al. isolated a rearranged product. After esterification with the taxotere side chain, a weak but significant microtubule depolymerisation activity was observed (Scheme 6.21).

After optimization of the side chain and slight modifications to the scaffold (hydrogenation of the double bond generated inactive β-methyl and active α-methyl

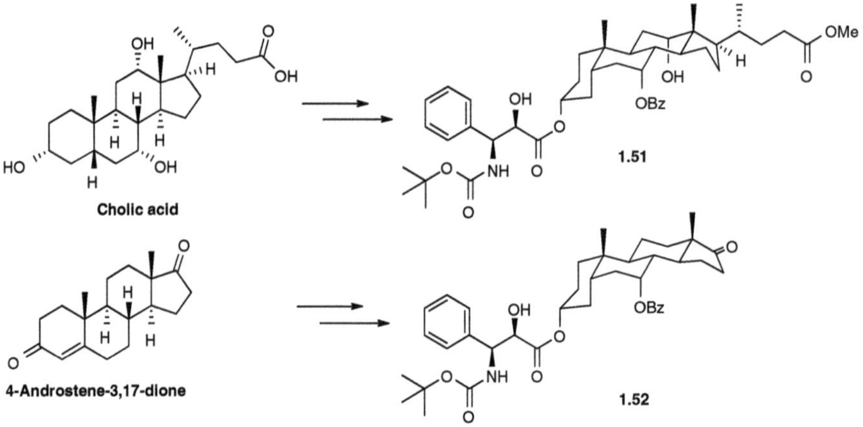

Scheme 6.22 Docetaxel mimetics by Roussi et al.

derivatives) strongly improved microtubule stabilizing compounds were obtained. Although compound **1.50** has even superior microtubule stabilizing activity than taxol itself, it does not inhibit tumor cell growth effectively. As seen in the cyclopropyl analog **1.29** the microtubule stabilization activity and the cytotoxic properties have been decoupled. Although these compounds may have limited application in tumor depression they might be interesting starting points for the treatment of Alzheimers disease. For example it has already been proven that taxol very effectively protects against β-amyloid toxicity in primary neurons (Michaelis et al. 1998). Compound **1.50** showed higher protective activity against β-amyloid-induced toxicity of primary neurons than taxol itself. The survival rates were as high as 89% with no reduced cell viability observed (Michaelis et al. 2005). The non-cytotoxic properties of compound **1.50** compared to members of the taxane family are a clear advantage for this application.

Despite the efficient semi-synthetic approaches to the taxane scaffold a much simpler scaffold reachable by synthesis with retained biological activity would still be desirable. The groups of Ojima, Kingston, Roussi, and many others were working on structurally simpler taxane mimetics. Starting from cholic acid and 4-androstene-3,17-dione Roussi and Guéritte synthesized two docetaxel mimetics with the parent phenyl isoserine and benzoate side chains. Molecular modeling studies showed a good overlay with the T-shape docetaxel conformation, thought to be representative of the active tubulin binding conformation (Roussi et al. 2005). Both compounds could be readily synthesized in a few steps starting from the natural products. Unfortunately they showed no microtubule disassembly inhibitory activity. Methyl ester **1.51** showed a weak activity on MCF7 cells together with a significant cytotoxicity on KB cells (Scheme 6.22).

Ojima developed a synthetic route to 5-7-6, 5-8-6, 5-6-6 tricyclic, and 5-6 bicyclic scaffolds which were decorated with the phenyl isoserine side chain of taxol and bearing the acetoxy and benzyloxy functionalities of the parent natural product

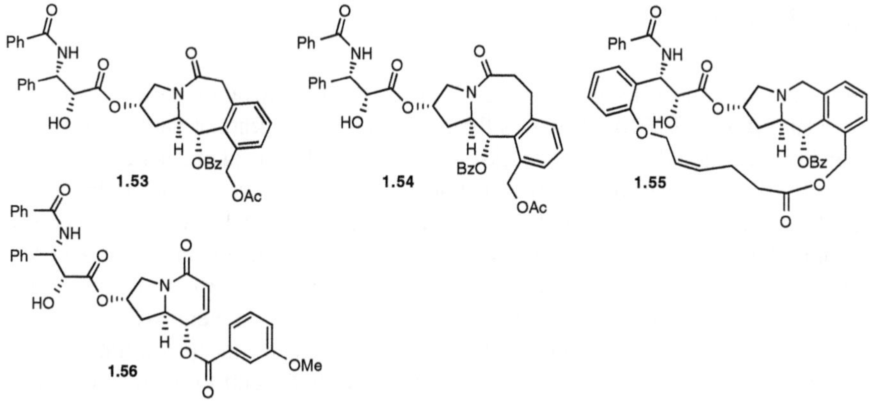

Scheme 6.23 Paclitaxel mimetics by Ojima et al.

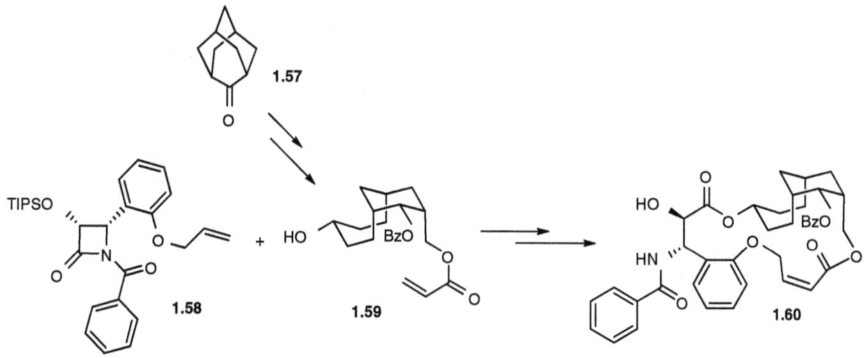

Scheme 6.24 Paclitaxel mimetics by Kingston et al.

(Sun et al. 2010). The 5-7-6 tricyclic paclitaxel mimic **1.53** was the most potent against drug-sensitive human breast cancer cell lines. Although less active than paclitaxel itself the decrease in synthetic complexity makes these mimetics very interesting starting points for optimization (Scheme 6.23).

Starting from commercially available 2-adamantanone Kingston et al. generated a macrocyclic paclitaxel mimicking the T-taxol bioactive conformation (Ganesh et al. 2006). The measurements of bioactivity and tubulin stabilization were complicated by the low solubility in DMSO water solutions. Although compound **1.60** is less cytotoxic as paclitaxel on ovarian cancer cell line A2780 this may well be due to its insolubility (Scheme 6.24).

While in its 40s it seems fair to say that taxol has many fascinating sides still to be discovered. There is something very special going on for this scaffold as the above incredible efforts are showing: in this case nature does not give up its secrets easily.

6.3 Eribulin

Esai announced on 15th of November 2011 the FDA approval of Halaven (Eribulin mesylate) (Ledford 2010). This was the celebration of 25 years of hard work in labs both in academia and in industry (Jackson et al. 2009). The parent molecule from which initially eribulin was developed is Halichondrin B. Halichondrin B was originally isolated from the marine sponge *Halichondria okadai* Kadota collected on the coast of Aburatsubo in the Miura Peninsula south of Tokio. The same organism okadaic acid had been isolated from. Okadaic acid is remarkably toxic, unfortunately in vivo not showing any antitumor activity if dosed in subtoxic amounts (Scheme 6.25).

H. okadai Kadota is a black-colored sponge living in the mediolittarol zone (intertidal zone in particular the middle shore). Initially 600 kg of material were collected (Uemura et al. 1985; Hirata and Uemura 1986). In a tedious procedure 12.5 mg of halichondrin B were isolated from the original 600 kg of biomaterial by Hirata and Uemura using a bioassay guided fractionation. Already in the first publication together with NMR and X-ray structural data, very exciting in vitro (B-16 melanoma cells) and in vivo data was published. The structure of the nearly related norhalicondrin A was unambiguously elucidated by X-ray crystallography of the corresponding *p*-bromophenacyl ester (Scheme 6.26).

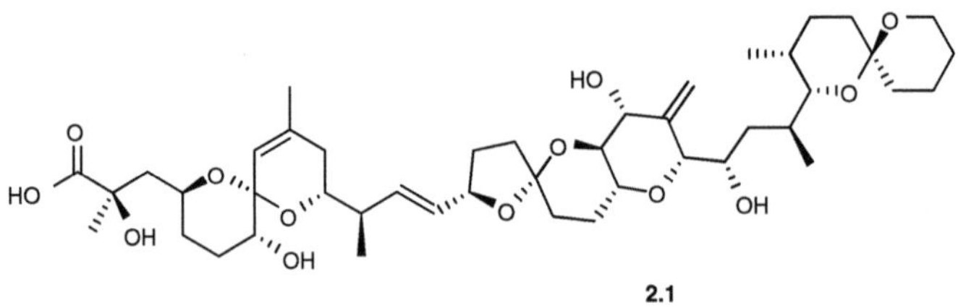

Scheme 6.25 Structure of okadaic acid

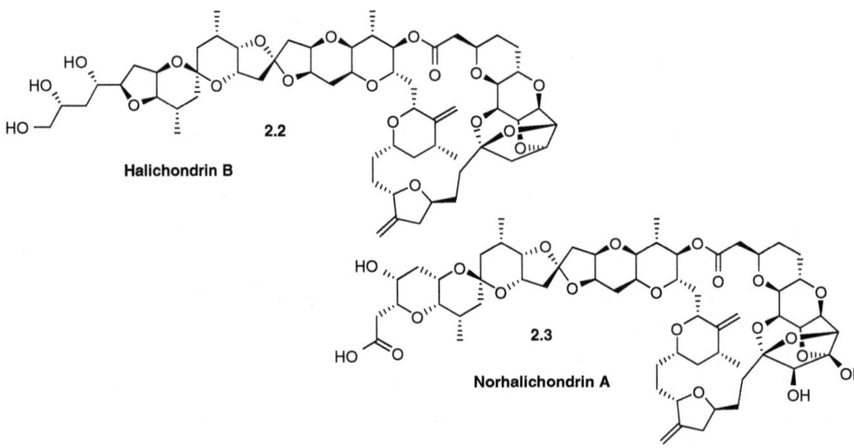

Scheme 6.26 Structures of halichondrin B and norhalichondrin A

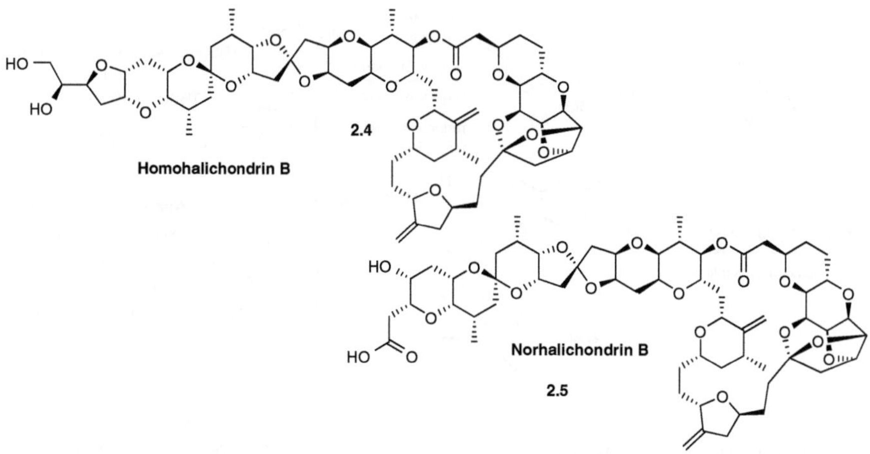

Scheme 6.27 Structures of homohalichondrin B and norhalichondrin B

Together with halichondrin B a whole class (seven further compounds) of structural-related compounds were isolated and their structure elucidated. In 1988, Blunt and Munro isolated the halichondrins from an unrelated New Zealand sponge the black shallow-water Axinellid *Raspalia agminata* (Internal research report Dr. R. J. Lake et al. 1988). From the bright yellow sponge *Lissodendoryx* sp. (family Myxillidae, order Poecilosclerida) collected from deep water off the Kaikoura Peninsula Munro and Blunt isolated isohomohalichondrin B (Litaudon et al. 1994) (Scheme 6.27).

From the Western Pacific marine sponge *Axinella* sp. collected in Palau Pettit et al. isolated homohalichondrin B and halichondrin B (Pettit et al. 1991). In spite of the diversity of locations and species it was speculated whether the halichondrins are secondary metabolites by a symbiont of the sponge rather than metabolites of the sponge itself (Scheme 6.28).

From all of the isolated compounds of the halichondrin structural family halichondrin B seems to be the most active presenting subnanomolar IC_{50} values in cancer cell tissue cultures (Bai et al. 1991).

Halichondrin B is binding to tubulin without interfering with colchicine. It binds to the vinca domaine of tubulin but not exactly to the vinca site. It inhibits vinca alkaloids (especially vinblastine) binding to tubulin in a non-competitive manner. In vitro it has been shown that halichondrine B disrupts normal mitotic spindle architecture and induces cell cycle arrest in the G_2/M phase. The in vivo data was complemented by very encouraging in vivo data as well. In several human solid tumor xenograft models halichondrin B showed exceptional results. Owing to the extraordinary bioactivity halichondrin B was recommended by the NCI for preclinical trials in 1992.

The amounts isolated from the original extraction by Hirata and Uemura were discouraging. A re-isolation project by Blunt and Munro from *Lissodendoryx* sp. did

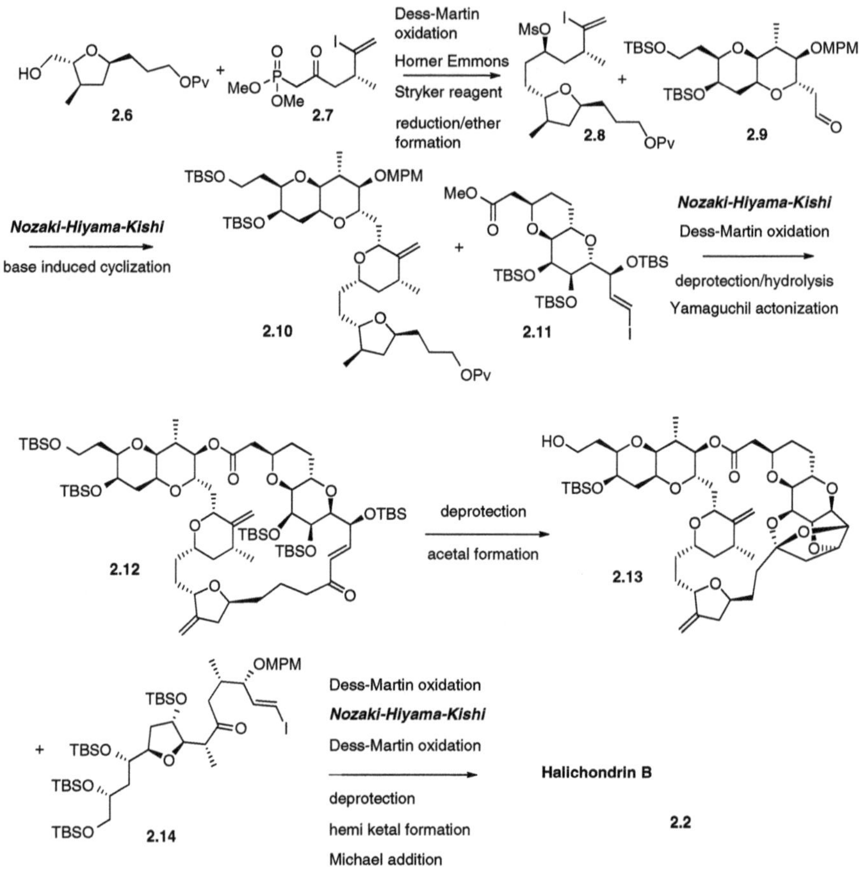

Scheme 6.28 Total synthesis of halichondrin B by Kishi et al.

also not produce halichondrin B in sufficient amounts. From 200 kg of sponge Blunt and Munro isolated 43 mg of halichondrin B (Litaudon et al. 1997). A pilot study to cultivate the sponge in aquaculture did reveal an even decreased amount of isolated halichondrin B relative to the harvested biomaterial (Munro et al. 1999). The only sustainable method for the procurement of halichondrin B emerged with the iconic hallmark total synthesis of Aicher et al. (1992).

The Kishi synthesis of halichondrin B is characterized by a highly convergent approach. Kishi's synthesis really mastered the Ni/Cr Nozaki–Hiyama–Kishi cross coupling reaction for bringing the convergent synthetic intermediates together. This together with the readily accessible carbohydrate starting materials made already Kishi's original synthesis highly efficient. As the success of halichondrin-derived compounds in the clinic was surfacing the importance of a synthetic access to the halichondrins became even more pressing. Accordingly the Kishi group continuously improved the synthetic route to halichondrin B (Duan and Kishi 1993; Stamos and

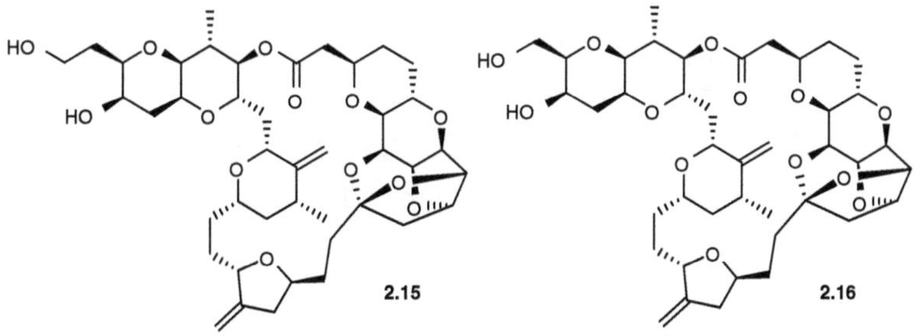

Scheme 6.29 Structure of the macrolactone diol part of halichondrin B

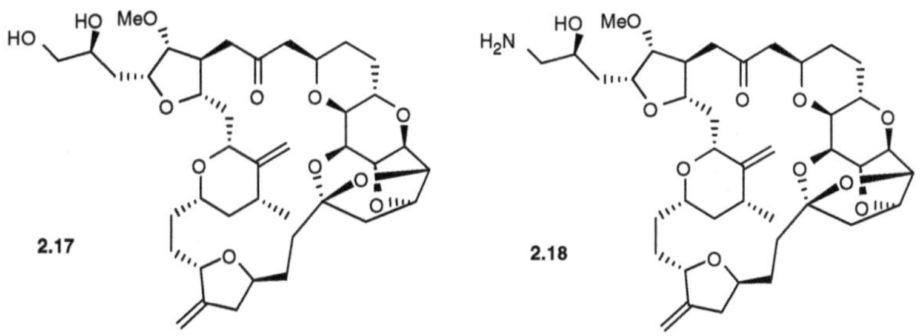

Scheme 6.30 Structures of the drug candidates ER-077349 and ER-086526 (ER-7389, Eribulin)

Kishi 1996; Stamos et al. 1997; Guo et al. 2009; Kim et al. 2009; Dong et al. 2009). Of equal importance as the access to halichondrin B was the synthetic access to unnatural analogs being generated using the same synthetic protocol. Although there were numerous natural products isolated showing attractive bioactivity, they also had the molecular complexity with halichondrin B in common (Scheme 6.29).

In 1992 samples of the synthetic halichondrin B and several synthetic intermediates were send to the Esai Research Institute by the Kishi group for cell growth inhibitory activity evaluation. The most remarkable discovery was the activity of the macrocyclic ketone **2.17** comprising the C1-C38 fragment of halichondrin B (Wang et al. 2000; Kishi et al. 1995). The in vitro results of compound **2.17** and halichondrin B, although a lot less complex in structure, were within a magnitude comparable. With this observation a synthetic access to halichondrin-related compounds became in reach of economic feasibility. Compound **2.15** was also prepared by the Kishi group and showed in vivo antitumor activity in the LOX human melanoma xenograft model. With these two structures in hand a viable starting point for a drug discovery program was reached (Scheme 6.30).

Although it is fair to say that Esai's decision to start lead optimization program from here was a remarkably brave one, this status would have never been reached without the exceptional total synthesis option provided by the Kishi group. It is by far the most impressive example of efficient contemporary organic synthesis and its impact on health of humankind. One of the major hurdles in the development of the

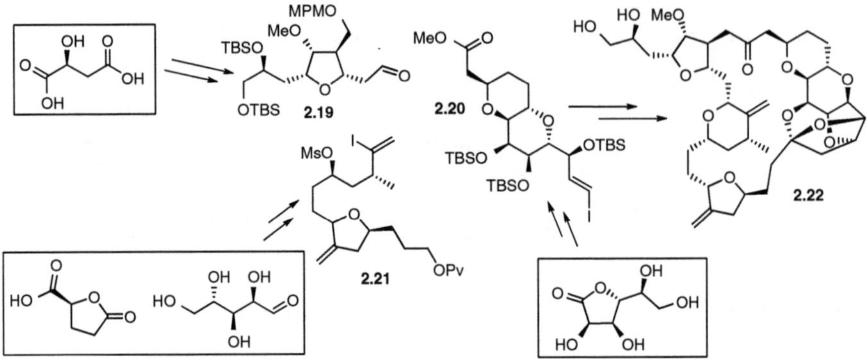

Scheme 6.31 Convergent key intermediates for the synthesis of **2.22**

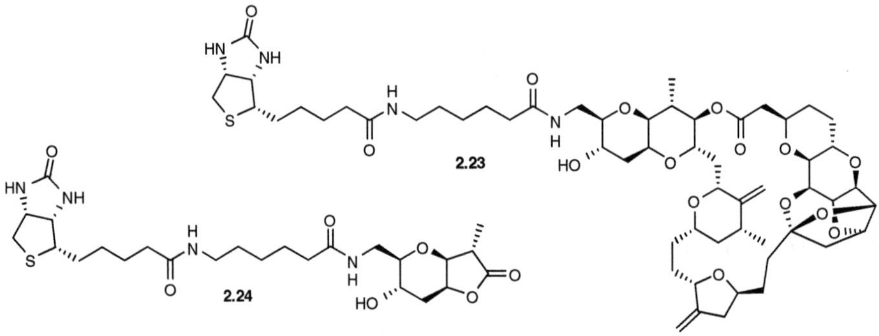

Scheme 6.32 Structure of the biotinylated macrolactone ER-040798 (**2.23**) and a negative control compound (**2.34**)

halichondrin-derived compounds is probably the instability of the macrocyclic lactone (Scheme 6.31).

At the Esai Research Institute over 180 analogs of halichondrin B were generated and tested. From this tremendous effort two macrocyclic ketone analogs were developed up to candidate status (Zheng et al. 2004; Littlefield et al. 1999). The major differences between halichondrin B and the two candidates were the transformation of the C1 lactone to a ketone functionality, a substitution of the C31 methyl with a methoxy group and the substitution of the C29-C38 tricyclic system with a single THF. In contrast to the macrolactone these macrocyclic ketones are stable even in plasma for a sufficient period of time underpinning the theory of a ketone being a more stable bioisostere to a lactone. ER-077349 and ER-086526 (ER-7389, eribulin) showed both a very exciting bioactivity in vitro (eribulin has an IC_{50} of 0.07 nM on MDA-MB-435 human breast cancer cells compared to 0.54 nM for vinblastine and 2.6 for paclitaxel). Eribulin and ER-077349 were also active on several other human cancer cell lines in contrast they did not show significant toxicity on MR-90 human fibroblast cell indicating excellent selectivity. The in vitro activities were in average very comparable to the activities of halochondrin B.

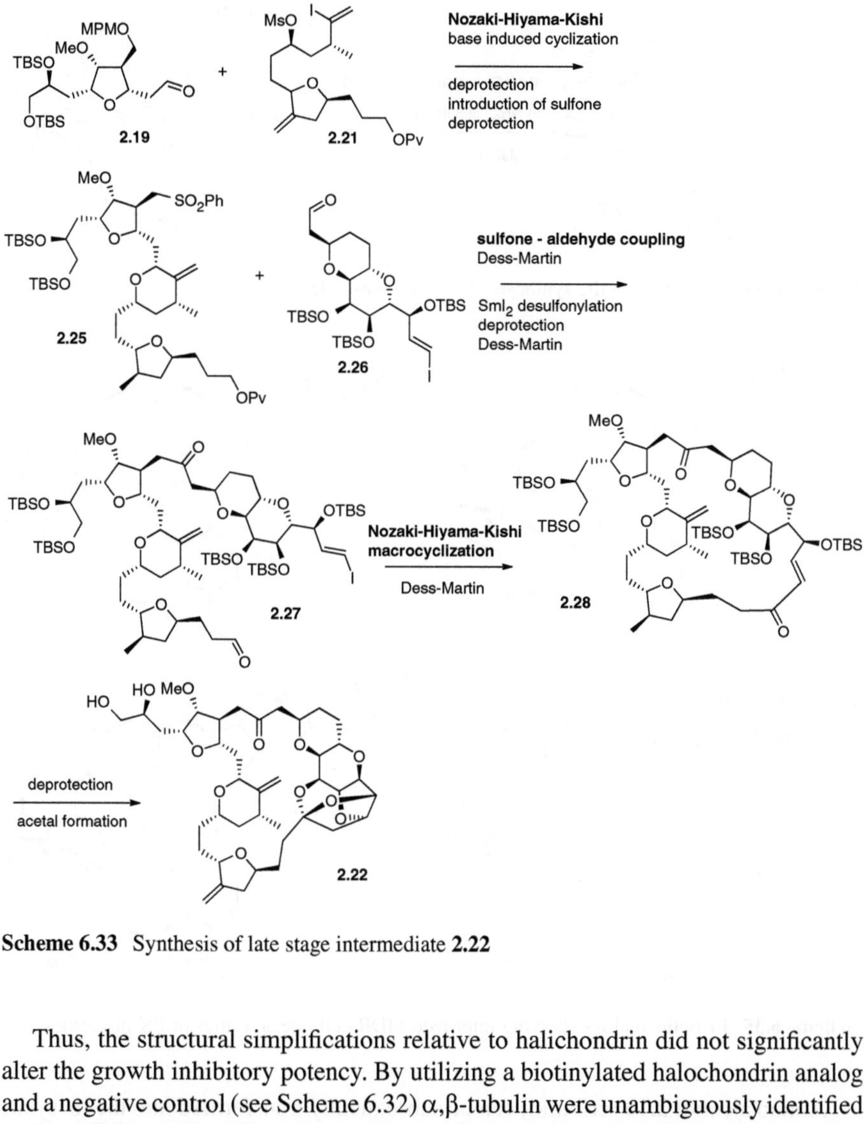

Scheme 6.33 Synthesis of late stage intermediate **2.22**

Thus, the structural simplifications relative to halichondrin did not significantly alter the growth inhibitory potency. By utilizing a biotinylated halochondrin analog and a negative control (see Scheme 6.32) α,β-tubulin were unambiguously identified as the cellular targets. By using cell flow cytometry the cell cycle arrest was localized in the G_2/M Phase.

As already the total synthesis of halichondrin B the Esai synthesis of eribulin is highly convergent and starting from carbohydrate synthons (Smith et al. 1995; Stamos and Kishi 1996).

The three convergent fragments were coupled by a Nozaki–Hiyama–Kishi reaction followed by a sulfone aldehyde addition and a subsequent Nozaki–Hiyama–Kishi macrocyclization (Scheme 6.33).

This late stage intermediate diol was furthermore transformed into the amino alcohol **2.18** (Eribulin).

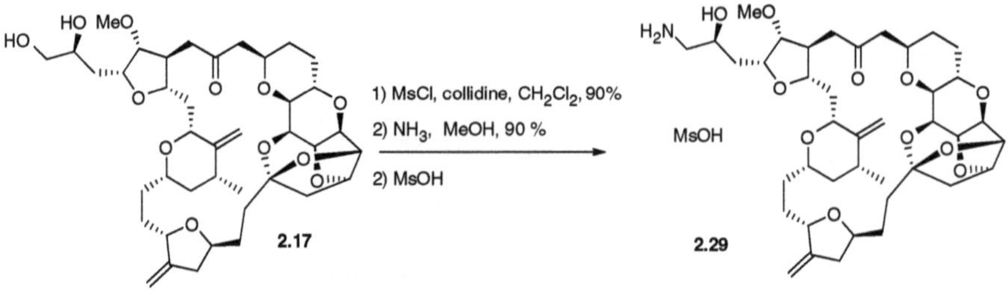

Scheme 6.34 Final steps in the synthesis of eribulin mesylate

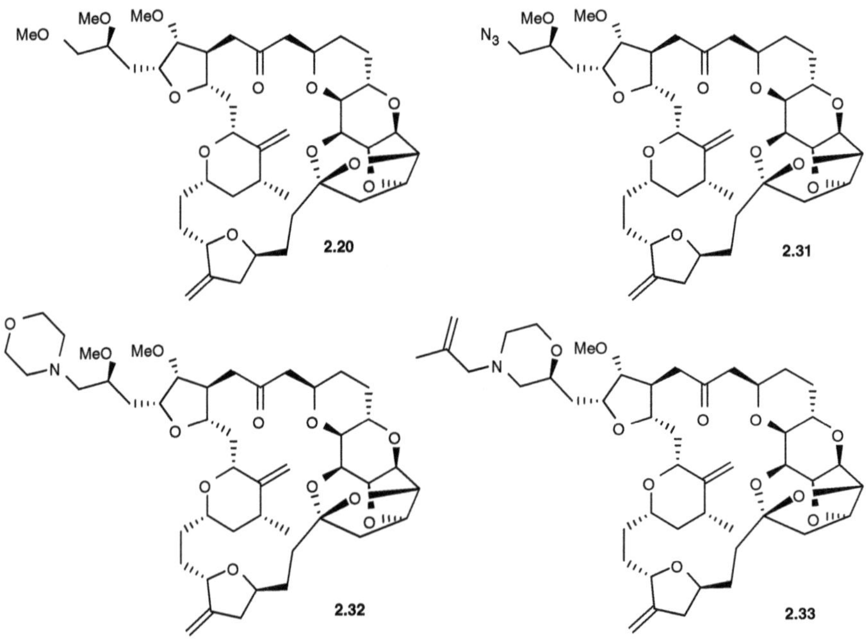

Scheme 6.35 Eribulin analogs showing improved MDR cell line activities or PK properties

Amino alcohol **2.18** (eribulin) can be subsequently transformed into the mesylate salt being the active ingredient in Esai's Halaven now on the market. After successful phase III clinical trials halaven got FDA approval in 2010 for metastatic breast cancer (Towle et al. 2001; Cortes et al. 2011; Cortes et al. 2010; Twelves et al. 2010).

Apart from being a key synthetic intermediate and showing interesting activities on its own right diol **2.17** can also function as a late stage intermediate for the generation of novel analogs with improved PK properties. MDR tumors are one major reason for shortages in clinical outcomes especially in the case of refractory tumors. One mechanism by which cancer cells can develop MDR is the overexpression of efflux pumps such as P-gp. Eribulin is a substrate for P-gp, as are paclitaxel and doxorubicin. For the diol **2.17** similar in vitro and in vivo activities compared to eribulin were detected but diol **2.17** was less susceptible to P-gp than eribulin (Scheme 6.35).

By comparison of the ratio of IC_{50} values on doxorubicin-resistant human sarcoma cells (MES-SA Dx5/Rx1) to IC_{50} values on the parental MES-SA cells a good indicator for P-gp susceptibility is obtained (fold resistance ratio (FR)). By correlating the FR value with calculated $logD_{7.4}$ values a decision can be made which compounds can be further pursued showing better activity on multidrug resistance cell lines (Narayan et al. 2011a).

Compounds **2.30** and **2.31** bearing more lipophilic side chains compared to eribulin showed a highly improved fold resistance ratio and displayed extremely low P-gp susceptibility. Unfortunately **2.30** and **2.31** showed also decreased activity on mouse xenograft models. By keeping the amine functionality in morpholine **2.32** with an decreased basicity the analogs comprised a better oral bioavailability and while keeping excellent activity on MDR cell lines displayed efficacy in DLD-1 xenograft mouse models (Narayan et al. 2011b). By further increasing the lipophilicity of the amine like in compound **2.33** high exposure levels to the brain and CSF could be achieved while retaining excellent in vitro activities (Narayan et al. 2011c).

6.4 Geldanamycin

Geldanamycin is a benzoquinone ansamycin which binds to the heat shock protein HSP90. Geldanamycin was first isolated by De Boer et al. from Upjohn by fermentation of *Streptomyces hygropicus* in 1970 (De Boer et al. 1970). The molecular structure of Geldanamycin was published shortly after by NMR and IR analysis (Sasaki et al. 1970). Antiparasitic, antifungal, and antibacterial activities were reported for geldanamycin. From the beginning the high reactivity of the 17-methoxy group made it unsuitable for antibiotic drug discovery. The 17-methoxy group is reactive towards nucleophiles commonly found in biological systems. There are several natural products described comprising similar bioactivity and structural relationship to geldanamycin (e.g. herbimycin and macbecin I and II) (Scheme 6.36).

In 1995 several analogs of geldanamycin were reported blocking the expression of the ErbB-2 oncoprotein in human breast cancer cell lines (Schnur et al. 1995a, b). While geldanamycin has been postulated to have cell growth inhibition capabilities by kinase (v-Src, Raf-1, ErbB2) inhibition (Fukasawa et al. 1991), subsequently inhibition of HSP90 could be shown in vitro (Whitesell et al. 1994). HSP90 is a 90 kDa member of a protein super family and does exist in two isoforms. HSP90α is the inducible form overexpressed in cancer cells, while HSP90β is the constitutive form (Biamonte et al. 2010). HSP90 is one of the most abundant proteins in unstressed cells (Wegele et al. 2004). It is ubiquitous molecular chaperone found in eubacteria and all branches of eukarya. HSP90 is an elongated dimer with a dissociation constant in the low nanomolar range. HSP90 consists of three major domains: a highly conserved amino-terminal ATPase domain, a middle domain, and a carboxy-terminal dimerization domain. As a response to cellular stress HSP90 gets overexpressed, predominantly in cancer cells. HSP90 helps its client proteins to fold in the bioactive conformation. Many of the HSP90 client proteins are

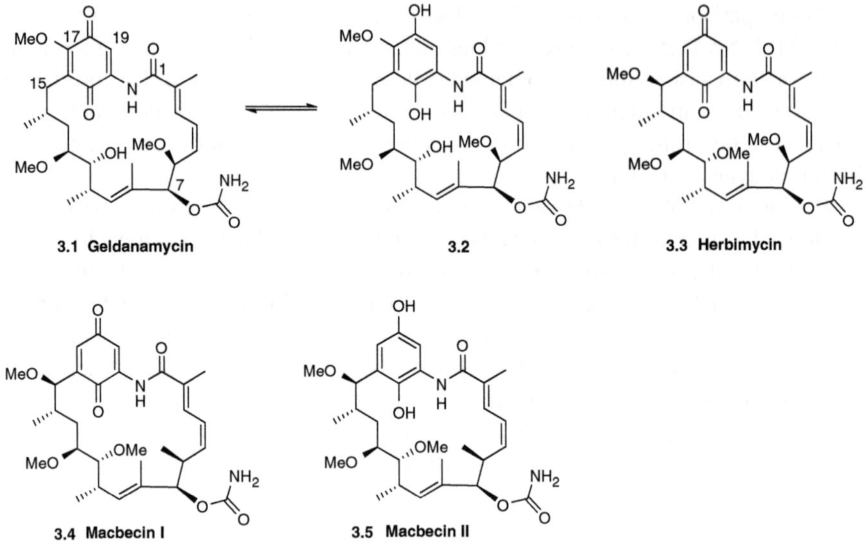

Scheme 6.36 Structures of ansamycins geldanamycin, herbimycin, and macbecin I and II

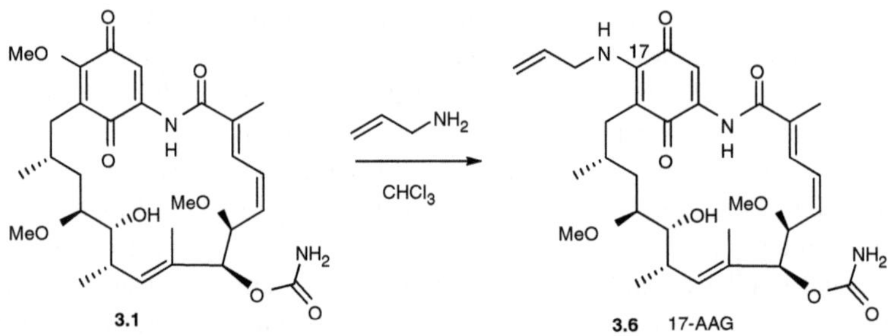

Scheme 6.37 Synthesis of 17-AAG (17-allyl amino geldanamycin)

notorious oncogenes. Some of these are furthermore validated cancer targets (HER-2/neu (Herceptin (trastuzumab)), Bcr-Abl (Gleevec (imatinib mesylate)) or VEGF (Sutent (sunitinib))). By inhibition of HSP90 the client proteins undergo degradation as a result of misfolding very effectively processed by an ubiquitinylation proteasome pathway.

Herbimicin and geldanamycin were the first ligands described to bind to HSP90. Due to the highly reactive 17-methoxy group geldanamycin showed unfortunately a high liver toxicity and was not suitable as a development candidate for drug discovery.

On the other hand this high reactivity could be used in a replacement reaction by amines at room temperature. The in vivo anti-cancer activity and the synthesis of these derivatives was first described in a patent application from Kaken Chemical Company (Sasaki and Inoue 1980) (Scheme 6.37).

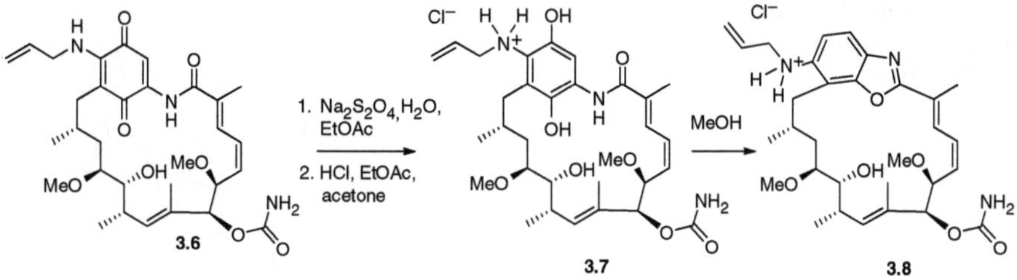

Scheme 6.38 Synthesis of the 17-AAG hydroquinone salt

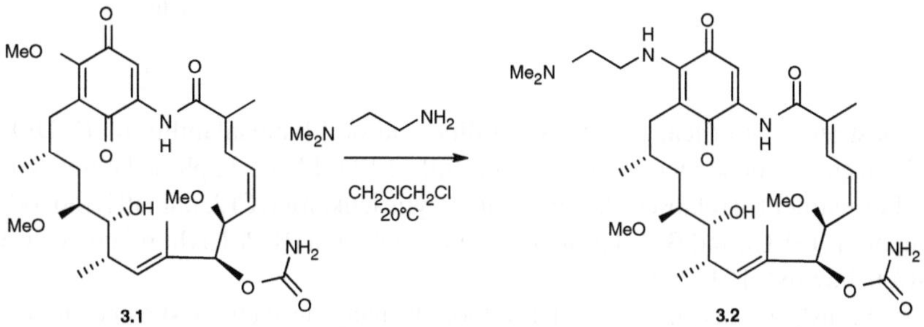

Scheme 6.39 Synthesis of 17-DMAG

Only after 1994 17-AAG was recognized to inhibit HSP90. In 1999 17-AAG entered clinical trials. Although to a lesser extent 17-AAG showed as the parent compound hepatoxicity. Furthermore 17-AAG is very insoluble and difficult to formulate. As shown in Scheme 6.36 geldanamycin (**3.1**) can be converted in vivo into the hydroquinone form **3.2**. Infinity Pharmaceuticals reduced 17-AAG with sodium dithionite and used the HCl salt of the hydroquinone a much better water soluble compound than the parent one (Adams et al. 2005). Unfortunately the hydro chloride salt **3.7** is only stable at −20°C not at room temperature. Under acidic conditions the benzoxazole analog formed under this condition and showed diminished bioactivity (Ge et al. 2006) (Scheme 6.38). The hydroquinone 17-AAG entered clinical trials in 2005.

By introducing an amino functionality into geldanamycin Kosan Inc. did also increase the solubility compared to the parent compound (Tian et al. 2004). More than 60 derivatives were synthesized and tested. From these 17-DMAG showed the most promising in vitro and in vivo profile. By probing the 7-carbamate position for superior substituents no residue with improved bioactivity could be identified (Rastelli et al. 2005). Previously it was shown that although many derivatives showed improved in vitro activity only simple alkyl amine or alkoxy substituents showed improved in vivo activities compared to geldanamycin (Scheme 6.39).

In 2006 17-DMAG entered clinical trials, which were discontinued in 2008 as a result of focussing the resources on a 17-AAG formulation. Independently the NCI

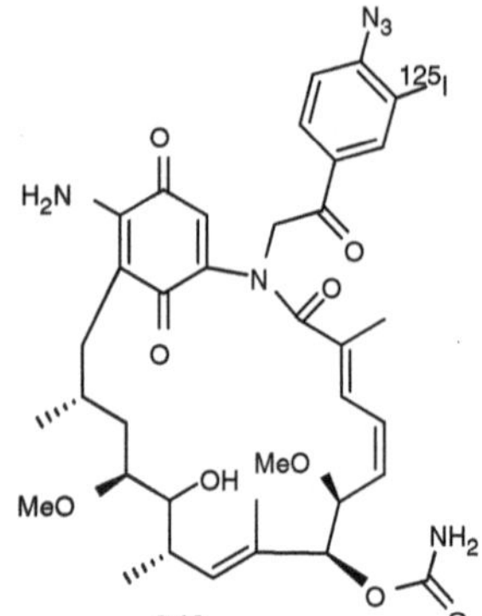

3.10

Scheme 6.40 Photoaffinity labeled 17-amino geldanamycin

demonstrated the efficacy, water solubility, and oral bioavailability of 17-DMAG and selected it for advanced studies. Currently 17-DMAG is in phase I clinical trials for the treatment of relapsed chronic lymphocytic leukemia (CLL), small lymphocytic lymphoma (SLL), and B-cell prolymphocytic leukemia (B-PLL) (http://clinicaltrials.gov/ct/show/NCT01126502).

Previously it was published that although many derivatives showed improved in vitro activity only simple alkyl amine or alkoxy substituents show improved in vivo activities. For photo affinity labeling studies the azido iodo phenacyl amide **3.10** could be synthesized and showed sufficient in vitro activity (Schnur and Corman 2006) (Scheme 6.40). The compound could be used for photoaffinity labeling proteins which interact with geldanamycin.

Geldanamycin, 17-AAG, hydroxyquinone 17-AAG, and 17-DMAG share a common unfavorable feature, the benzoquinone toxophore which undergoes redox cycles involving a NAD(P)H/quinine oxidoreductase (NQO1) catalyzed 2-electron reduction of quinone to dihydroquinone, leading to superoxide radical generation, which has been suggested as a component of their cytotoxicity (Kelland et al. 1999; Dikalov et al. 2002). 17-AAG and analogs have been shown to undergo Michael additions with glutathione at C17 leading to cellular depletion. There are at least two compounds in the literature bearing a phenol substitute instead of the quinone moiety. Kyowa Hakko Ltd. prepared geldanamycin derivatives devoid of quinones by reacting 17-AAG and anologs with methyl Grignard followed by reduction to obtain 4-methyl phenol derivatives **3.11** (Yamaguchi et al. 2007) (Scheme 6.41).

During the screening of extracts from fermentation of *Streptomyces* sp. from an Indonesian soil at Kyowa Hakko Ltd. four nonquinone HSP90 inhibitors were discovered inheriting interesting binding affinities to HSP90 and in vitro activities against a variety of cancer cell lines (Onodera et al. 2008) (Scheme 6.42).

By biosynthetic engineering of the macbecin I synthase cluster Biotica Ltd. developed nonquinone HSP90 inhibitors (Zhang et al. 2008). The macbecin biosynthetic gene-cluster was identified and its sequence obtained. Biotica's strategy was based on the thorough understanding of the quinone biosynthetic pathway for

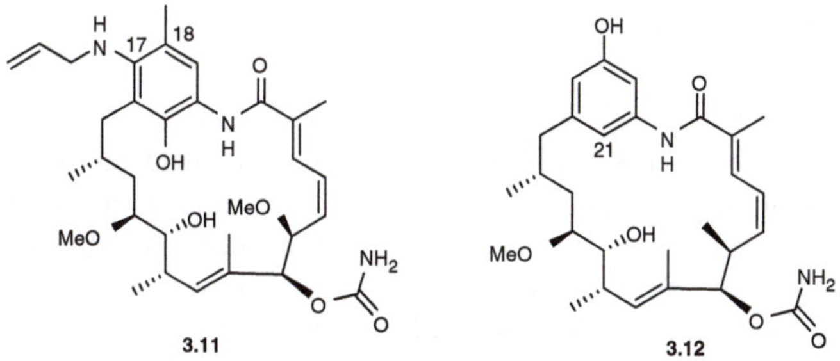

Scheme 6.41 Phenol analogs of geldanamycin and macbecin I

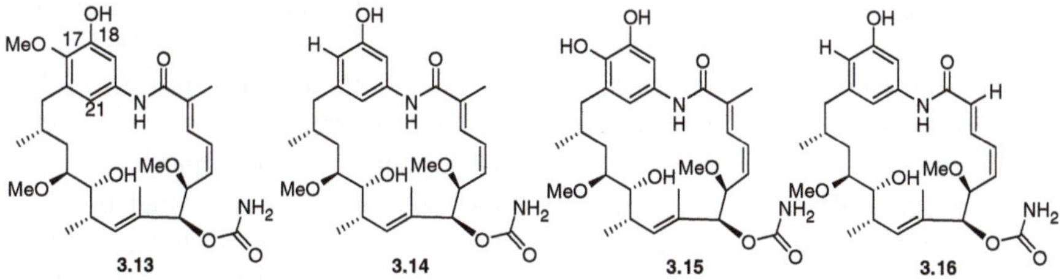

Scheme 6.42 Geldanamycin analogs isolated by Kyowa Hakko

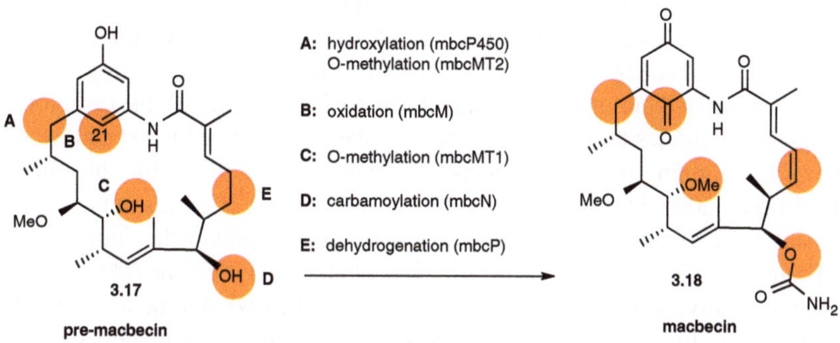

Scheme 6.43 Post-PKS modifications in the macbecin biosynthesis

ansamycin polyketides. The hydroxylation of a phenol precursor followed by auto oxidation occurred with molecular oxygen. Molecular genetic manipulation of the producing organism to delete mbcM (responsible for introducing oxygen at C21) yielded a novel strain producing in the fermentation broth **3.12**. C21-oxidation is the second post-polyketide synthase (PKS) step in the biosynthesis of macbecin I (Martin et al. 2011). To produce **3.12** as a single product the organism was further modified. Cultivation of the modified strain Biot-3982 led to the exclusive production of **3.12** in more than 200 mg/L fermentation titer. Mice dosed i.p. at 250 mg/kg/day for 5 days showed no apparent toxicity in contrast to previous studies with macbecin did (dosing 10 or 20 mg/kg) (Scheme 6.43).

6.5 Camptothecin

Camptothecin **4.1** was discovered by Wani and Wall (the discoverers of taxol) in 1966 during an extensive screening program of random plants products at the Cancer Chemotherapy National Service Center (Wall et al. 1966). Initially it was isolated from *Camptotheca acuminata* (Nyssaceae), later it has also been detected in several other non-related plant species (e.g. *Nothapodytes foetida* (Syn. *Mappia foetida*), Icacinaceae, *Ophiorrhiza pumila*, Rubiaceae). Despite its quinoline structure, camptothecin belongs biogenetically to a family of modified monoterpenoid indole alkaloids (Yamazaki et al. 2004) (Scheme 6.44).

Very soon after its discovery it was found, that camptothecin showed extremely promising results as an antitumor agent in animal models. The camptothecin sodium salt was then taken to clinical trials (Moertel et al. 1972; Hutchinson 1981; Cai and Hutchinson 1983), which were abandoned due to unfavorable properties like insufficient water solubility, substantial toxicity (Erickson-Miller et al. 1997; Giovanella et al. 1989), and—maybe most important of all—its rapid inactivation via lactone hydrolysis at physiological pH yielding the hydroxy acid which is devoid of any antitumor activity (Gottlieb and Luce 1972). In addition the acid binds very effectively to human serum albumine and is thus inaccessible for cellular uptake. The sodium salt is cleared by the kidneys, causing hemorrhagic cystis and myelotoxicity. The interest in the compound was renewed only several years later, when it became clear that camptothecin acts via a completely new mode of action interacting with a prime target hitherto not related to cancer therapy, topoisomerase I (Pommier 2006; Hsiang and Liu 1988). Thus the impetus was given for great efforts in the synthesis and biological profiling of less toxic and more soluble camptothecin analogs. All activities around camptothecin have been covered in several excellent reviews highlighting different aspects of research efforts (Venditto and Simanek 2010; Li et al. 2006; Du 2003), camptothecin is a member of a class of natural products sharing a highly conjugated and mostly planar polycyclic quinoline core (11H-indolizino[1,2-b]quinoline-9-one) (Das et al. 2001) (Scheme 6.45).

Due to relatively detailed structural models (see below) and the accessibility of biological data for a large number of structural analogs the rationale basis for the generation of new, improved camptothecin derivatives is strong. For the optimization there are several major objectives standing in the focus of attention like improved

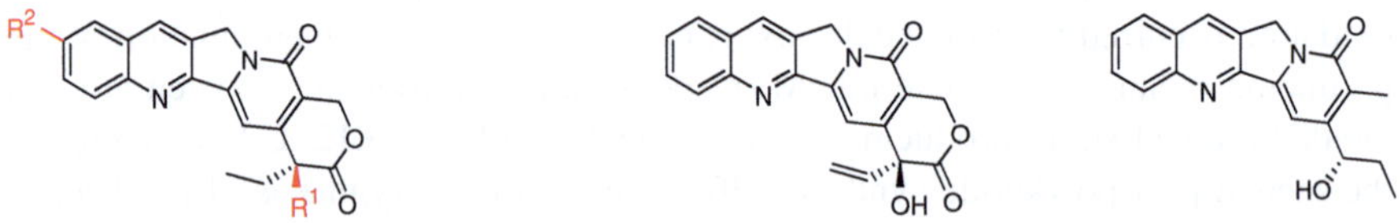

4.1 R^1 = OH, R^2 = H **Camptothecin**
4.2 R^1 = H, R^2 = H **Deoxycamptothecin**
4.3 R^1 = OH, R^2 = OMe **10-Methoxycamptothecin**
4.4 R^1 = OH, R^2 = OH **10-Hydroxycamptothecin**

4.5 18, 19-Dehydrocamptothecin

4.6 Mappicine

Scheme 6.44 Structural representation of naturally occurring camptothecins and analogs

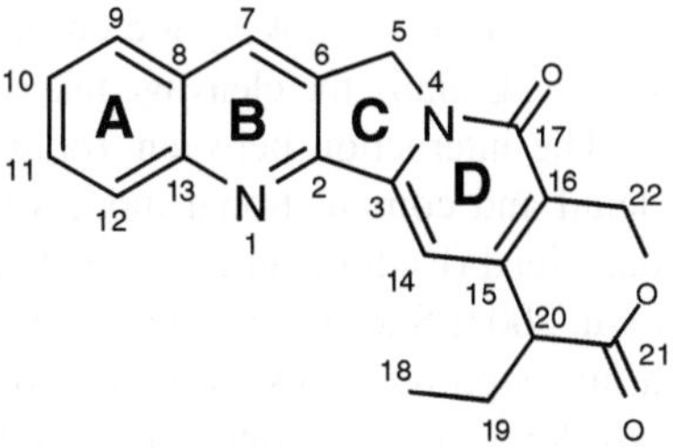

Scheme 6.45 The core structure of therapeutically relevant camptothecin derivatives

water solubility, improved plasma stability, reduced plasma protein binding and acute toxicity and susceptibility to defined enzymatic cleavage.

Important features of the SAR are:

- Planarity and conjugation of the ABCD ring system is mandatory. Disruption of planarity leads to a substantial loss of activity. The quinoline unit can be replaced by other ring systems.
- Modifications particularly in 9- and 10-position are well tolerated (see Topotecan and Irinotecan). Mono- and disubstitution with electron rich substituents in positions 9 and 10 often increase the biological activity. Modification at C7 is often used for the improvement of solubility and to improve the affinity to DNA.
- Substitutions at C11 are sometimes helpful.
- No modifications at C12 are tolerated.
- Substitution at C5 and C14 (CD-system) usually results in a loss of activity (exemption 14-aza-camptothecin). This is reflected by findings that the pyridone-amide plays a central role in the stabilization of the DNA/Topoisomerase/drug interaction.
- The pyridone-amide unit is essential.
- The lactone unit is essential and plays a key role for biological activity. Extension of the 6-membered ring into a 7-membered β-hydroxylactone improves stability and biological activity. The *S*-stereochemistry is an essential feature for the activity, the *R*-isomer is almost inactive. Replacement of the OH-group or its omittance leads to diminished activity. Esterification often leads to reduced toxicity and can improve the in vitro antitumor activity.

Concerning resistance P-gp plays a role, even though the effect is much lower than for other MDR substrates like doxorubicin or vinblastine.

Prime target for camptothecin is Topoisomerase I. The physiological function of this essential nuclear enzyme is the release of torsional strain in supercoiled DNA through the introduction of transient breaks in the DNA-backbone. In contrast to Topoisomerase II which causes a complete break of the double strand Topoisomerase I breaks only one strand thus allowing rotation around the remaining phospodiester linkage of the intact strand and consequently reducing torsional strain. After relaxation religation takes place, a reaction that is much faster than the cleaving step which ensures that the concentration of cleaved DNA remains low. This uncoiling function is a central DNA processing step for events like replication, transcription, recombination, repair, chromatin assembly or chromosome segregation. Due to the sheer size of the eukaryotic chromosome supercoiling can only be accomplished locally. Usually the cleavage takes place between T at −1 and G at +1.

The interaction between Topoisomerase I and dsDNA has been elucidated in detail and computational models for its interaction with camptothecin have been described (Redinbo et al. 1998; Padlan and Kabat 1991; Laco et al. 2002; Thomas et al. 2004; Staker et al. 2005). On a molecular level the interaction takes place via a nucleophilic attack of a catalytic Tyr on a phosphodiester linkage generating a covalently bound Topoisomerase-DNA-conjugate, the so-called binary complex. The broken DNA-strand allows torsional relaxation. The role of camptothecin lies in its ability to stabilize the covalent binary complex thus hindering religation. The newly formed ternary complex interferes with the moving replication fork, which leads to the induction of replication arrest and lethal double-stranded breaks in DNA. As mammalian cells can not efficiently repair these double strand breaks, the formation of this ternary complex eventually leads to apoptosis.

Structural data point to an intercalation of camptothecin into the DNA part of the binary complex, occupying the space of the +1 base pair. The so generated stacking interactions stabilize the complex. Moreover a hydrogen-bond between camptothecin and the enzyme is formed and conformational changes of the phosphodiester link are induced with the consequence that religation is unlikely to happen. No data could be generated concerning a possible role of the hydrolyzed lactone in the binding pocket, a possible scenario including the activation of camptothecin via hydrolysis is still under discussion.

Currently two camptothecin derivatives are therapeutically used—topotecan **4.7** and irinotecan **4.8**—though a wide variety of analogs (including camptothecin itself) are in different phases of clinical investigation. In both cases the problem of insolubility of initial camptothecin was overcome by the introduction of solubilising groups which also tend to have a stabilizing effect on the lactone. One major drawback of the structures and is the interpatient variability which continues to hamper the wide spread use of camptothecins (Scheme 6.46).

Topotecan (Hycamtin) entered clinical trials in 1989 under National Cancer Institute sponsorship in collaboration with SmithKline Beecham. In 1996, it was approved for use by the United States FDA for previously treated patients with advanced ovarian cancer. It is a semi-synthetic analog of camptothecin (Dennis et al. 1997; Kollmannsberger et al. 1999) (Scheme 6.47).

Compared to camptothecin topotecan contains additionally a C9 tertiary amine side chain and a C10 hydroxyl group both enhancing aqueous solubility. Topotecan is not a prodrug. It is used for the treatment of advanced ovarian cancer in patients with disease that has recurred or progressed following therapy with platinum-based regimens. It is also used as a second-line therapy for treatment-sensitive small cell lung cancer, as well as in combination with *cis*-platin for the treatment of stage IV-B, recurrent, or persistent cervical cancer not amenable to curative treatment with surgery and/or radiation therapy (http://drugbank.ca/drugs/DB01030).

Irinotecan (camptosar) was initially developed in Japan. It is a prodrug formulation of the decarboxylated metabolite SN-38 (7-ethyl-10-hydroxy-camptothecin) which plays at least a major role in the antitumor activity of irinotecan in vivo (Scheme 6.48).

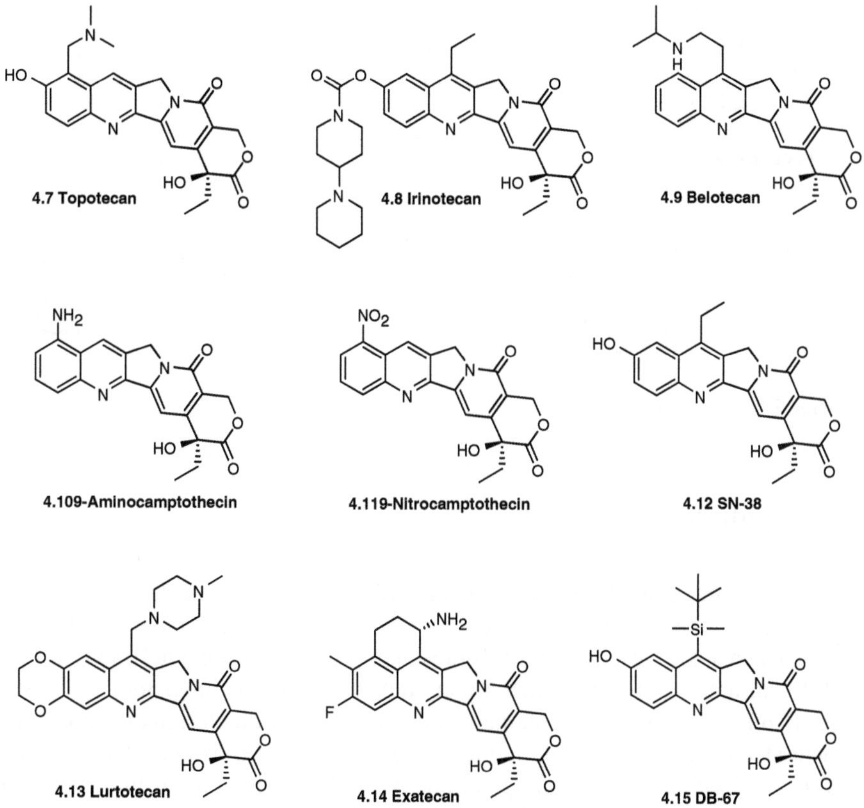

Scheme 6.46 Structural representation of camptothecin analogs in therapeutical use or clinical testing

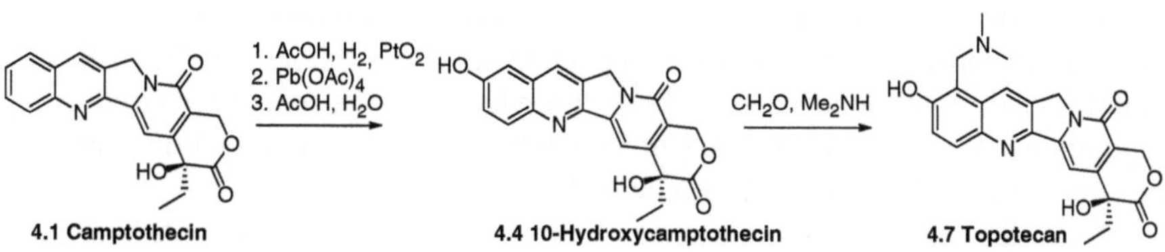

Scheme 6.47 Synthesis of topotecan

It is used for the treatment of metastatic colorectal cancer (first-line therapy when administered with 5-fluorouracil and leucovorin) and in combination with *cis*-platin for the treatment of extensive small cell lung cancer. Irinotecan is currently under investigation for the treatment of metastatic or recurrent cervical cancer.

Belotecan (camptobell) is the third drug in clinical use based on the camptothecin scaffold, which is applied mainly for the treatment of small cell lung cancer.

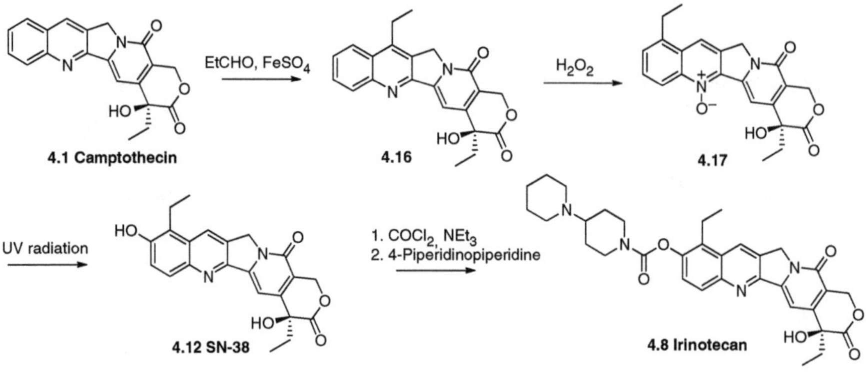

Scheme 6.48 Synthesis of irinotecan

6.6 Epothilones

Epothilones are a small structural family of 16-membered macrolactones first isolated from the gliding myxobacterium *Sorangium cellulosum* by Höfle and Reichenbach at the Gesellschaft für Biotechnologische Forschung in Braunschweig in 1987 during a study on the activity of natural products as fungicides and pesticides (Höfle et al. 1996). The name is a combination of the designations of three structural elements contained in the compound class: *Epo*xide, *Thi*azol*e*, and Ket*one*. In 1995 the activity of epothilone A and B as inhibitors of tumor growth was discovered during a screening at the NCI (Bollag et al. 1995). Of particular interest is the fact, that epothilones retain their biological activity in tumors insensitive to paclitaxel. Moreover, despite the much reduced structural complexity, compared to paclitaxel the binding affinity of epothilone B is higher and the solubility is 30-fold increased (Altmann 2003; Chou et al. 1998a).

Up to now six members of the family are known, epothilone A to F. Structural differences consist in the occurrence of an epoxide (epothilones A, B, E, F) or a double bond (epothilones C, D), the hydroxylation of the thiazole methyl group (epothilones E, F) and the presence of a methyl group instead of a hydrogen atom in position 12 (epotilones B, D, F) (Scheme 6.49).

Due to the outstanding importance of the structural class gains by its biological activity the visibility was high and consequently much effort has been made to investigate the epothilones. Furthermore the biosynthesis of the epothilones was investigated in detail. It starts with the formation of the unique 2-methyl-4-carboxythiazole, which is transferred to a type I PKS to generate the 16-membered ring, followed by post-PKS modifications like oxidations and introduction of further methyl groups (Molnar et al. 2000). Several total synthesis approaches have been published (Luduvico et al. 2006; Balog et al. 1996; Su et al. 1997; Yang et al. 1997; Schinzer et al. 1997; Mulzer et al. 2000; Bode and Carreira 2001; Nicolaou et al. 1997a).

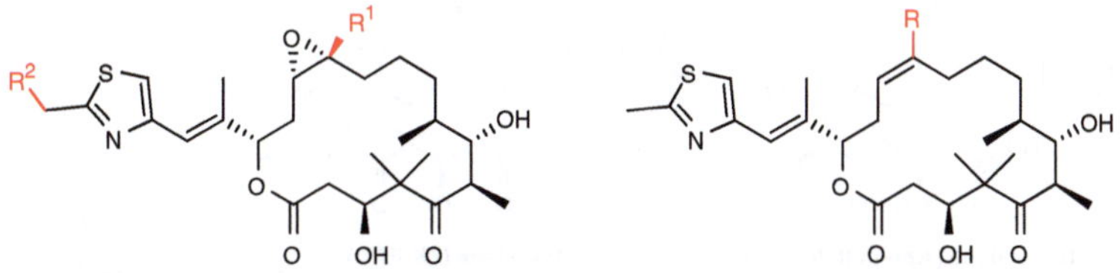

5.1 R^1 = H, R^2 = H **Epothilone A**
5.2 R^1 = H, R^2 = Me **Epothilone B**
5.3 R^1 = H, R^2 = OH **Epothilone E**
5.4 R^1 = Me, R^2 = OH **Epothilone F**

5.5 R = H **Epothilone C**
5.6 R = Me **Epothilone D**

Scheme 6.49 Structural representation of naturally occurring epothilones

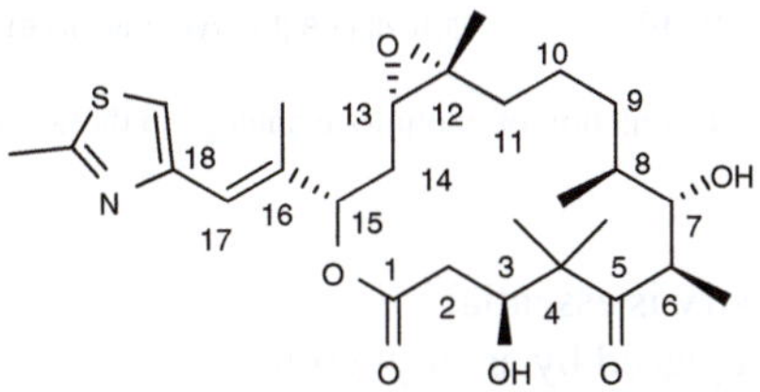

Scheme 6.50 The core structure of therapeutically relevant epothilone derivatives

The mechanism of action is similar to that of taxanes and discodermolide and consists in the stabilization of the structure of microtubuli which are essential for cell division (Bollag 1995). This stabilization causes the restriction of the dynamic processes (polymerization-depolymerization of tubuline heterodimers) mainly during mitosis (Jordan and Wilson 2004). The consequence is mitotic arrest and subsequently apoptosis or—in lower concentrations—aneupliodic cell division (Hoffmann et al. 2008). Due to this concentration-dependant dichotomy of effects the term biphasic mode of action was used (Ganguly et al. 2010; Chen et al. 2003). Since the mode of action strongly affects the cytoskeleton also other cells with a distinct microtubuli network are affected as well leading to inherent side effects. Dosing of epothilones is limited due to the occurrence of peripheral neuropathy, a typical side effect of these compounds. Compared to taxanes the epothilones tendency for the development of resistances is greatly reduced. Epothilones share the same binding site to the α,β-tubulin heterodimer subunit as the taxanes, though the binding contacts are different (Nettles et al. 2004; Ojima et al. 1999), even though epothilone B efficacy against tubulin-mutated cell lines suggests different interactions. Discussion about whether the two classes of natural products also share a common pharmacophore is still ongoing (Giannakakou et al. 2000; Forli et al. 2010; Reese et al. 2007).

The structure activity relationship has been investigated in detail. Important features are (Scheme 6.50):

- Stereochemistry at C3 is essential
- Gem dimethyl group at C4 can be replaced with cyclopropyl
- Keto group at C5 is essential

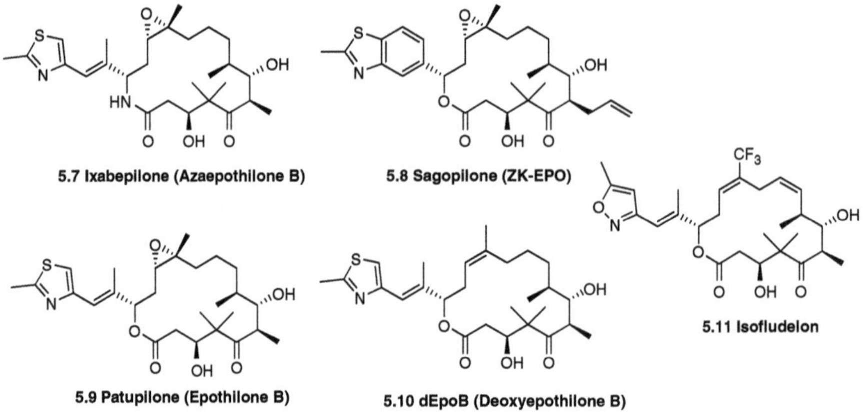

Scheme 6.51 Structural representation of epothilone analogs in therapeutical use or clinical testing

- C6-C7 stereochemistry is essential
- C6-Methyl can be replaced by an allyl group
- C8-Methyl is essential (can not be replaced by nor or dimethyl)
- 9,10-Unsaturation increases activity
- Presence of an *R*-configurated substituent at C12 is important (Me, Et, PR, Hex, CF_3, CN are tolerated)
- Epoxide is not essential (episulfide, alkene, aziridine, cyclopropane are tolerated)
- C15-Stereochemistry is essential
- C16-Methyl is not essential
- C16-C17 *E*-configuration is important
- Position of hetero aromatic nitrogen is important (thiazole, oxazole, and pyridyl tolerated)
- Small groups at C26 are tolerated (e.g. amine)

The outstanding biological activity of the epothilones inevitable gave the pharmaceutical industry an enormous impetus for the development of new cancer drugs on the structural basis of the epothilones. One drug (Ixabepilone) is approved by the FDA, several others are or were in clinical trials (Scheme 6.51).

Ixabepilon (Ixempra, BMS-247550), a product of Bristol-Myer-Squibb was approved by the FDA in 2007 for the treatment of aggressive metastatic or locally advanced breast cancer no longer responding to currently available chemotherapies (anthracyclines, taxanes) and is in clinical use in the USA under the name Ixempra either in monotherapy or in combination with capecitabin. The EMEA refused marketing authorization in 2008 (European Medicines Agency 2009). Structurally Ixabepilon is an aza-analogon of epothilone B and is semi synthetically produced from it. The lactam moiety slows down enzymatic and spontaneous hydrolysis, unfortunately the lactam moiety causes the recognition of the compound by efflux pumps. The synthesis consists of a Pd-catalyzed attack of an azide ion with subsequent ring opening, the reduction of the resulting open chain epothilone analog

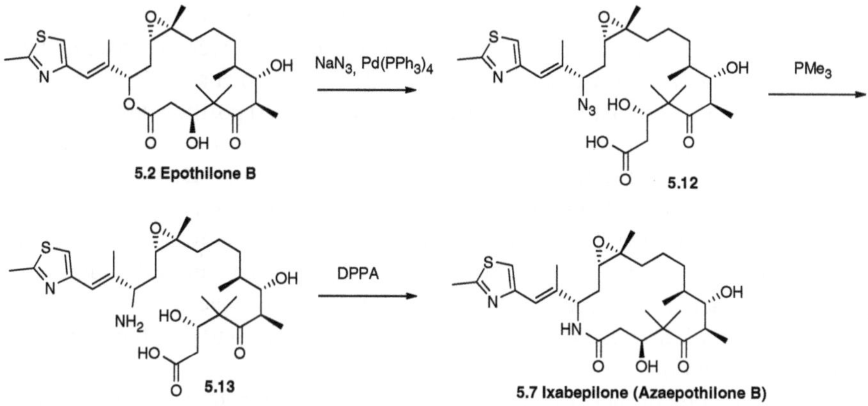

Scheme 6.52 Semi-synthesis of ixabepilone starting from epothilone B

under Staudinger conditions and a ring closure reaction with DPPA (Diphenyl phosphoryl azide) as a coupling reagent (Borzilleri et al. 2000) (Scheme 6.52).

Sagopilone (ZK-Epo) is an epothilone B derivative, developed by Schering AG in which the methylthiazole moiety is replaced by a methyl benzthiazole and the methyl group in C6 is substituted by an allyl group. Due to the relatively grave alterations as compared to epothilone B a semi-synthetic approach was not economical and a total synthetic route was developed. The route established by Schering consisted of 39 steps by which about 350 compounds were synthesized. Sagopilone proved to have the best preclinical properties out of these (Klar et al. 2006). Apart from its outstanding biological activity two features make sagopilone especially interesting. First it is able to cross the blood brain barrier giving rise to the prospect of a possible agent for the treatment of brain tumors. Secondly sagopilone seems not to be recognized by efflux pumps which makes the fast development of resistances improbable.

Patupilone (EPO906) from Novartis is the natural product epothilone B itself (Rothermel et al. 2003). In vitro data suggest that patupilone binds to β-tubulin with higher affinity than taxanes or other members of the epothilones family (Buey et al. 2004). Although the in vitro affinity constant (K_i) for patupilone was 0.71 μM, the IC_{50} for cell proliferation ranged from 0.13 to 2.92 nM (Kowalski et al. 1997; Nicolaou et al. 1997b). This may be due to increased intracellular patupilone concentrations. Patupilone inhibited tumor growth in paclitaxel-resistant human lung carcinoma A549, where resistance to paclitaxel was not mediated by P-gp overexpression (Martello et al. 2003). Patupilone went to Phase III clinical trials, but was stopped since it did not show a significant overall survival advantage in a phase III trial of patients with advanced ovarian cancer, refractory or resistant to platinum-based therapy. Patupilone failed a phase III trial for ovarian cancer.

In cooperation with the MSKCC Kosan Biosciences Inc. started a synthetic program to improve the pharmaceutical properties of the epothilones, developing derivatives called second generation epothilones. Based on the consideration, that the high cytotoxicity of epothilone B might stem from the epoxide, this group was replaced by a simple double bond which led to dEpoB with lower cytotoxicity but a reduced potency as well (Chou et al. 1998b, c, 2001; Rivkin et al. 2005).

By introduction of an additional double bond in C9-C10 position and thus introduced enhanced rigidity of the molecule the potency and serum stability could be significantly improved. KOS-1584 (9,10 didehydroepothilone D, Dehydelone) at 6 nM effectively induced tubulin polymerization in a cell-based assay using MCF7, a breast cancer cell line. Potent anti-proliferative activity against a broad range of tumor cell lines was observed in vitro, including hematologic and solid tumor cell lines, such as leukemia, breast, colon, lung, and ovary (average IC_{50} 3.8 ± 1.4 nM) and is currently being evaluated in phase II clinical trials and combination trials (Yiqing et al. 2005).

The observation of increased toxicity and a narrow therapeutic window of KOS-1584 led to the development of so-called third generation epothilones. By substituting the C12-methyl group by CF_3 the therapeutic index of the new compound Fludelone could be improved (Rivkin et al. 2004). By substitution of the methylthiazole moiety of Fludelone by an isoxazole ring the potency could be further improved and a remarkable metabolic stability could be achieved (Chou et al. 2008). The compound Isofludelone (KOS-1803) currently is in clinical phase II for solid tumors.

6.7 Podophyllotoxin

The cyclolignan Podophyllotoxin has been isolated for the first time from *Podophyllum peltatum* (Berberidaceae) native to eastern North America. To prevent overcollection of the slow-growing wild *Podophyllum* plants, cell cultures of lignan-producing plant species were systematically investigated for their podophyllotoxin production. The most promising production of podophyllotoxin was achieved by *Linum album* (Linaceae). To optimize productivity, the biosynthesis of podophyllotoxin has also been studied in detail in cell cultures. The biosynthetic sequence starts with a stereospecific dimerization of two coniferyl alcohols, followed by modification of the aromatic substitution, cyclization to the cyclolignan skeleton and the final oxidation of 7-deoxpodophyllotoxin to podophyllotoxin (Seidel et al. 2002). Although lignans are typically produced by plants, there are some reports of podophyllotoxin producing endophytic fungi, which could be promising candidates for large scale production of podophyllotoxin (Kour et al. 2008).

Podophyllotoxin is an aryltetralin lignan that can extracted from the roots and rhizomes of a variety of different *Podophyllum* species. The structure was elucidated in 1932 by Borsche and Niemann (1932) and later confirmed by Gensler via total synthesis (Gensler et al. 1954). The use of Podophyllotoxin containing plants has a long tradition in traditional medicine—mainly in Asia—for an impressive variety of ailments like gout, tuberculosis, gonorrhoea, syphilis, psoriasis, cough, venereal wart any many more. In 1942 it was introduced by Kaplan (1942) for the treatment of venereal warts and is still in use today (Liu et al. 2007). Even though

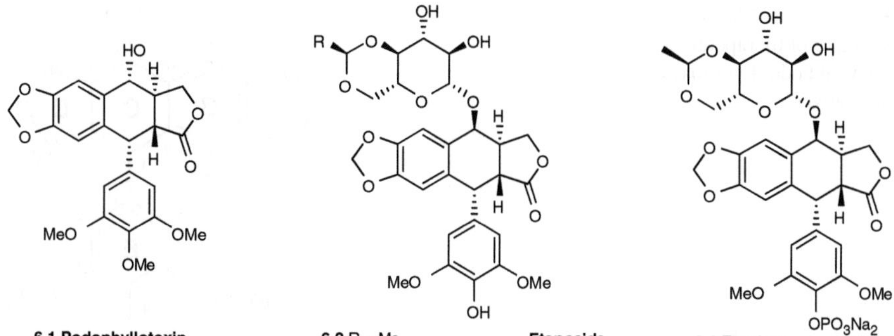

6.1 Podophyllotoxin — **6.2** R = Me **Etoposide**; **6.3** R = 2-Thiophenyl **Teniposide** — **6.4 Etophos**

Scheme 6.53 Structural representation of podophyllotoxin and clinically relevant derivatives

Podophyllotoxin has very promising properties as a antineoplastic agent, its clinical development as antitumor agent was abandoned due to severe side effects like gastrointestinal toxicity (Canel et al. 2000). However, the remarkably activity of Podophyllotoxin inspired an intense research effort in the search for less toxic analogs. Very successful actors in this area were researchers from Sandoz whose efforts led to the development of acetalized glycosylated Podophyllotoxin analogs like Etoposide and Teniposide (Stoll et al. 1954; Emmenegger et al. 1961; Kuhn and von Wartburg 1969; Stähelin and von Wartburg 1991) which are still in clinical use today even though they suffer of side effects like myelosuppression, gastrointestinal toxicity, or hypersensitivity.

Roughly 20 years after the introduction of these two structures as therapeutic agents the mode of action was clarified (Ross et al. 1984; Long and Brattain 1984; Long and Minocha 1983; Long et al. 1984) (Scheme 6.53).

Podophyllotoxin has an impressive range of biological activities (Gordaliza et al. 2004). Aside of the antineoplastic activity which has led to the development of two antitumor drugs (Etoposid, Teniposid) and the antiviral potency which is the reason for its use against venereal warts (caused by the papilloma virus) Podophyllotoxin analogs show immunomodulatory activity which in principle opens the possibility of the development of candidates to be used in the clinic for organ transplantations. Moreover anti-leishmanial activity, LOX-inhibition, and some others were reported. The reason for the antiviral activity seems to be due to an interaction with tubulin which disrupts the build up of the cytoskeleton und thus interferes with the viral replication (Bedows and Hatfiled 1982; Hammonds et al. 1996; Charlton 1998). This binding affinity to tubilin is completely absent for glycosylated derivatives like Etoposide or Teniposide, the reason is thought to be the sterical hindrance imposed by the carbohydrate residues.

The mode of action for etoposide and teniposide is based on the interaction with Topoisomerase II which is a nuclear enzyme with the function of the introduction of a break in dsDNA in order to deal with DNA tangles and supercoils. Unlike Topoisomerase I—the target protein of Camptothecin and its derivatives—it is

Scheme 6.54 The core structure of therapeutically relevant podophyllotoxin derivatives

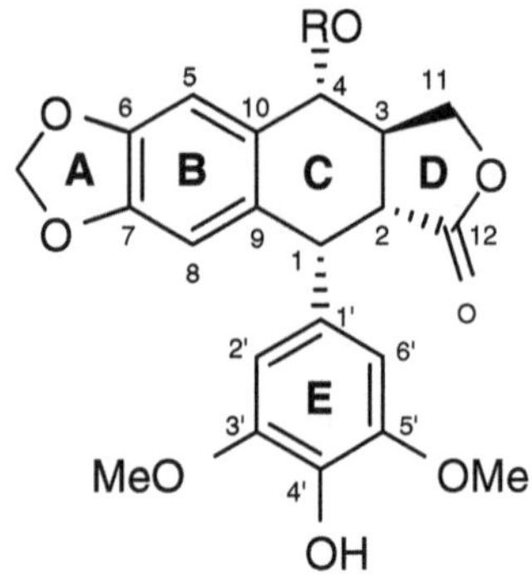

ATP-dependant. During the catalytic process the enzyme changes the linking number of circular DNA by ±2.

The exact details of the mechanism of action are not well understood, though it is widely accepted that Podophyllotoxin, Etoposide, and Teniposide bind to the Topoisomerase-DNA complex and thus stabilize the temporary break caused by the enzyme. Religation is inhibited, DNA unwinding and replication are stopped and eventually apoptosis occurs. The drugs are specific for the late S and early G_2 stage (Imbert 1998; Hande 1998; Burden and Osheroff 1998; Pommier 1997). The fact that Etoposide can form stable radicals has led to a discussion whether this factor may add to the biological activity of Podophyllotoxin derivatives with a free 4′-OH group (Haim et al. 1987; Sinha 1989).

Based on molecular modeling a common binding mode for intercalating and non-intercalating Topoisomerase II inhibitors was published. It was suggested, that inhibitory activity is based on three structural domains of the molecule. The intercalation is thought to be accomplished by the almost planar rings A and B, while minor groove binding and maybe also interaction with Topoisomerase II takes place via interaction with the dimethoxyphenol proportion of Etoposide and Teniposide. C4 is the position where variable substituents are thought to be able to bind to the minor groove as well (Cho et al. 1996).

A plethora of Podophyllotoxin analogs and their biological have been published and extensively reviewed. Key features of the SAR concerning the cyctotoxic activity are (Liu et al. 2007; Xu et al. 2009; Lv and Xu 2011) (Scheme 6.54):

- The 4-β configuration is essential.
- The free 4′-OH is at least important. It is thought to be involved in groove binding and Topoisomerase association.
- The *trans*-lactone (2-α, 3-β) moiety is crucial.
- The dioxolane (ring A) is optimal, it is involved in the intercalation process during the formation of the ternary complex.
- The free rotation of the dimethoxyphenol ring (ring E) is an essential feature for biological activity.
- *R* is variable.

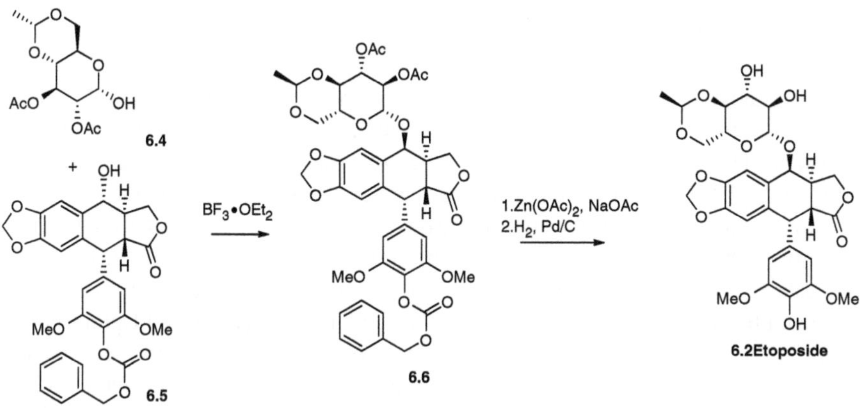

Scheme 6.55 Semi-synthesis of etoposide starting from podophyllotoxin

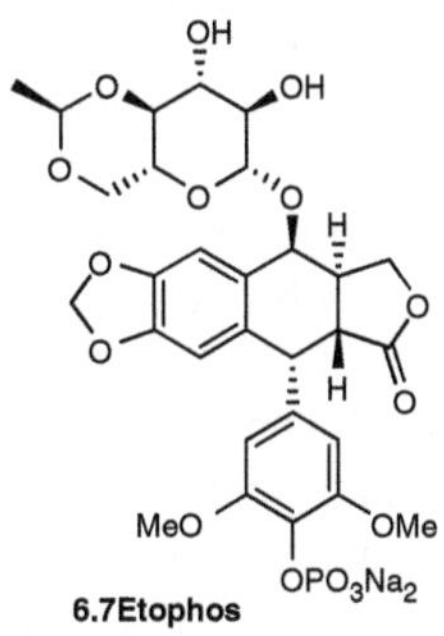

Scheme 6.56 Structural representation of etopophos

Etoposide **6.2** (Eposin, Etopophos, Vepesid, VP-16) is a semi-synthetic glycosylated analog of Podophyllotoxin and is in use in combination with other chemotherapeutic agents in the treatment of refractory testicular tumors and as first-line treatment in patients with small cell lung cancer. It is also used to treat other malignancies such as lymphoma, non-lymphocytic leukemia, and glioblastoma multiforme. Even though it is a valuable drug which is in clinical use it suffers from several disadvantages like moderate potency, poor solubility, fast metabolic inactivation, side effects due to toxicity and the development of drug resistances (Scheme 6.55).

A substantial improvement was the introduction of a phosphate group to the free phenolic group of etoposide forming a prodrug which can be administered in higher doses than etoposide as a short intravenous injection, whereafter it is rapidly converted to the parent compound by plasma phosphatases, and thus constitutes an improved formulation of etoposide. Thus water solubility as well as bioavailability (0.04–50%) was greatly improved. The drug was launched in 1986 by BMS under the name Etophos (Scheme 6.56).

Teniposide (Vumon) is another glycosylated Podophyllotoxin derivative and is mainly used for the treatment of refractory acute lymphoblastic leukemia.

6.8 Anthracyclines

The Anthracyclines are a group of compounds which are produced by soil bacteria and proved to be a source of very potent agents for the treatment of different kinds of cancer. The first compounds of this structural class of natural products were isolated in the 1950s by Brockmann (Brockmann and Bauer 1950; Brockmann 1963). The cytotoxic potential of the anthracyclines was discovered soon after but the high toxicity of the known compounds hindered the use in cancer therapy at that time. A breakthrough was made, when research groups from Farmitalia (Grein et al. 1963) and from Rhone-Poulenc (Dubost et al. 1963) independently isolated Daunorubicin from *Streptomyces* strains. Later scientists from Farmitalia performed random mutagenesis with the daunorubicin producing strain *Streptomyces peceutius* which for the first time yielded doxorubicin (Arcamone et al. 1969) which was shown to have improved pharmacological properties as compared to daunorubicin (Blum and Carter 1974; Weiss 1992; http://en.wikipedia.org/wiki/Daunorubicin). Anthracyclines belong to the group of type II aromatic polyketides. They are produced by bacteria of the genus *Streptomyces* and some closely related genera. During the last 50 years, about 500 different natural anthracyclines have been isolated (Laatsch and Fotso 2008). Common structural fragment of all anthracyclines is a tetracyclic ring system with a cyclohexyl unit linear attached to an anthraquinone (7,8,9,10-tetrahydro-5,12-naphthacenequinone). Most of the anthracyclines bear sugar moieties as O-glycosides (usually attached at O-7 position in the cyclohexane ring). Sugar moieties often contain quite unique aminosugars like L-rhodosamine and L-daunosamine. The anthracycline skeleton is produced by a type II PKS. The biosynthesis of anthracyclines is well known in detail including most of the enzymes and genes, predominantly causing this structure class to be the first target for combinatorial biosynthesis (Salas 2008) (Scheme 6.57).

Important from a therapeutical point of view are four closely related structures: doxorubicin **7.1**, daunorubicin **7.2**, Idarubicin **7.4**, and Epirubicin **7.3**. Structural variations concentrate on C1 (OMe or H), C8 (acetyl or hydroxyacetyl) and C5 of

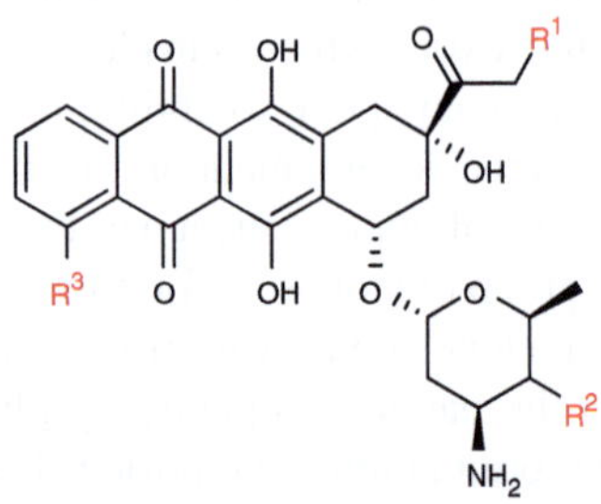

7.1 Doxorubicin	R^1=OH,	R^2=α–OH, R^3=OMe
7.2 Daunorubicin	R^1=H,	R^2=α–OH, R^3=OMe
7.3 Epirubicin	R^1=OH,	R^2=β–OH, R^3=OMe
7.4 Idarubicin	R^1=H,	R^2=α–OH, R^3=H

Scheme 6.57 Clinically important anthracyclines

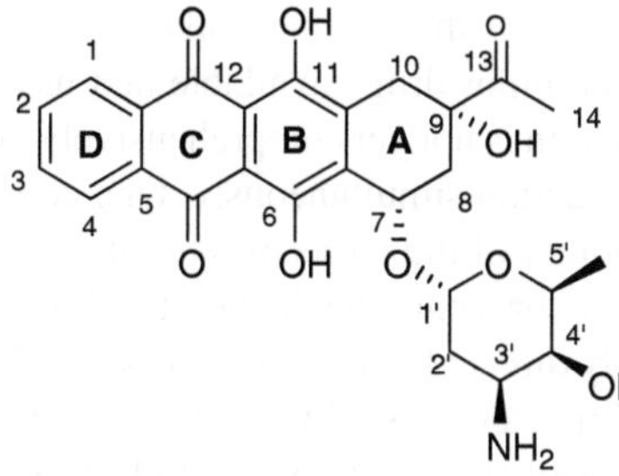

Scheme 6.58 The core structure of therapeutically relevant anthracyclines

the sugar moiety (daunosamin, α or β) while the anthracycline scaffold is untouched in all these structures (Scheme 6.58).

These structural differences have direct consequences for the selectivity of the anthracyclines. Compounds with a primary alcohol in the side chain (doxorubicin and epirubicin) show good activity against solid tumors and hematologic malignancies, while structures lacking this functional group (dauno- and idarubicin) are primarily used for the treatment of acute myeloblastic leukemia and AIDS-related Kaposi Sarcoma. The biological activity depends crucially on the natural absolute configuration in ring A. Due to their later discovery, Epi- and Idarubicin are described as second generation antitumor anthracyclines.

Naturally as for a class of natural products with interesting biological activity a plethora of methods for total synthesis, semi-synthesis, or bioengineering have been published. These efforts mainly were triggered in order to improve the pharmacological profile (efficacy vs. cardiotoxicity). Due to the complex structures this research proved extremely laborious.

One of the main drawbacks of the structural class of anthracyclines is the inherent dose-related and cumulative cardiotoxicity of the compounds which significantly reduces their therapeutic index. Upon treatment patients can experience arrhythmias, hypotension, and a mild depression of myocardial contractibility, in few cases (1%) acute myocarditis and pericardial effusions (acute cardiotoxicity) are observed, conditions which are in most cases reversible and usually do not lead to the termination of the treatment. Moreover anthracyclines may cause dilative cardiomyopathy and congestive heart failure. Responsible for these side effects are reductive events at the quinone and the carbonyl group which are accompanied by iron and free radical reactions (Menna et al. 2008). Other side effects include bone marrow toxicity, gastrointestinal disorders, stomatitis, or alopecia.

Tumor resistance against anthracyclines follows the line of classical multidrug resistance which is due to the presence of P-glycoprotein (P-gp) in the plasma membrane. No other mechanism for the development of resistances against anthracyclines is known to date. This fact combined with the fluorescent abilities of doxorubicin can be used for the visualization of the effusion of the drug from the cell. The susceptibility of anthracylines to resistance is one of the major reasons for the ongoing effort in the research in this area.

Even though the anthracycline compound class is among the most important antitumor drugs available and used in cancer therapy for many years, the mode of action is not yet comprehensively understood, also because different mechanisms are in action simultaneously. One component of the therapeutic effect originates in the ability of the molecules to intercalate into dsDNA, preferably between two adjacent G/C basepairs flanked on the 5′ side by an A/T base pair with the daunosamine moiety directed to the minor groove (DeVita et al. 1993). Moreover the complex inhibits Topoisomerase II with the consequence of double strand fission during the super coiling of the DNA. Not yet clear are the effects at the outside of the cell membrane, effects caused by covalent DNA binding and the induction of apoptosis (Keizer et al. 1990). Well known is the fact, that doxorubicin causes the generation of free radicals (Cummings et al. 1991) which is at least in part the cause for side effects like cardiomyopathy (Seifert et al. 1994; Siveski-Iliskovic et al. 1995; Dorr 1996).

Daunorubicin is the member of the structure family first isolated in the 1950s from *Streptomyces peucetius* and also was the first anthracycline in clinical use. It is among the most effective agents for the treatment of leukemia (AML and ALL), usually in combination with other cytostatics.

Doxorubin (Hydroxydaunorubicin, trade name Adriamycin) which was approved in 1974 is the most prominent and most frequently used member of this class of cytotoxic compounds. To date no analog has shown activity that is clearly superior to that of doxorubicin which has a broad spectrum of activity and remains one of the most effective drugs for the treatment of solid tumors. It is used for the treatment of various kinds of cancer like Hodgkin's disease, non-Hodgkin's lymphoma, leukemia, neuroblastoma, sarcoma, Wilms' tumor, bladder cancer, breast cancer, lung cancer, ovarian cancer, stomach cancer, and thyroid cancer. The drug commonly is used in combination with other anti-cancer agents like cyclophosphamide, taxotere, bleomycin, vinblastine, vincristine, or 5-fluorouracil. Several prodrugs of doxorubicin are momentarily in clinical trials.

Epirubicin (traded name Farmorubicin/Pharmorubicin and Ellence (USA)) is a very close semi-synthetic analog of doxorubicin with the only structural difference of an epimerized hydroxyl group in position 4 of the aminosugar moiety. This results in an improved toxicological profile with good antitumor activity and reduced cardiotoxicity though a similar activity spectrum as compared to Doxorubicin is retained.

Idarubicin (trade names Zavedos (UK) and Idamycin (USA)) which was synthesized in order to improve the pharmacological profile of daunorubicin lacks a methoxy group in the aglycon moiety of the latter, which results in enhanced lipophilicity and an altered pharmacological profile. The most striking difference compared to daunorubicin is a shift in the modes of action, since the ability to induce Topoisomerase II mediated DNA-cleavage and the intracellular drug accumulation is enhanced which leads to increased antitumor potency. Moreover in contrast to daunorubicin which is not orally available at all, so the drug is orally administered. However, idarubicin suffers from a drawback common to all analogs of daunorubicin—the low activity against solid tumors. Its application is thus restricted to leukemia. Idarubicin was approved in the USA in 1990.

Scheme 6.59 Structural representation of aclacinomycin A

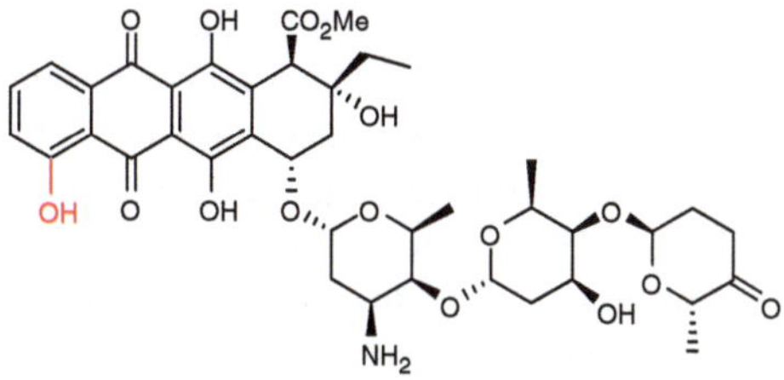

7.5AclacinomycinA

Aclacinomycin A (aclarubicin, aclacin) is the latest development in the field of approved antitumor active anthracyclines. The new trisaccharide-anthracycline derivative has been isolated from *S. galilaeus*. It is employed for the treatment of leukemias that are resistant to standard therapy. The compound shows less cardiotoxicity compared to other agents within the anthracycline class (Oki 1980) (Scheme 6.59).

6.9 Perspective

This review is far from being a comprehensive overview about the field of natural products in cancer therapy. The sheer size of this sector of medicinal chemistry makes it impossible to be covered within the limits of a book chapter. However, it is amazing to see the variety of molecules nature creates which are useful to tackle so many facets of cancer. In fact natural products and derivatives still make up for the predominant part of agents used for cancer therapy (also valid for antibiotics). Even more astonishing is the fact, that most of the compounds used in this context are synthesized by plants (Taxol by *T. brevifolia*, Camptothecin by *C. acuminata* etc.) or bacteria (Anthracyclines by *Streptomyces* strains, Epothilones by *myxobacteria*) for which the uncontrolled cell proliferation follows entirely different mechanisms or does not exist at all. The question that logically arises is why do these organisms invest substantial amounts of energy and devote considerable parts of their genetic information (for the setup of the enzymatic machinery) in the production of secondary metabolites which are seemingly useless for the organism itself.

One explanation for this intriguing question might be a hypothesis according to which plants, devoid of possibilities for escape, had to develop defence mechanism against herbivores, or other plants competing for light, space, or nutrition. Plants do not possess an immune system of their own which would cause fatal vulnerability to microorganisms. In order to tackle these vital threats plants (and also microorganisms) had to evolve effective defence mechanisms. This line of defence seems to be established by secondary metabolites (Wink 2010; Williams et al. 1989).

Naturally these compounds would have unpleasant taste, causing unstable conditions or even death to the attacking organism which would require biological activity based on targets present in these organisms. Given the fantastic multitude of structures identified to be secondary metabolites from plants alone it is not unlikely that a large number of target classes can be addressed on a multitude of different mechanisms of action. The astonishing number of natural products with biological activity against cancer-related targets might be a hint for the account of this theory.

If there is some sense in these reflections an interesting outlook might be drawn. Only 20–30% of all higher plants have been investigated thoroughly yielding already tens of thousands of microorganisms of secondary metabolites. Given the impressive multitude of different modes of action working on cancer-related targets already discovered by now one can expect more spectacular discoveries in this field.

References

Adams J, Gao Y, Georges Evangelinos AT, Grenier L, Pak R, Porter JR, Wright JL (2005) WO2005063714
Aicher TD, Buszek KR, Fang FG, Forsyth CJ, Jung SH, Kishi Y, Matelich MC, Scola PM, Spero DM, Yoon SK (1992) J Am Chem Soc 114:3162
Ajikumar PK, Xiao W-H, Tyo KEJ, Wang Y, Simeon F, Leonard E, Mucha O, Phon TH, Pfeifer B, Stephanopoulos G (2010) Science 330:70
Altmann KH (2003) Mini Rev Med Chem 3:149
Arcamone F, Cassinelli G, Fantini G, Grein A, Orezzi P, Pol C, Spalla C (1969) Biotechnol Bioeng 11:1101
Bai R, Pauli KD, Herald CL, Malspeis L, Pettit GR, Hamel E (1991) J Biol Chem 266:15882
Balog DM, Meng D, Kamanecka T, Bertinato P, Su D-S, Sorensen EJ, Danishefsky SJ (1996) Angew Chem Int Ed Engl 35:2801
Bedows E, Hatfiled EM (1982) J Nat Prod 45:725
Biamonte MA, Van de Water R, Arndt JW, Scannevin RH, Perret D, Lee W-C (2010) J Med Chem 53:3
Bissery M-C. (2002) US 6,403,634
Blum R, Carter S (1974) Ann Intern Med 80:249
Bode JW, Carreira EM (2001) J Am Chem Soc 123:3611
Bollag DM (1995) Cancer Res 55:2325
Bollag DM, McQueney PA, Zhu J, Hensens O, Koupal L, Leisch J, Goetz M, Lazarides E, Woods CM (1995) Cancer Res 55:2325
Borchardt JK (2002) Drug News Perspect 15:187
Borsche W, Niemann J (1932) Liebigs Ann Chem 494:126
Borzilleri RM, Zheng X, Schmidt RJ, Johnson JA, Kim S-H, DiMarco JD, Fairchild CR, Gougoutas JZ, Lee FYF, Long BH, Vite GD (2000) J Am Chem Soc 122:8890
Brockmann H (1963) Fortschr Chem Org Naturst 21:121
Brockmann H, Bauer K (1950) Naturwissenschaften 337:492
Buey R, Diaz J, Andreu J (2004) Chem Biol 11:225
Burden DA, Osheroff N (1998) Biochim Biophys Acta 1400:139
Caesar GJ. Commentarii de Bello Gallico, (52 bc) Book VI, 32
Cai J-C, Hutchinson CR (1983) In: Brossi A (ed) The alkaloids, vol 21. Academic, New York, p 101
Canel C, Moraes RM, Dayan FE, Ferreira D (2000) Phytochemistry 54:115
Charlton JL (1998) J Nat Prod 61:1447
Chen S-H, Huang S, Wie J, Farina V (1993) J Org Chem 58:4520
Chen JG, Yang CP, Cammer M, Horwitz SB (2003) Cancer Res 63:7891
Cho SJ, Tropsha A, Suffness M, Cheng YC, Lee KH (1996) J Med Chem 39:1383
Chou TC, Zhang X-G, Balog A, Shu D-S, Meng D, Savin KA, Bertino JR, Danishefsky SJ (1998a) Proc Natl Acad Sci U S A 95:9642

Chou T-C, Zhang X-G, Balog A, Su D-S, Meng D, Savin K, Bertino JR, Danishefsky SJ (1998b) Proc Natl Acad Sci U S A 95:9642

Chou T-C, Zhang X-G, Harris CR, Kuduk SD, Balog A, Savin K, Bertino JR, Danishefsky SJ (1998c) Proc Natl Acad Sci U S A 95:15798

Chou T-C, O'Connor OA, Tong WP, Guan Y, Zhang X-G, Stachel SJ, Lee C, Danishefsky SJ (2001) Proc Natl Acad Sci U S A 98:8113

Chou T-C, Zhang X, Zhong Z-Y, Li Y, Feng L, Eng S, Myles DR, Johnson R, Wu N, Yin JI, Wilson RM, Danishefsky SJ (2008) Proc Natl Acad Sci U S A 105:13157

Colin M, Guènard D, Guèritte-Voegelein F, Potier P. (1989) FR 8610401

Collin M, Guènard D, Guèritte-Voegelein F, Potier P. (1990) EP 0253738

Cortes J, Vahdat L, Blum JL, Twelves C, Campone M, Roche H, Bacelot T, Awada A, Paridaens R, Goncalves A, Shuster DE, Wanders J, Fang F, Gurnani R, Richmond E, Cole PE, Ashworth S, Allison MA (2010) J Clin Oncol 28:3922

Cortes J, O'Shaughnessy J, Loesch D, Blum JL, Vahdat LT, Petrakova K, Chollet P, Manikas A, Diéras V, Delozier T, Vladimirov V, Cardoso F, Koh H, Bougnoux P, Dutcus CE, Seegobin S, Mir D, Meneses N, Wanders J, Twelves C (2011) Lancet 377:914

Cragg GM, Grothaus PG, Newman DJ (2009) Chem Rev 109:3012

Croteau R, Ketchum REB, Long RM, Kaspera R, Wildung MR (2006) Phytochem Rev 5:75

Cummings J, Willmott N, Smyth JF (1991) Eur J Cancer 27:532

Danishefsky SJ, Masters JJ, Young WB, Link JT, Snyder LB, Magee TV, Jung DK, Isaacs RCA, Bornmann WG, Alaimo CA, Coburn CA, Di Grandi MJ (1996) J Am Chem Soc 118:2843

Das B, Madhusudhan P, Reddy PV, Anitha Y (2001) Indian J Chem 40B:453

DE 4416 374 (1995); DE 195 13 040 (1996)

De Boer C, Meulman PA, Wnuk RJ, Peterson DH (1970) J Antibiot 23:442

Denis J-N, Greene AE, Serra AA, Luche M-J (1986) J Org Chem 51:46

Denis J-N, Greene AE, Guénard D, Guéritte-Voegelein F, Mangatal L, Potier P (1988) J Am Chem Soc 110:5917

Dennis MJ, Beijnen JH, Grochow LB, van Warmerdam LJ (1997) Semin Oncol 24(1 suppl 5): S5-12–S5-18

DeVita VT Jr, Hellman S, Rosenberg SA (1993) Cancer—principles and practice of oncology. J.B. Lippincott, New York, p 377

Didier E, Oddon G, Panze D, Leion P, Piquet D. (1999) US 5,962,702

Diéras V, Limentani S, Romieu G, Tubiana-Hulin M, Lortholary A, Kaufman P, Girre V, Besenval M, Valero V (2008) Ann Oncol 19:1255

Dikalov S, Landmesser U, Harrison DG (2002) J Biol Chem 277:25480

Dong C-G, Henderson JA, Kaburagi Y, Sasaki T, Kim JT, Urabe D, Guo H, Kishi Y (2009) J Am Chem Soc 131:15642

Dorr RT (1996) Semin Oncol 23(4 suppl):23

Du W (2003) Tetrahedron 59:8649

Duan JJW, Kishi Y (1993) Tetrahedron Lett 34:7541

Dubost M, Ganter P, Maval R, Ninet L, Pinneret S, Preud'Homme J, Werner GH (1963) C R Acad Sci 257:1813

Dumontet C, Jordan MA (2010) Nat Rev Drug Discov 9:790

Emmenegger H, Stähelin H, Rutschmann J, von Wartburg A (1961) Arzneimittelforschung 11:327

Erickson-Miller CL, May RD, Tomaszewski J (1997) Cancer Chemother Pharmacol 39:467

European Medicines Agency. Withdrawal Assessment Report for Ixempra, Apr 2009

Forli S, Manetti F, Altmann K-H, Botta M (2010) ChemMedChem 5:35–40

Fukasawa H, Li P-M, Yamamoto C, Murakami Y, Mizuno S, Uehara Y (1991) Biochem Pharmacol 42:1661

Ganesh T, Norris A, Sharma S, Bane S, Alcaraz AA, Synder JP, Kingston DGI (2006) Bioorg Med Chem 14:3447

Ganesh T, Yang C, Norris A, Glass T, Bane S, Ravindra R, Banerjee A, Metaferia B, Thomas SL, Giannakakou P, Alcaraz AA, Lakdawala AS, Synder JP, Kingston DGI (2007) J Med Chem 50:713

Ganguly A, Yang H, Cabral F (2010) Mol Cancer Ther 9:2914
Ge J, Normant E, Porter JR, Ali JA, Dembski MS, Gao Y, Georges AT, Grenier L, Pak RH, Patterson J, Sydor JR, Tibbitts TT, Tong JK, Adams J, Palombella VJ (2006) J Med Chem 49:4606
Gelderblom H, Verweij J, Nooter K, Spareboom AE (2001) Eur J Cancer 37:1590
Gensler WJ, Samour CM, Wang SY (1954) J Am Chem Soc 76:315
Giannakakou P, Gussio R, Nogales E, Downing KH, Zaharevitz D, Bollubuck B, Poy G, Sackett D, Nicolaou KC, Fojo T (2000) Proc Natl Acad Sci U S A 97:2904
Giovanella BC, Stehlin JS, Wall ME (1989) Science 246:1046
Gordaliza M, Garcia PA, Miguel de Coral JM, Castro MA, Gomez-Zurita MA (2004) Toxicon 44:441
Gottlieb JA, Luce JK (1972) Cancer Chemother Rep 56:103
Grein A, Spalla C, DiMarco A, Canevazzi G (1963) Giorn Microbiol 11:109
Guénard D, Guèritte-Voegelein F, Potier P (1993) Acc Chem Res 26:160
Guèritte-Voegelein F, Guènard D, Lavelle F, Le Goff M-T, Mangatal L, Potier P (1991) J Med Chem 34:992
Guo H, Dong C-G, Kim D-S, Urabe D, Wang J, Kim JT, Liu X, Sasaki T, Kishi Y (2009) J Am Chem Soc 131:15387
Haim N, Nemec J, Roman J, Sinha BK (1987) Cancer Res 47:5835
Hammonds TR, Denyer SP, Jackson DE, Irving WL (1996) J Med Microbiol 45:167
Hande KR (1998) Eur J Cancer 34:1514
Hirata Y, Uemura D (1986) Pure Appl Chem 58:701
Hoffmann J, Vitale I, Buchmann B, Galluzzi L, Schwede W, Senovilla L, Skuballa W, Vivet S, Lichtner RB, Vicencio JM, Panaretakis T, Siemeister G, Lage H, Nanty L, Hammer S, Mittelstaedt K, Winsel S, Eschenbrenner J, Castedo M, Demarche C, Klar U, Kroemer G (2008) Cancer Res 68:5301
Höfle G, Bedorf N, Steinmetz H, Schomburg D, Gerth K, Reichenbach H (1996) Angew Chem Int Ed Engl 35:1567
Holton RA (1991). US 5,015,744
Holton RA (1992). US 5,175,315
Holton RA (1993). US 5,274,124
Holton RA, Somoza C, Kim H-B, Liang F, Biediger RJ, Boatman D, Shindo M, Smith CC, Kim S, Nadizadeh H, Suzuki Y, Tao C, Vu P, Tang S, Zhang P, Murthi KK, Gentile LN, Liu JH (1994a) J Am Chem Soc 116:1597
Holton RA, Kim H-B, Somoza C, Liang F, Diediger RJ, Boatman PD, Shindo M, Smith CC, Kim S, Nadizadeh H, Suszuki Y, Tao C, Vu P, Tang S, Zhang P, Murthi KK, Gentile LN, Liu JH (1994b) J Am Chem Soc 116:1599
Hsiang YH, Liu LF (1988) Cancer Res 48:1722
http://clinicaltrials.gov/ct/show/NCT01126502
http://drugbank.ca/drugs/DB01030
http://en.wikipedia.org/wiki/Daunorubicin
Hutchinson CR (1981) Tetrahedron 37:1047
Imbert TF (1998) Biochimie 80:207
Internal research report Dr. R. J. Lake, Post-Doctoral Fellow, Marine Chemistry Group, University of Canterbury, 1988
Itokawa H, Lee KH (2003) Taxus: the genus Taxus. Taylor and Francis, London, pp 1–18
Jackson KL, Henderson JA, Phillips AJ (2009) Chem Rev 109:3044
Johnson T (1999) CRC ethnobotany desk reference. CRC Press, Boca Raton
Jordan M, Wilson L (2004) Nat Rev Cancer 4:253
Kaplan IW (1942) Med Surg J 94:388
Keizer HG, Pinedo HM, Schuurhuis GJ, Joenje H (1990) Pharmacol Ther 47:219
Kelland LR, Sharp SY, Rogers PM, Myers TG, Workman P (1999) J Natl Cancer Inst 91:1940
Kim D-S, Dong C-G, Kim JT, Guo H, Huang J, Tiseni PS, Kishi Y (2009) J Am Chem Soc 131:15636
Kishi Y, Fang FG, Forsyth CJ, Scola PM, Yoon SK (1995) US 5,436,23
Klar U, Graf H, Schenk O, Röhr B, Schulz H (1998) Bioorg Med Chem Lett 8:1397

Klar U, Buchmann B, Schwede W, Skuballa W, Hoffmann J, Lichtner RB (2006) Angew Chem Int Ed Engl 45:7942
Kollmannsberger C, Mross K, Jakob A, Kanz L, Bokemeyer C (1999) Oncology 56:1
Kour A, Shawl AS, Rehman S, Sultan P, Qazi PH, Suden P, Khajuria RK, Verma V (2008) World J Microbiol Biotechnol 24:1115
Kowalski R, Giannakakou P, Hamel E (1997) J Biol Chem 272:2534
Kuhn M, von Wartburg A (1969) Helv Chim Acta 52:948
Laatsch H, Fotso S (2008) Top Curr Chem 282:3
Laco GS, Collins JR, Luke BT, Kroth H, Sayer JM, Jerina DM, Pommier Y (2002) Biochemistry 41:1428
Ledford H (2010) Nature 468:608
Leighton J, Kalla RL (1958) Ann N Y Acad Sci 76:513
Li QY, Zu YG, Shi RZ, Yao LP (2006) Curr Med Chem 13:2021
Litaudon M, Hart JB, Blunt JW, Lake RJ, Munro MHG (1994) Tetrahedron Lett 35:9435
Litaudon M, Hickford SJH, Lill RE, Lake RJ, Blunt JW, Munro MHG (1997) J Org Chem 62:1868
Littlefield BA, Palme M, Seletsky BM, Towle MJ, Yu MJ (1999) WO 99/65894
Liu YQ, Yang L, Tian X (2007) Curr Bioact Compd 3:37
Long BH, Brattain MG (1984) In: Issell BF, Muggia FM, Carter SA (eds) Etoposide: current status and new developments. Academic, New York, p 223
Long BH, Minocha A (1983) Proc Am Assoc Cancer Res 24:1271
Long BH, Musial ST, Brattain MG (1984) Biochemistry 23:1183
Luduvico I, Hyaric ML, Almeida MV, Da Silva AD (2006) Mini Rev Org Chem 3:49
Lv M, Xu H (2011) Mini Rev Med Chem 11:901
Mann J (1994) Nature 367:594
Martello L, Verdier-Pinard P, Shen H (2003) Cancer Res 63:1207
Martin C, Zhang M, Gaisser S, Coates N, Wilkinson B (2011) US 2011/00991452
Menna P, Salvatorelli E, Gianni L, Minotti G (2008) Top Curr Chem 283:21
Michaelis ML, Ranciat N, Chen Y, Bechtel M, Ragan R, Hepperle M, Liu Y, Georg G (1998) J Neurochem 70:1623
Michaelis ML, Ansar S, Chen Y, Reiff ER, Seyb KI, Himes RH, Audus KL, Georg GI (2005) J Pharmacol Exp Ther 312:659
Moerman DE (1986) Medicinal plants of native America. Museum of Anthropology, University of Michigan, Ann Arbor
Moertel CG, Schutt AJ, Reitemeier RJ, Hahn RG (1972) Cancer Chemother Rep 56:95
Molnar I, Schupp T, Ono M, Zirkle RE, Milnamow M, Nowak-Thompson B, Engel N, Toupet C, Stratmann A, Cyr DD, Gorlach J, Mayo JM, Hu A, Goff S, Schmid J, Ligon JM (2000) Chem Biol 7:97
Mukaiyama T, Shiina S, Iwadare H, Saitoh M, Nishimura T, Ohkawa N, Sakoh H, Nishimura K, Tani Y-I, Hasegawa M, Yamada K, Saitoh K (1999) Chem Eur J 5:121
Mulzer J, Mantoulidis A, Öhler E (2000) J Org Chem 65:7456
Munro MGH, Blunt JW, Dumdei EJ, Hickford SJH, Lill RE, Battershill CN, Duckworth AR (1999) Prog Ind Microbiol 35:155
Narayan S, Carlson EM, Cheng H, Du H, Hu Y, Jiang Y, Lewis BM, Seletsky BM, Tendyke K, Zhang H, Zheng W, Littlefield BA, Towle MJ, Yu MJ (2011a) Bioorg Med Chem Lett 21:1630
Narayan S, Carlson EM, Cheng H, Condon K, Du H, Eckley S, Hu Y, Jiang Y, Kumar V, Lewis BM, Saxton P, Schuck E, Seletsky BM, Tendyke K, Zhang H, Zheng W, Littlefield BA, Towle MJ, Yu MJ (2011b) Bioorg Med Chem Lett 21:1634
Narayan S, Carlson EM, Cheng H, Condon K, Du H, Eckley S, Hu Y, Jiang Y, Kumar V, Lewis BM, Saxton P, Schuck E, Seletsky BM, Tendyke K, Zhang H, Zheng W, Littlefield BA, Towle MJ, Yu MJ (2011c) Bioorg Med Chem Lett 21:1639
Nettles JH, Li H, Cornett B, Krahn JM, Snyder JP, Downing KH (2004) Science 305:866–869
Newman DJ (2008) J Med Chem 51:2589
Newman DJ, Cragg GM (2010) In: Buss AD, Buttler MS (eds) Natural product chemistry for drug discovery, RSC biomolecular sciences no. 18. RSC, London, pp 3–27

Nicolaou KC, Yang Z, Liu JJ, Ueno H, Natermet PG, Guy RK, Claiborne CF, Renaud J, Couladouros EA, Paulvannan K, Sorensen EJ (1994) Nature 367:630
Nicolaou KC, Ninkovic S, Sarabia F, Vourloumis D, He Y, Vallberg H, Finlay MRV, Yang Z (1997a) J Am Chem Soc 119:7974
Nicolaou K, Winssinger N, Pastor J (1997b) Nature 387:268
Ojima I, Habus I, Zhao M, Georg GI, Jayasinghe LR (1991) J Org Chem 56:1681
Ojima I, Chakravarty S, Inoue T, Lin S, He L, Horwitz SB, Kuduk SD, Danishefsky SJ (1999) Proc Natl Acad Sci U S A 96:4256
Ojima I, Chen J, Sun L, Borella CP, Wang T, Miller ML, Lin S, Geng X, Kuznetsowa L, Qu C, Gallager D, Zhao X, Zanardi I, Xia S, Horwitz SB, Mallen-St. Clair J, Guerriero JL, Bar-Sagi D, Veith JM, Pera P, Bernacki RJ (2008) J Med Chem 51:3203
Oki T (1980). In: Crooke ST, Reich SD (eds) Anthracyclines (Proc. workshop). Academic, New York, pp 323–342
Onodera H, Kaneko M, Takahashi Y, Uochi Y, Funahashi J, Nakashima T, Soga S, Suzuki M, Ikeda S, Yamashita Y, Rahayu ES, Kanda Y, Ichimura M (2008) Bioorg Med Chem Lett 18:1577 (wrong page in database search online for 1588!)
Padlan EA, Kabat EA (1991) Methods Enzymol 203:3
Pettit G, Herald CL, Boyd MR, Leet JE, Dufresne C, Doubek DL, Schmidt JM, Cerney RL, Hooper JNA, Rützler KC (1991) J Med Chem 34:3339
Pommier Y (1997) In: Teacher BA (ed) Cancer therapeutic: experimental and clinical agents. Humana Press, Taotowa, p 153
Pommier Y (2006) Nat Rev Cancer 6:789
Querolle O, Dubois J, Thoret S, Roussi F, Guéritte F, Guénard D (2004) J Med Chem 47:5937
Rastelli G, Tian Z-Q, Wang Z, Myles DC, Liu Y (2005) Bioorg Med Chem Lett 15:5016
Redinbo MR, Stewart L, Kuhn P, Champoux JJ, Hol WGJ (1998) Science 279:1504
Reese M, Sanchez-Pedregal VM, Kibicek K, Meiler J, Blommers MJJ, Griesinger C, Carlomagno T (2007) Angew Chem Int Ed Engl 46:1864
Rivkin A, Yoshimura F, Gabarda AE, Cho YS, Chou T-C, Dong H, Danishefsky SJ (2004) J Am Chem Soc 126:10913
Rivkin A, Chou T-C, Danishefsky SJ (2005) Angew Chem Int Ed 44:2838
Ross W, Rowe T, Glisson B, Yalowich J, Liu L (1984) Cancer Res 44:5857
Rothermel J, Wartmann M, Chen T (2003) Semin Oncol 30(S6):51
Roussi F, Ngo QA, Thoret S, Guéritte F, Guénard D (2005) Eur J Org Chem:3952
Salas J (2008) Nat Prod Rep 26:628
Samaranayake G, Magri NF, Jitrangsri C, Kingston DGI (1991) J Org Chem 56:5114
Sasaki K, Inoue Y (1980) DE 3006097
Sasaki K, Rinehart KL, Slomp G, Grostic MF, Olson EC (1970) J Am Chem Soc 92:7591
Schiff PB, Fant J, Horwitz SB (1979) Nature 277:665
Schinzer D, Limberg A, Bauer A, Böhm OM, Cordes M (1997) Angew Chem Int Ed Engl 36:523
Schnur RC, Corman ML (2006) J Lab Comp Radiopharm 34:529
Schnur RC, Corman ML, Gallaschun RJ, Cooper BA, Dee MF, Doty ML, Muzzi ML, Moyer JD, DiOrio CI, Barbacci EG, Miller PE, O'Brien AT, Morin MJ, Foster BA, Pollack VA, Savage DM, Sloan DE, Pustilnik LR, Moyer MP (1995a) J Med Chem 38:3806
Schnur RC, Corman ML, Gallaschun RJ, Cooper BA, Dee MF, Doty JL, Muzi ML, DiOrio CI, Barbacci EG, Miller PE, Pollack VA, Savage DM, Pustilnik LR, Moyer JD, Moyer MP (1995b) J Med Chem 38:3813
Seidel V, Windhövel J, Eaton G, Alfermann AW, Arroo RRJ, Medarde M, Petersen M, Woolley JG (2002) Planta 215:1031
Seifert CF, Nesser ME, Thompson DF (1994) Ann Pharmacother 28:1063
Sènilh V, Blechert S, Colin M, Guènard D, Picot F, Potier P, Varenne P (1984) J Nat Prod 47:131
Sinha BK (1989) Chem Biol Interact 69:293
Siveski-Iliskovic N, Hill M, Chow DA, Singal PK (1995) Circulation 91:10
Smith AB III, Chen SS-Y, Nelson FC, Reichert JM, Salvatore BA (1995) J Am Chem Soc 117:12013

Sonnendecker G, Kremers E (1986) Urdangs history of pharmacy. American Institute of the History of Pharmacy, Madison, p 17
Stähelin HF, von Wartburg A (1991) Cancer Res 51:5
Staker BL, Feese MD, Cushman M, Pommier Y, Zembower D, Stewart L, Burgin AB (2005) J Med Chem 48:2336
Stamos DP, Kishi Y (1996) Tetrahedron Lett 37:8643
Stamos DP, Sean SC, Kishi Y (1997) J Org Chem 62:7552
Stephenson F (2002) Fla State Univ Res Rev 12(3):12–37
Stoll A, von Wartburg A, Angliker E, Renz J (1954) J Am Chem Soc 76:5004
Su D-S, Meng D, Bertinato P, Balog DM, Sorensen EJ, Danishefsky SJ, Zheng Y-H, Chou T-C, He L, Horwitz SB (1997) Angew Chem Int Ed Engl 36:757
Sun L, Veith JM, Pera P, Bernacki RJ, Ojima I (2010) Bioorg Med Chem 18:7101
Synder JP, Nettles JH, Cornett B, Downing KH, Nogales E (2001) Proc Natl Acad Sci U S A 98:5312
The New York Times, p 29 (3 May 1987)
Thomas CJ, Rahier NJ, Hecht SM (2004) Bioorg Med Chem 12:1585
Tian Z-Q, Liu Y, Zhang D, Wang Z, Dong SD, Cerreras CW, Zhou Y, Rastelli G, Santi DV, Myles DC (2004) Bioorg Med Chem 12:5317
Towle MJ, Salvato KA, Budrow J, Wels BF, Kuznetsov G, Aalfs KK, Welsh S, Zheng W, Seletsky BM, Palme MH, Habgood GJ, Singer LA, DiPietro LV, Wang Y, Chen JJ, Quincy DA, Davis A, Yoshimatsu K, Kishi Y, Yu MJ, Littlefield BA (2001) Cancer Res 61:1013
Twelves C, Cortes J, Vahdat LT, Wanders J, Akerele C, Kaufman PA (2010) Clin Breast Cancer 10:160
Uemura D, Takahashi K, Yamamoto T, Katayama C, Tanaka J, Okumura Y, Hirata Y (1985) J Am Chem Soc 107:4796
Venditto VJ, Simanek EE (2010) Mol Pharm 7(2):307
Wall ME, Wani MC, Cook CE, Palmer KH, McPhail AI, Sim GA (1966) J Am Chem Soc 88:3888
Wang Y, Habgood GJ, Christ WJ, Kishi Y, Littlefield BA, Yu MJ (2000) Bioorg Med Chem Lett 10:1029
Wani MC, Taylor HL, Wall ME, Coggon P, McPhail AT (1971) J Am Chem Soc 93:2325
Wegele H, Müller L, Buchner J (2004) Rev Physiol Biochem Pharmacol 151:1
Weiss RB (1992) Semin Oncol 19:670
Wender PA, Mucciaro TP (1992) J Am Chem Soc 114:5878
Wender PA, Badham NF, Conway SP, Floreancig PE, Glass TE, Houze JB, Krauss NE, Lee D, Marquess DG, McGrane PL, Meng W, Natchus MG, Shuker AJ, Sutton JC, Taylor RE (1997) J Am Chem Soc 119:2757
Whitesell L, Mimnaugh EG, De Costa B, Myers CE, Neckers LM (1994) Proc Natl Acad Sci U S A 91:8324
Williams DH, Stone MJ, Hauck PR, Rahman SK (1989) J Nat Prod 52:1189–1208
Wilson CR, Sauer J-M, Hooser SB (2001) Toxicon 39:175
Wink M (2010) Encyclopedia of life sciences. Wiley, Chichester
Xu H, Lv M, Tian X (2009) Curr Med Chem 16:327
Yamaguchi S, Nakashima T, Kanda Y (2007) WO 2007/001049
Yamazaki Y, Kitajima M, Arita M, Takayama H, Sudo H, Yamazaki M, Aimi N, Saito K (2004) Plant Physiol 134:161
Yang Z, He Y, Vourloumis D, Vallberg H, Nicolaou KC (1997) Angew Chem Int Ed Engl 36:166
Yiqing Z, Zhong Z, Liu F, Sun M, Craig D, Eng S, Feng L, Sherrill M, Cropp GF, Yu K, Hannah AL, Johnson RG Jr (2005) Proc Am Assoc Cancer Res:46
Yukimune Y, Tabata H, Higashi Y, Hara Y (1996) Nat Biotechnol 14:1129
Zhang M-Q, Gaisser S, Nur-E-Alam M, Sheehan LS, Vousden WA, Gaitatzis N, Peck G, Coates NJ, Moss SJ, Radzom M, Foster TA, Sheridan RM, Gregory MA, Roe SM, Prodromou C, Pearl L, Boyd SM, Wilkinson B, Martin CJ (2008) J Med Chem 51:5494
Zheng W, Seletsky BM, Palme MH, Lydon PJ, Singer LA, Chase CE, Lemelin CA, Shen Y, Davis H, Tremblay L, Towle MJ, Salvato KA, Wels BF, Aalfs KK, Kishi Y, Littlefield BA, Yu MJ (2004) Bioorg Med Chem Lett 14:5551

Chapter 7
The Role of Genetic Engineering in Natural Product-Based Anticancer Drug Discovery

Claudia Eva-Maria Unsin, Scott R. Rajski, and Ben Shen

Abstract Genetic engineering is the process of altering, in a premeditated fashion, the genetic makeup of an organism and has started to play an increasing role in the production and development of clinically significant antitumor compounds. Genes can be introduced into the microbial producer of medicinally relevant secondary metabolites or inactivated to achieve changes in the metabolic profile. The pharmaceutical use of natural products with anticancer activity is most often limited by factors such as low production titers or poor solubility. Genetic engineering has provided an alternative to circumvent such difficulties by affording recombinant strains capable of high titers, as well as, recombinant strains able to produce novel analogs with characteristics superior to those of the parent natural product. This chapter highlights recent genetic engineering advances that have been successfully applied to the development of natural product and natural product-based anticancer agents. Titer improvement and combinatorial biosynthesis in actinomycetes to produce new compounds will be discussed in detail.

C. Eva-Maria Unsin • S.R. Rajski
School of Pharmacy, Division of Pharmaceutical Sciences, University of Wisconsin, 777 Highland Avenue, Madison, WI 53705-2222, USA
e-mail: claudia_unsin@yahoo.de; srrajski@pharmacy.wisc.edu

B. Shen (✉)
Departments of Chemistry and Molecular Therapeutics, Natural Products Library Initiative at The Scripps Research Institute, The Scripps Research Institute, 130 Scripps Way, #3A1, Jupiter, FL 33458, USA
e-mail: shenb@scripps.edu

F.E. Koehn (ed.), *Natural Products and Cancer Drug Discovery*, Cancer Drug Discovery and Development, DOI 10.1007/978-1-4614-4654-5_7,

7.1 Introduction

Actinomycetes are a major source of bioactive natural products. More than 10,000 substances with bioactivity have been isolated so far from terrestrial and marine actinomycetes (Berdy 2005, 2012) and many are clinically used as antitumor agents, antibiotics, or immunosuppressants. Natural products such as bleomycin, doxorubicin, rapamycin, mithramycin, and C-1027 constitute, or have inspired, important anticancer therapeutics. The significance of natural product to human health is by no means, however, restricted to anticancer agents. Regardless of their uses, drug intolerance, severe organ toxicity, and pathogen resistance often provide significant driving forces for analog design efforts. Genetic engineering is an important approach to devise new compounds that lack the aforementioned undesirable characteristics. Under the umbrella of genetic engineering, combinatorial biosynthesis is applied to "mix and match" biosynthetic genes from different gene clusters or gene environments. The resultant gene products often catalyze the production of hybrid substances possessing novel features relating to structure, activity, and/or solubility. Mutagenesis is another means of producing novel natural products via genetic manipulation although mutagenesis does not generally permit the precision of engineering approaches. Inactivation of enzymes from a biosynthetic pathway through genetic ablation of specific genes leads in many cases to the generation of a drug derivative, especially if the respective enzyme acts at the end of the biosynthetic pathway. Titers can be elevated by heterologous expression of the biosynthetic genes in a suitable host or by manipulation of the producer's genetic system. Regulatory genes can be altered to optimize expression levels. Most often, the overexpression of biosynthetic genes leads to profound increases in natural product titer. Alternatively, the expression of resistance factors or exporters can help stimulate production or increase natural product export from the producing cells (Fig. 7.1).

7.2 Titer Improvement

Some antitumor substances are highly active but are produced in quantities insufficient to support necessary development efforts. Exacerbating this difficulty is that, in many cases, the structural complexity of these natural products vastly complicates or prohibits their practical construction via total synthesis. It is therefore highly desirable to enhance the ability of the producing organism to produce such compounds in substantially greater quantities.

7.2.1 Titer Improvement in Native Producing Strains

An organism's primary and secondary metabolic pathways are linked via the metabolic burden that each places on the other. Although antitumor natural products result

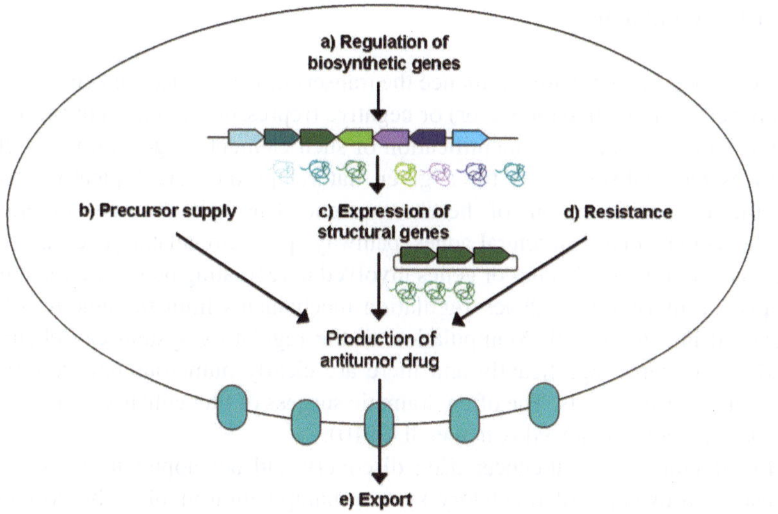

Fig. 7.1 Important steps targeted by genetic engineering approaches to improve production titers of natural antitumor substances. (**a**) Derepression or activation of regulatory genes leads to overexpression of structural genes; amplification of genes for (**b**) precursor supply or (**c**) structural genes overcomes pathway bottlenecks; (**d**) enhancement of self-resistance increases production levels; (**e**) enhanced expression of exporter proteins and efflux pumps increases the cell's ability to produce more compound by circumventing possible cell death and/or feedback mechanisms ordinarily used to limit compound production

from an organisms' secondary metabolic machinery there remains an inextricable link between natural product titers and the absolute metabolic burden with which the producing organism needs to cope. As the principal producer of anticancer natural products, the actinomycetes have therefore evolved strict regulatory mechanisms crucial to the maintenance of primary metabolism but that often limit natural product titers. Indeed, these regulatory mechanisms are at the root of many obstacles encountered in the industrial production of anticancer drugs. Such obstacles have been largely addressed through a combination of approaches based on genetic engineering (Fig. 7.1). Targeted inactivation of repressor genes and overexpression of transcriptional activators both constitute strategies that have been very effective in increasing natural product titers from their wild-type strains (Fig. 7.1a). The overexpression of structural genes has also served as a useful approach to overcome natural product-based feedback inhibition (Fig. 7.1b, c). Additionally, gene dosage plays an important role in the improvement of production titers. Relevant genes are multiplied or overexpressed in order to increase natural product production. Self-resistance can also contribute significantly to low titers. Increased resistance of the producer and amplified expression of structural genes can both lead to improved natural product titers (Fig. 7.1d). Additionally, exporters, either as part of the self-defense system or simply to transport the drug outside the cell, can be overexpressed to enhance natural product synthesis (Fig. 7.1e). Importantly, these strategies can be used individually or in combination to circumvent the problem of development- and production-limiting natural product titers.

7.2.1.1 Regulation

Pathway-specific regulators influence the transcription of structural genes and these can have either positive (activator) or negative (repressor) effects on the expression of biosynthetic genes. The identification of such elements is generally straightforward as biosynthetic genes for a given natural product are typically clustered together in specific regions of the chromosome. This is in contrast to pleiotropic regulators that control structural genes, pathway-specific regulator genes, and others such as morphological genes or genes involved in regulating precursor and cofactor supply. Furthermore, feedback regulation mechanisms limit the amount of drug produced inside the cell. Manipulation of the regulatory system can change the production profile significantly and there are clearly numerous opportunities for such engineering efforts. The often dramatic success of titer enhancement strategies has been recently reviewed (Chen et al. 2010).

In the context of anticancer drug discovery and development an extensively investigated example of regulatory system manipulation involves biosynthesis of the anthracycline-type antitumor antibiotic doxorubicin (Fig. 7.2), the production of which is tightly regulated in *Streptomyces peucetius*. Overexpression of transcriptional activators has been shown to increase doxorubicin production up to 4.3-fold relative to the wild-type strain (Malla et al. 2010a). *S. peucetius* strains bearing plasmids with an increasingly larger number of the regulatory genes *dnrN*, *dnrI*, *afsR*, and *metK1-sp* produced increasingly greater amounts of doxorubicin. Consistent with this work, when the global regulatory gene *afsR*, an established transcriptional activator, was overexpressed in *S. peucetius*, doxorubicin production was enhanced up to eightfold (Maharjan et al. 2009). Similar experiments with the pikromycin producer *Streptomyces venezuelae* and the actinorhodin producer *Streptomyces lividans* yielded similar results; *afsR* overexpression in *S. venezuelae* led to a ~5-fold improvement in pikromycin (Fig. 7.2) production relative to the wild-type strain and *afsR* overexpression in *S. lividans* led to a 1.5-fold improvement in actinorhodin (Fig. 7.2) production relative to wild-type. Similar titer improvements have been achieved in the fredericamycin producer *Streptomyces griseus* ATCC 49344. Overexpression of *fdmR1*, which codes or a SARP family member pathway-specific activator, led to a 5.6-fold improved production of fredericamycin A (FDM A, Fig. 7.2) by *S. griseus*. The recombinant strain produces FDM A with a titer of ~1.5 g/L sufficient to support future development efforts for this antitumor pentadecaketide (Chen et al. 2008a). These recent examples of regulatory manipulation to improve anticancer agent titers from fermentation compliment many earlier examples.

Just as overexpression of activators can lead to improved natural product titers so too can the repression of negative regulators. An excellent example of this comes in the form of efforts to improve the production of the antibacterial compounds platensimycin and platencin (Fig. 7.2). The production of both compounds is encoded by a single gene cluster in *Streptomyces platensis* MA7327. Within this cluster, the *ptmR1* gene was found to encode for what appears to be a member of the GntR family of transcriptional repressors. Consistent with its role as a repressor of platensimycin

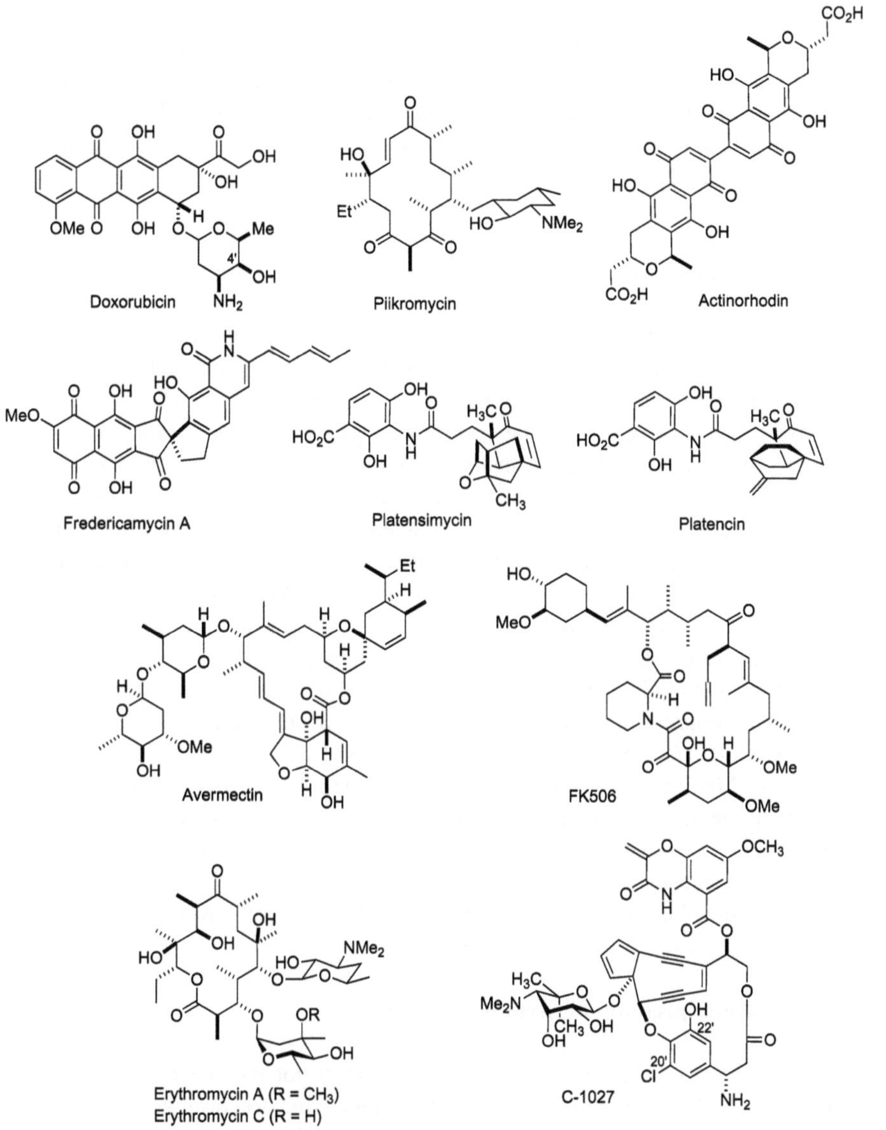

Fig. 7.2 Clinically significant natural products whose development and application were enabled by titer improvements accomplished by altered expression of regulatory genes (doxorubicin, pikromycin, platensimycin, platencin, fredericamycin), structural genes (erythromycin A, doxorubicin), resistance genes (actinorhodin, C-1027) and export genes (doxorubicin) or by enhancements in biosynthetic precursor availability (actinorhodin, avermectin, pikromycin, FK506) during fermentation, or by a combination of these approaches

and platencin biosynthesis inactivation of *ptmR1* afforded *S. platensis* strains that produced both natural products in titers ~100-fold superior to those of the wild-type strain (Smanski et al. 2009). Clearly, repressors of biosynthetic machineries constitute excellent targets of regulatory modification and complement, perhaps in multiple ways, the effectiveness of targeting activators to achieve improved natural product titers (Chen et al. 2010).

7.2.1.2 Precursor Supply

Bottlenecks in biosynthetic pathways restrict natural product titers and are often associated with the limitations of primary metabolism in providing key precursors necessary for secondary metabolic pathways. Such limitations can be overcome by amplifying the gene or genes that code for enzymes associated with such bottlenecks; increased enzyme levels translate to diminished bottleneck effects and hence, improved titers. Examples correlating precursor supply to improve natural product titers have focused on carbohydrate metabolism, fatty acid precursors, and intracellular cofactor supplies (Olano et al. 2008). Such examples include heterologous overexpression of the *S*-adenosyl-L-methionine (SAM) synthetase *metK*, which improved production of a number of anticancer agents and antibiotics, such as actinorhodin, avermectin, and pikromycin (Fig. 7.2), by providing SAM as a substrate or cofactor (Huh et al. 2004). More recently, it has been demonstrated that titers of the immunosuppressant drug FK506 (Fig. 7.2) can be improved by enhancing the intracellular pool of methylmalonyl-CoA, a critical building block for FK506 production. Medium supplementation with methyl oleate and simultaneous introduction of the methylmalonyl-CoA mutase pathway into the FK-506 producer *Streptomyces clavuligerus* CKD1119 led to a threefold improvement in FK-506 titers through transient intracellular increases in acetyl-CoA, which is rapidly converted via the endogenous machinery to methylmalonyl-CoA (Mo et al. 2009).

7.2.1.3 Overexpression of Structural Genes

Structural genes code for any RNA or protein product not associated with regulation and resistance. In the context of secondary metabolism, structural genes are necessary for providing functional platforms on which natural products are constructed. An absence or limited expression of these genes correlates to the production of intermediates or shunt metabolites that are co-isolated with intact natural products. Such impurities can be minimized by structural gene overexpression and this strategy has also been shown to induce enhanced natural product titers. The biosynthesis of the antibiotic erythromycin A (Fig. 7.2) has been studied intensively in the past and serves as a model for polyketide biosynthesis in actinomycetes. A large number of natural anticancer agents are polyketides. The tailoring genes *eryK* (P450 hydroxylase) and *eryG* (SAM-dependent *O*-methyltransferase) were overexpressed in the producer strain *Saccharopolyspora erythraea*. Larger quantities of these enzymes led to titer improvement and purity of the antibiotic erythromycin A by conversion of its intermediates (Chen et al. 2008b). A good example of titer improvement of antitumor substances is engineering of tailoring genes in doxorubicin production. Glycosylation is considered to be a rate-limiting step in doxorubicin biosynthesis. Heterologous expression of structural sugar biosynthesis and glycosyltransferase genes elevated doxorubicin levels in *S. peucetius* up to 5.6-fold (Malla et al. 2009). Structural gene deletion or inactivation also can be used to increase natural product titers. This is particularly applicable when applied to systems in which one natural product is biosynthetically transformed to another, less useful, compound. However, by and large, structural genes have been the target of overexpression in efforts to improve anticancer natural product titers.

7.2.1.4 Overexpression of Resistance Genes

Enhanced levels of natural product resistance in the producer strain often correlate to enhanced production of compound. One example of natural product resistance involves CagA, the apoprotein of the enediyne antitumor antibiotic C-1027 (Fig. 7.2). CagA binds to the toxic chromophore of C-1027, stabilizing the ordinarily labile enediyne and playing an important role in C-1027 resistance to the producer *Streptomyces globisporus* (Beerman et al. 2009; Kennedy et al. 2007; Liu et al. 2002). Disruption of the *cagA* gene showed that the apoprotein is important but not essential for resistance and that other factors seem to be required for self-resistance. Though not the sole source of resistance to C-1027, overexpression of *cagA* leads to increased C-1027 production (Cui et al. 2009). Such a strategy to enhance titer is limited only by the resistance mechanism; overexpression of resistance genes can impact more than just one compound at a time. For instance, overexpression of streptomycin resistance (*rpsL* ribosomal protein mutation) is known to increase the production of the polyketide antibiotic actinorhodin (Hesketh and Ochi 1997). Additionally, multiple drug resistance mutations have been used to obtain a strain with 180-fold increased productivity. Using the same idea of enhanced resistance to multiple antibiotics a new approach called "ribosome engineering" has been developed (Wang et al. 2008). By screening for a *Streptomyces coelicolor* strain that is resistant to eight different antibiotics a strain capable of producing 1.63 g/L actinorhodin was identified. The high degree of resistance was attributed to mutant ribosomes which sustain a high level of protein synthesis, even at a late stage of growth when antibiotic production starts. Furthermore, the mutants showed an increased ability to accumulate ppGpp, which is an important signaling molecule for the onset of antibiotic production. This example illustrates that resistance is tied to the regulatory network within the cell and can contribute to improved natural product titers.

7.2.1.5 Overexpression of Export Genes

Natural product biosynthesis is often subjected to negative feedback regulation and a critical component of such mechanisms is the expression of exporter genes.

Indeed, many anticancer natural products are removed from the cell by export proteins. Intracellular accumulation of products and thus inhibition of biosynthetic enzymes can be modulated by efflux systems. For instance, doxorubicin production has been enhanced 2.2-fold by overexpression of the export genes *drrA* and *drrB*. These genes encode an ATP-binding cassette (ABC) transporter that pumps the antibiotic out of the cell thereby conferring resistance (Malla et al. 2010b). Conversely, deletion of *drrA* and *drrB* resulted in a dramatic decrease of antitumor antibiotic production in *S. peucetius* (Srinivasan et al. 2010). For the purpose of titer improvement it is important to note that enhancement of drug export is not limited to product-specific transporters. For instance, it is now well established that the nonspecific multi-drug resistance (MDR) exporters can be modulated so as to increase drug production (Adrio and Demain 2006).

7.2.2 Heterologous Expression

Many natural products with complex structures and anticancer activity have been identified from marine and terrestrial actinomycetes, but large-scale production is out of reach. Strains often grow very slowly or only under conditions that are not feasible in the laboratory. Heterologous expression, the expression and exploitation of biosynthetic gene clusters introduced into non-native host strains, has been very effectively used to overcome these limitations and has been very rigorously reviewed previously (Galm and Shen 2006; Wenzel and Muller 2005). Complete gene clusters for antitumor antibiotics can be transferred into a more suitable organism that grows faster, is genetically characterized and stable, and is amenable to manipulation. Cosmids, BACs (bacterial artificial chromosomes), or YACs (yeast artificial chromosomes) are generally used as vector systems for biosynthetic gene clusters. These genetic elements carry large regions of DNA inserts (up to 1,000 kb) and are easily transferable into other organisms, the "heterologous hosts." It is important for antitumor antibiotic biosynthesis that all structural genes, regulatory genes, and resistance genes of a gene cluster be present and active. Activating enzymes, cofactors, or strain-specific chaperones are often needed but are not located in the gene cluster. If the host, for instance *Escherichia coli*, does not have these required factors, the respective genes have to be transferred separately into the heterologous host. Sometimes promoters have to be exchanged in order to achieve efficient expression in the host. The genetic code for the translation of mRNA into amino acids differs in some organisms; expression of specific tRNAs overcomes this limitation.

Since an organism's endogenous secondary metabolism represents an unproductive carbon sink relative to the introduced machinery, it is often advisable to inactivate endogenous antibiotic gene clusters present in the heterologous host so that the vast majority of the host's metabolic flux can be dedicated to production of the desired natural product. This strategy has been very successfully exploited and numerous anticancer products have been successfully expressed in heterologous systems

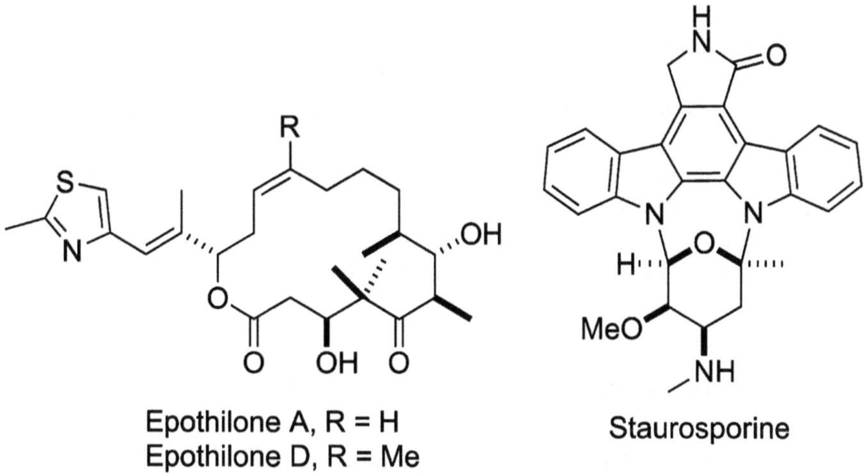

Fig. 7.3 Selected examples of natural products successfully produced in heterologous host expression systems

including but not limited to the epothilones (Fig. 7.3) in *S. coelicolor* (Tang et al. 2000) and in *E. coli* (Mutka et al. 2006), staurosporine (Fig. 7.3) in *S. lividans* (Onaka et al. 2002), FDM A in *Streptomyces albus* (Wendt-Pienkowski et al. 2005), and bleomycins in *Streptomyces flavoviridis* (Huang et al. 2012). Not surprisingly, heterologous expression could emerge as promising alternative for the industrial production and development of many natural anticancer agents (Galm and Shen 2006; Wenzel and Muller 2005).

7.3 Generation of New Analogs by Genetic Engineering

Two major enzymatic machineries are commonly found in natural antitumor agents: polyketide synthases (PKS) and non-ribosomal peptide synthetases (NRPS). Both systems are multi-enzyme complexes with modular organizations. Small precursors, acyl units or amino acids, are activated and attached to a carrier protein. Repetitive condensations of activated units lead to a polyketide or polypeptide scaffold, respectively. Some modifications, such as reductions, can take place during or after the assembly of the backbone. Tailoring enzymes catalyze finely tuned and structurally discrete transformations to the core scaffold.

The high degree of modularity of PKS and NRPS enzymes makes them easy to manipulate. Genes from different gene clusters can be readily mixed and matched to produce novel hybrid natural compounds and this means of obtaining structural diversity ideally represents the field of combinatorial biosynthesis. In addition to diversification of a core scaffold or molecular framework, the introduction of exogenous tailoring enzymes to a specific producer's genome allows another level of structural complexity to be achieved. The promise and potential of combinatorial biosynthetic methods as applied to drug discovery have been previously reviewed (Van Lanen and Shen 2006).

Just as combinatorial biosynthesis can exploit the introduction or modification of genes en route to new structures, directed mutagenesis is also a useful tool to manipulate functional genes within a biosynthetic gene cluster. Gene deletions or insertions result in the inactivation of biosynthetic genes for a given natural product. As a result, the gene is no longer translated into a functional enzyme and the corresponding catalytic step is omitted in the biosynthetic pathway leading ultimately to new, albeit truncated, natural products.

7.3.1 Combinatorial Biosynthesis

7.3.1.1 Exchanging Domains Within Modules

Some organisms contain elements from different biosynthetic gene clusters. This phenomenon is found in nature and has proven to be a very effective means of enhancing structural diversity in natural products. Both enzymatic machineries work together via protein–protein interactions to catalyze the biosynthesis of a hybrid compound, such as the non-ribosomal peptide/polyketide bleomycin (Du et al. 2000) that results from a gene cluster bearing both PKS and NRPS modules. The combination of elements from different origins can be applied to genetic engineering and this often calls for the use of linkers, short peptides at the end of enzymes which allow compatible domains to communicate by protein–protein interaction. It is possible to exchange enzyme domains if the linkers are still matching and the exchanged domain has relaxed substrate specificity. For instance, an acyltransferase (AT) domain of the erythromycin PKS has been successfully combined with a rapamycin AT domain to yield a novel triketide (Oliynyk et al. 1996). An alternative strategy has been to use thioesterases (TE) to release polyketide or polypeptide chains from their synthases by hydrolysis or macrocyclization, thereby making the basic scaffold available for subsequent tailoring (Schwarzer et al. 2001). With the approach of swapping domains, libraries of hybrid PKS and NRPS can be generated and screened for novel anticancer agents (Keller and Schauwecker 2003; Kim et al. 2004).

7.3.1.2 Variation of the Starter Unit

The use of alternative starter units can be achieved by the exchange of loading modules. Initial experiments exploiting this approach to molecular diversity involved the actinorhodin PKS system. When coexpressed with the R1128 initiation module, the actinorhodin minimal PKS produced novel hexaketides with propionyl and isobutyryl primer units (Tang et al. 2004). This principle of mix and match can be applied to enzyme domains from anticancer biosyntheses. Prodiginine antibiotics show anticancer and immunosuppressive activity. The initiation enzyme RedP from undecylprodiginine biosynthesis in *S. coelicolor* was replaced by its homologous enzyme from fatty acid biosynthesis, FabH. Besides undecylprodiginine, two new analogs, methylundecylprodiginine and methyldodecylprodiginine, were produced (Fig. 7.4) (Mo et al. 2005).

In contrast to the exchange of loading domains, utilization of synthetic *N*-acetylcysteamine starter units or precursor analogs also leads to novel analogs with altered starter molecules. Synthesis of unnatural polyketides with *N*-acetylcysteamine (SNAc) thioesters as starter substrates and malonyl-coenzyme A (CoA) or methylmalonyl-CoA as extender substrates provided novel products that target the HER signaling pathway in human breast cancer cells (Kwon et al. 2010). In vitro reconstitution of the first three epothilone PKS/NRPS modules results in epothilone precursor biosynthesis; if serine was utilized as a starter molecule instead of cysteine, an oxazole instead of thiazole moiety was found in the epothilone structure (Fig. 7.4) (Schneider et al. 2002). Feeding of unnatural starter units therefore has been shown to provide access to new natural product analogs in the future.

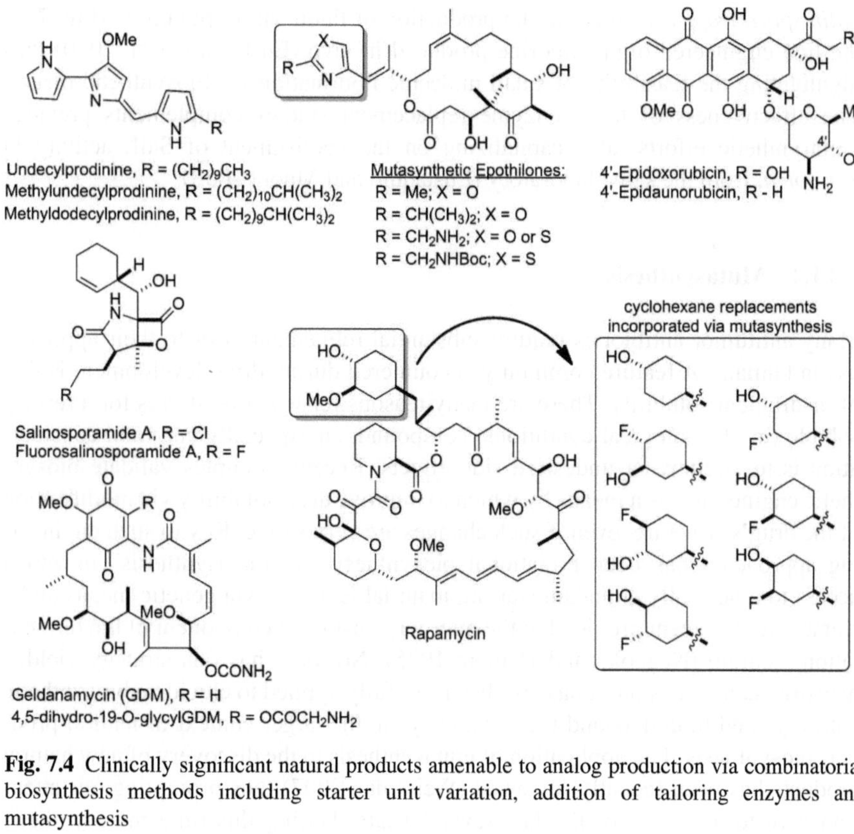

Fig. 7.4 Clinically significant natural products amenable to analog production via combinatorial biosynthesis methods including starter unit variation, addition of tailoring enzymes and mutasynthesis

7.3.1.3 Addition of Tailoring Enzymes

Glycosylation significantly alters the bioactivity of compounds and therefore is a very important modification exploitable for the generation of novel antitumor substances. Genes for aglycon biosynthesis can be coupled with heterologous genes for sugar biosynthesis and glycosyltransferases. The genetic engineering of *S. peucetius* bypasses the classical low-yielding semisynthesis of 4′-epidoxorubicin and 4′-epidaunorubicin (Fig. 7.4); both are important cancer chemotherapy drugs. These 4′-epimeric anthracyclines are produced when the *S. peucetius dnmV* gene for daunosamine biosynthesis is replaced with the *Streptomyces avermitilis avrE* or *Saccharopolyspora eryBIV* genes. By introducing further mutations that increase daunorubicin and doxorubicin titers in the wild-type producer strain, analog production can be improved significantly (Madduri et al. 1998). This example, though now dated, illustrates the effectiveness of this engineering strategy to produce new hybrid antibiotics.

A more recent approach to product diversity is the modification of natural products by halogenation. The unique tailoring enzyme fluorinase FlA from *Streptomyces cattleya* was used to engineer fluorometabolite production, which is rarely found in natural product biosyntheses. Replacement of the chlorinase gene *salL* with *flA* in *Salinispora tropica* resulted in the production of fluorosalinosporamide (Fig. 7.4), the first engineered organofluorine produced in vivo (Eustaquio et al. 2010) thus highlighting the feasibility of small molecule flourination via biosynthetic means. The effectiveness of these enzyme replacement studies complements previous mutasynthetic efforts, also capitalizing on the abolishment of SalL activity in *S. tropica*, from the same laboratory (Eustaquio and Moore 2008).

7.3.1.4 Mutasynthesis

Many antitumor antibiotics require substantial refinement prior to their approved use in humans. A feature commonly encountered during drug development is that of insufficient solubility. There are many reasons for why one strives for a readily soluble (in physiological conditions) compound; an especially common consideration is to circumvent undesired side effects. Recent examples validate biosynthetic engineering as a means by which to improve drug solubility via modification of the drug's structure, even if such changes are very subtle. Key to such engineering approaches has been mutational biosynthesis or "mutasynthesis" in which access to a normally abundant starting material is denied via genetic means and a suitable replacement provided to the microorganism as a component of the fermentation medium (Nagaoka and Demain 1975). Not only has this strategy yielded new drug candidates but it has also been carefully applied to elucidate biosynthetic pathways and to understand these pathways in the larger context of natural products biosynthesis. The application of mutasynthesis to the discovery of new natural products has been recently reviewed (Weissman 2007) thus prompting our attention here to advances of the last several years. During this time mutasynthetic approaches have resulted in the production of new antibacterial agents belonging to the aminocoumarin (Lutz 2009a, b), pikromycin (Gupta et al. 2008), and uridyl peptide antibiotic (Gruschow et al. 2009) classes.

Anticancer drug discovery has benefited substantially from the application of mutasynthetic methods to new natural products biosynthesis. Numerous examples of the successful application of mutasynthetic approaches to new anticancer drug candidates have appeared in the last several years. For instance, fluorinated rapamycin analogs have been produced recently by feeding mutants of the producing strain *Streptomyces hygroscopicus*, which cannot produce the natural starter acid required to initiate polyketide synthesis fluorohydroxycyclohexanecarboxylic acids and their ester derivatives (Fig. 7.4). The result of these mutasynthetic experiments was the production of six fluorinated rapamycin analogs with titers ranging from 4 to 28 mg/L (Goss et al. 2010). Though classically viewed as an immunosuppressive agent, rapamycin, and by association, rapamycin analogs are now being carefully examined as cancer therapeutics; Torisel® and Afinitor®, both rapamycin analogs, have recently been approved for the treatment of advanced renal cell carcinoma. Mutasynthetic efforts have also involved geldanamycin (GDM), an antitumor antibiotic also produced by *S. hygroscopicus* (Fig. 7.4). Motivated by dose limiting hepatotoxicity the production of GDM analogs displaying fewer and less severe side effects has been achieved via a mutasynthetic approach. Gene inactivation of the carbamoyltransferase gene *gdmM* yielded 4,5-dihydro-7-*O*-descarbamoyl-7-hydroxy-19-*O*-glycylgeldanamycin, which was isolated and subsequently converted into 4,5-dihydro-19-*O*-glycylgeldanamycin by a PKS-deficient *S. hygropicus* strain. The resulting analog showed lower cytotoxicity against HepG2 cancer cells and significantly greater water-solubility than GDM (Li et al. 2008). Another class of anticancer agents, the salinasporamides, which operate via proteasome inhibition have also been the subject of mutasynthetic efforts (Eustaquio and Moore 2008). Salinosporamide A, a chlorinated natural product from *S. tropica*, is believed to get its chloride moiety from 5′-chloro-5′-deoxyadenosine (ClDA) produced by virtue of the enzyme SalL and its ability to convert S-adenosylmethionine (SAM) to ClDA in the presence of chloride. Ablation of SalL activity and administration of synthetically produced 5′-fluoro-5′-deoxyadenosine (FDA) afforded a fluorinated analog of salinosporamide A. Although the fluorinated analog is not as potent at proteasome inhibition as the chlorinated natural product, these experiments reveal the amenability of the salinosporamide biosynthetic machinery to mutasynthetic methods while revealing a previously underappreciated SAM-dependent halogenation pathway. Mutasynthesis is thus a powerful tool for answering biosynthetic questions in addition to providing new natural products.

Our work and that of others on the glutarimide-containing polyketides migrastatin, *iso*-migrastatin, and lactimidomycin support the significance of cell migration inhibitors as potential antimetastasis agents (Fig. 7.5) (Ju et al. 2008, 2009; Perez and Danishefsky 2007). Tumor cell migration inhibitors such as migrastatin, iso-migrastatin, and lactimidomycin, studied extensively in our laboratories, constitute excellent antimetastasis candidates for the control and treatment of cancers. It is therefore significant that mutasynthesis of the natural product beauvericin, a cyclic depsipeptide inhibitor of tumor cell migration isolated from *Beauveria bassiana*,

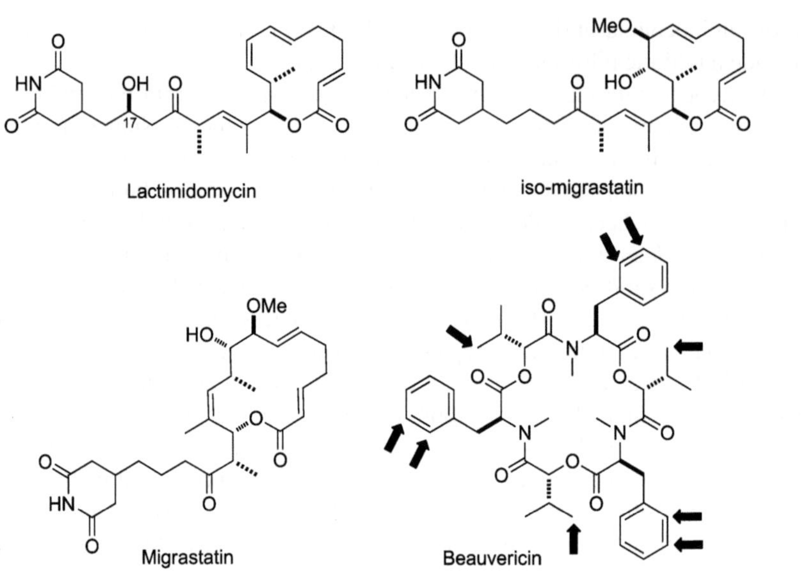

Fig. 7.5 Natural product inhibitors of tumor cell migration processes currently being investigated as antimetastasis agents. A combination of precursor supply manipulations and mutasynthesis has been effectively used to produce 14 new beauvericin analogs some of which have antiproliferative activities superior to beauvericin. Sites of possible structural departure from beauvericin are indicated by *darkened arrows*

has been used to produce new beauvericin analogs (Fig. 7.5). This was accomplished first by abolishing the function of KIVR, the biosynthetic enzyme responsible for production of D-hydroxyisovalerate, a common precursor for depsipeptide biosynthesis in *B. bassiana*. To the *kivr* knockout strain were then provided pairs of non-natural amino acids whose use by the remaining biosynthetic machinery resulted in 14 new cyclic depsipeptides (Xu et al. 2009). Testing of the wide assortment of new compounds has enabled an enhanced understanding of discrete molecular features and their correlation to the assortment of biological activities, such as cytotoxicity and antiproliferative activity as well as cell migration inhibition, displayed by beauvericin and its related congeners.

7.4 Conclusion

It is impossible to overstate the importance of natural products to the advent and evolution of human medicine particularly in the realm of cancer treatment. However, the microorganisms that produce these substances rarely produce them in quantities sufficient to support large-scale applications. It is therefore critically important to understand the optimization of natural product titer through genetic manipulation, alteration of regulatory pathways and precursor supply, and modulation of resistance, export and heterologous expression. Furthermore, it is often necessary to derivatize natural products to produce analogs with preferential physicochemical properties. In addressing these needs it is clear that the application of genetic engineering, combinatorial biosynthesis, precursor-directed biosynthesis, and tailoring enzymes will continue to enable advances in cancer therapeutics and medicinal improvements for the human condition.

Acknowledgments Studies on natural product biosynthesis, engineering, and drug discovery in the Shen laboratories are supported in part by National Institutes of Health grants AI051689, CA078747, CA094426, CA106150, CA113297, and GM086184.

References

Adrio JL, Demain AL (2006) Genetic improvement of processes yielding microbial products. FEMS Microbiol Rev 30:187–214

Beerman TA, Gawron LS, Shin S, Shen B, McHugh MM (2009) C-1027, a radiomimetic enediyne anticancer drug, preferentially targets hypoxic cells. Cancer Res 69:593–598

Berdy J (2005) Bioactive microbial metabolites. J Antibiot 58:1–26

Berdy J (2012) Thoughts and facts about antibiotics: where we are now and where we are heading. J Antibiot 65:385–395

Chen Y, Wendt-Pienkowski E, Shen B (2008a) Identification and utility of FdmR1 as a Streptomyces antibiotic regulatory protein activator for fredericamycin production in *Streptomyces griseus* ATCC 49344 and heterologous hosts. J Bacteriol 190:5587–5596

Chen Y, Deng W, Wu J, Qian J, Chu J, Zhuang Y, Zhang S, Liu W (2008b) Genetic modulation of the overexpression of tailoring genes *eryK* and *eryG* leading to the improvement of erythromycin A purity and production in *Saccharopolyspora erythraea* fermentation. Appl Environ Microbiol 74:1820–1828

Chen Y, Smanski MJ, Shen B (2010) Improvement of secondary metabolite production in Streptomyces by manipulating pathway regulation. Appl Microbiol Biotechnol 86:19–25

Cui Z, Wang L, Wang S, Li G, Hong B (2009) Disruption of *cagA*, the apoprotein gene of chromoprotein antibiotic C-1027, eliminates holo-antibiotic production, but not the cytotoxic chromophore. FEMS Microbiol Lett 301:57–68

Du L, Sanchez C, Chen M, Edwards DJ, Shen B (2000) The biosynthetic gene cluster for the antitumor drug bleomycin from *Streptomyces verticillus* ATCC15003 supporting functional interactions between nonribosomal peptide synthetases and a polyketide synthase. Chem Biol 7:623–642

Eustaquio AS, Moore BS (2008) Mutasynthesis of fluorosalinosporamide, a potent and reversible inhibitor of the proteasome. Angew Chem Int Ed 47:3936–3938

Eustaquio AS, O'Hagan D, Moore BS (2010) Engineering fluorometabolite production: fluorinase expression in *Salinispora tropica* yields fluorosalinosporamide. J Nat Prod 73:378–382

Galm U, Shen B (2006) Expression of biosynthetic gene clusters in heterologous hosts for natural product production and combinatorial biosynthesis. Expert Opin Drug Discov 1:409–437

Goss RJM, Lanceron S, Roy AD, Sprague S, Nur-e-Alam M, Hughes DL, Wilkinson DL, Moss SJ (2010) An expeditious route to fluorinated rapamycin analogues by utilizing mutasynthesis. Chembiochem 11:698–702

Gruschow S, Rackham EJ, Elkins B, Newill PLA, Hill LM, Goss RJM (2009) New pacidamycin antibiotics through precursor-directed biosynthesis. Chembiochem 10:355–360

Gupta S, Lakshmanan V, Seok KB, Fecik R, Reynolds KA (2008) Generation of novel pikdromycin antibiotic products through mutasynthesis. Chembiochem 9:1609–1616

Hesketh A, Ochi K (1997) A novel method for improving *Streptomyces coelicolor* A3(2) for production of actinorhodin by introduction of *rpsL* (encoding ribosomal protein S12) mutations conferring resistance to streptomycin. J Antibiot 50:532–535

Huang S-X, Feng Z, Wang L, Galm U, Wendt-Pienkowski E, Yang D, Tao M, Coughlin JM, Duan Y, Shen B (2012) A designer bleomycin with significantly improved DNA cleavage activity. J Am Chem Soc 134:13501–13509

Huh JH, Kim DJ, Zhao XQ, Li M, Jo YY, Yoon TM, Shin SK, Yong JH, Ryu YW, Yang YY, Suh JW (2004) Widespread activation of antibiotic biosynthesis by S-adenosylmethionine in streptomycetes. FEMS Microbiol Lett 238:439–447

Ju J, Rajski SR, Lim S-K, Seo J-W, Peters NR, Hoffman FM, Shen B (2008) Evaluation of new migrastatin and dorrigocin congeners unveils cell migration inhibitors with dramatically improved potency. Bioorg Med Chem Lett 18:5951–5954

Ju J, Rajski SR, Lim S-K, Seo J-W, Peters NR, Hoffmann FM, Shen B (2009) Lactimidomycin, iso-migrastatin and related glutarimide-containing 12-membered macrolides are extremely potent inhibitors of cell migration. J Am Chem Soc 131:1370–1371

Keller U, Schauwecker F (2003) Combinatorial biosynthesis of non-ribosomal peptides. Comb Chem High Throughput Screen 6:527–540

Kennedy DR, Gawron LS, Ju J, Liu W, Shen B, Beerman TA (2007) Single chemical modifications of the C-1027 enediyne core, a radiomimetic antitumor drug, affect both drug potency and the role of ataxia-telangiectasia mutated in cellular responses to DNA double-strand breaks. Cancer Res 67:773–781

Kim BS, Sherman DH, Reynolds KA (2004) An efficient method for creation and functional analysis of libraries of hybrid type I polyketide synthases. Protein Eng Des 17:277–284

Kwon SJ, Kim MI, Ku B, Coulombel L, Kim JH, Shawky JH, Linhardt RJ, Dordick JS (2010) Unnatural polyketide analogues selectively target the HER signaling pathway in human breast cancer cells. Chembiochem 11:573–580

Li Y, He W, Wang Y, Wang Y, Shao R (2008) A new post-PKS modification process in the carbamoyltransferase gene inactivation strain of *Streptomyces hygroscopicus* 17997. J Antibiot 61:347–355

Liu W, Christenson SD, Standage S, Shen B (2002) Biosynthesis of the enediyne antitumor antibiotic C-1027. Science 297:1170–1173

Lutz H (2009a) Genetic engineering of antibiotic biosynthesis for the generation of new aminocoumarins. Biotechnol Adv 27:1006–1014

Lutz H (2009b) Aminocoumarins: mutasynthesis, chemoenzymatic synthesis, and metabolic engineering. Methods Enzymol 459:437–455

Madduri K, Kennedy J, Rivola G, Inventi-Solari A, Filippini S, Zanuso G, Colombo AL, Gewain KM, Occi JL, MacNeil DJ, Hutchinson CR (1998) Production of the antitumor drug epirubicin (4′-epidoxorubicin) and its precursor by a genetically engineered strain of *Streptomyces peucetius*. Nat Biotechnol 16:69–74

Maharjan S, Oh TJ, Lee HC, Sohng JK (2009) Identification and functional characterization of an *afsR* homolog regulatory gene from *Streptomyces venezuelae* ATCC 15439. J Microbiol Biotechnol 19:121–127

Malla S, Niraula NP, Liou K, Sohng JK (2009) Enhancement of doxorubicin production by expression of structural sugar biosynthesis and glycosyltransferase genes in *Streptomyces peucetius*. J Biosci Bioeng 108:92–98

Malla S, Niraula NP, Liou K, Sohng JK (2010a) Improvement in doxorubicin productivity by overexpression of regulatory genes in *Streptomyces peucetius*. Res Microbiol 161:109–117

Malla S, Niraula NP, Liou K, Sohng JK (2010b) Self-resistance mechanism in *Streptomyces peucetius*: overexpression of *drrA*, *drrB* and *drrC* for doxorubicin enhancement. Microbiol Res 165:259–267

Mo S, Kim BS, Reynolds KA (2005) Production of branched-chain alkylprodiginines in *S. coelicolor* by replacement of the 3-ketoacyl ACP synthase III initiation enzyme, RedP. Chem Biol 12:191–200

Mo S, Ban YH, Park JW, Yoo YJ, Yoon YJ (2009) Enhanced FK506 production in *Streptomyces clavuligerus* CKD1119 by engineering the supply of methylmalonyl-CoA precursor. J Ind Microbiol Biotechnol 36(12):1473–1482

Mutka SC, Carney JR, Liu Y, Kennedy J (2006) Heterologous production of epothilone C and D in *Escherichia coli*. Biochemistry 45:1321–1330

Nagaoka K, Demain AL (1975) Mutational biosynthesis of a new antibiotic, streptomutin A, by an idiotroph of *Streptomyces griseus*. J Antibiot 28:627–635

Olano C, Lombó F, Méndez C, Salas JA (2008) Improving production of bioactive secondary metabolites in actinomycetes by metabolic engineering. Metab Eng 10(5):281–292

Oliynyk M, Brown MJ, Cortes J, Staunton J, Leadlay PF (1996) A hybrid modular polyketide synthase obtained by domain swapping. Chem Biol 3:833–839

Onaka H, Taniguchi S, Igarashi Y, Furumai T (2002) Cloning of the staurosporine biosynthetic gene cluster from *Streptomyces* sp. TP-A0274 and its heterologous expression in *Streptomyces lividans*. J Antibiot 55:1063–1071

Perez L, Danishefsky SJ (2007) Chemistry and biology in search of antimetastatic agents. ACS Chem Biol 2:159–162

Schneider TL, Walsh CT, O'Connor SE (2002) Utilization of alternate substrates by the first three modules of the epothilone synthetase assembly line. J Am Chem Soc 124:11272–11273

Schwarzer D, Mootz HD, Marahiel MA (2001) Exploring the impact of different thioesterase domains for the design of hybrid peptide synthetases. Chem Biol 8:997–1010

Smanski MJ, Peterson RM, Rajski SR, Shen B (2009) Engineered *Streptomyces platensis* strains that overproduce antibiotics platensimycin and platencin. Antimicrob Agents Chemother 53:1299–1304

Srinivasan P, Palani SN, Prasad R (2010) Daunorubicin efflux in *Streptomyces peucetius* modulates biosynthesis by feedback regulation. FEMS Microbiol Lett 305:18–27

Tang L, Shah S, Chung L, Carney J, Katz L, Khosla C, Julien B (2000) Cloning and heterologous expression of the epothilone gene cluster. Science 287:640–642

Tang Y, Lee TS, Khosla C (2004) Engineered biosynthesis of regioselectively modified aromatic polyketides using bimodular polyketide synthases. PLoS Biol 2:227–238

Van Lanen SG, Shen B (2006) Progress in combinatorial biosynthesis for drug discovery. Drug Discov Today Technol 3:285–292

Wang G, Hosaka T, Ochi K (2008) Dramatic activation of antibiotic production in *Streptomyces coelicolor* by cumulative drug resistance mutations. Appl Environ Microbiol 74:2834–2840

Weissman KJ (2007) Mutasynthesis—uniting chemistry and genetics for drug discovery. Trends Biotechnol 25:139–142

Wendt-Pienkowski E, Huang Y, Zhang J, Li B, Jiang H, Kwon H, Hutchinson CR, Shen B (2005) Cloning, sequencing, analysis, and heterologous expression of the fredericamycin biosynthetic gene cluster from *Streptomyces griseus*. J Am Chem Soc 127:16442–16452

Wenzel SC, Muller R (2005) Recent developments towards the heterologous expression of complex bacterial natural product biosynthetic pathways. Curr Opin Biotechnol 16:594–606

Xu Y, Wijeratne EMK, Espinosa-Artiles E, Gunatilaka AAL, Molnar I (2009) Combinatorial mutasynthesis of scrambled beauvericins, cyclooligomer depsipeptide cell migration inhibitors from *Beauveria bassiana*. Chembiochem 10:345–354

Chapter 8
Accessing Anti-cancer Natural Products by Plant Cell Culture

Lisa M. Leone and Susan C. Roberts

Abstract Natural products are a valuable part of drug discovery and represent key products in the marketplace. High-throughput screening of large, synthetic libraries has become a commonplace method in industrial drug development, partially due to past difficulties in sourcing large quantities of natural products. Plants have historically been key sources of blockbuster anti-cancer drugs, and new approaches to drug discovery and large-scale manufacturing are making innovative plant-derived natural products more available commercially and clinically. This chapter explores the growing capabilities of combinatorial biosynthesis in plants as well as technological developments in industrial plant cell culture, highlighting the stories of the plant anti-cancer agents paclitaxel and cyclopamine.

Abbreviations

BMS	Bristol-Myers Squibb
CMCs	Cambial meristematic cells
CRADA	Cooperative Research and Development Agreement
CYP450	Cytochrome P450 enzyme
DMAPP	Dimethylallyl diphosphate
FDA	Food and Drug Administration
HRC	Hairy root culture
IPP	Isopentenyl diphosphate
MIA	Monoterpene indole alkaloid
NCI	National Cancer Institute

L.M. Leone • S.C. Roberts (✉)
Department of Chemical Engineering, University of Massachusetts, Amherst, 686 North Pleasant Street, 159 Goessmann Laboratory, Amherst, MA 01003, USA
e-mail: sroberts@ecs.umass.edu

F.E. Koehn (ed.), *Natural Products and Cancer Drug Discovery*, Cancer Drug Discovery and Development, DOI 10.1007/978-1-4614-4654-5_8,

NP	Natural product
PCC	Plant cell culture
PCSC	Plant cell suspension culture
SM	Secondary metabolite
USDA	United States Department of Agriculture

8.1 Introduction

From 1981 to 2002, 74% of the anti-cancer agents approved by the Food and Drug Administration (FDA) were inspired by, derived from, or true natural products (NPs) (Wilson and Danishefsky 2006). Biologically derived compounds often exhibit "privileged structures" in terms of biological activity (Evans et al. 1988). Unfortunately, these structures are often chirally complex and pose unique challenges for large-scale commercial production. For a significant number of NPs, a route to chemical synthesis does not exist, and for the vast majority of NPs, total chemical synthesis methods are not commercially viable due to structural complexity, low yields, and environmental concerns (Kolewe et al. 2008). In the 1970s and 1980s, a lack of immediate production strategies for some NPs entering clinical trials led to a perceived "supply crisis" for certain drugs, including the anti-cancer compound paclitaxel. These situations led to reluctance in the pharmaceutical industry for investment in NPs.

The ease of chemical synthesis in simple, non-NP-derived molecules is perhaps one reason why synthetic combinatorial libraries became popular in drug development (Ortholand and Ganesan 2004). The total chemical synthesis of a synthetic combinatorial library's components as a starting point ensures that every compound will have a relatively simple (and known) route for eventual large-scale production. A disappointing result of this cost-saving methodology is that the systematic exploration of organic chemistry space has been limited to known and easily synthesizable structures, which are not necessarily biologically relevant (Lipkus et al. 2008). Among the millions of compounds screened in academia and industry, only a relatively small number of compounds from high-throughput screens have been shown to exhibit activity against drug targets, and even fewer have overcome toxicity issues and moved to clinical trials (Newman and Cragg 2007). Because of these difficulties, synthetic combinatorial libraries have shifted over the past decade from large collections of simple compounds to a diversity-centered exploration of a smaller chemistry space, aided by computational technologies (Schnur et al. 2011).

With this new focus, the drug industry has reverted back to a paradigm in which natural organisms have unique advantages over a chemist's bench. Plants, marine organisms, and microbes still represent an extremely diverse and relatively untapped space for lead discovery. The NP research of the past 30 years has made rapid progress in both increasing the diversity of leads through combinatorial biosynthesis and high-throughput screening of crude extracts, as well as alleviating possible supply issues through a variety of advanced technologies, including plant cell culture (PCC) and heterologous expression in microbial systems, as discussed below.

8.2 Plant-Derived Anti-cancer Agents

Naturally occurring plants, microbes, and fungi have been the source of a staggering amount of NP diversity throughout the history of drug discovery. In the case of plants, over 10,000 alkaloids and 23,000 terpenoids have been characterized, and these most certainly are an underestimation of the capabilities of plants, of which only a small portion of the world's species are estimated to have been sampled (Cheng et al. 2007). Many of these NPs (Fig. 8.1) have medicinal qualities, as summarized in Table 8.1.

Most plant NPs fall into the class of secondary metabolites (SMs), so called because they do not necessarily serve primary metabolic functions in the growth and maintenance of their native plant. Biosynthesis of plant SMs is complex, involving many different precursors from primary metabolism. Some of these pathways are interconnected, for example, all of the compounds within the alkaloid and terpenoid classes of NPs originate in part from the common precursor isopentenyl diphosphate (IPP) and its allylic isomer dimethylallyl diphosphate (DMAPP). Biosynthesis of many SMs often requires dozens of enzymes acting in concert, and as a result is under strict metabolic control. The abbreviated biosynthetic pathways of paclitaxel and vinblastine in *Taxus* spp. and *Catharanthus roseus*, respectively, are shown in Fig. 8.2. While SMs do not play a role in primary metabolism in that they do not directly pertain to plant growth or maintenance, many plant NPs have shown activity in nature as insecticides and anti-fungals, and are thought to confer evolutionary advantages to their native plant (Hartmann 2007).

The "screening" hypothesis of NP diversity postulates that the chances of any single NP possessing a potent bioactivity, as a fungicide or insecticide for example, is very low. Therefore, it is advantageous for plants to produce many slightly different

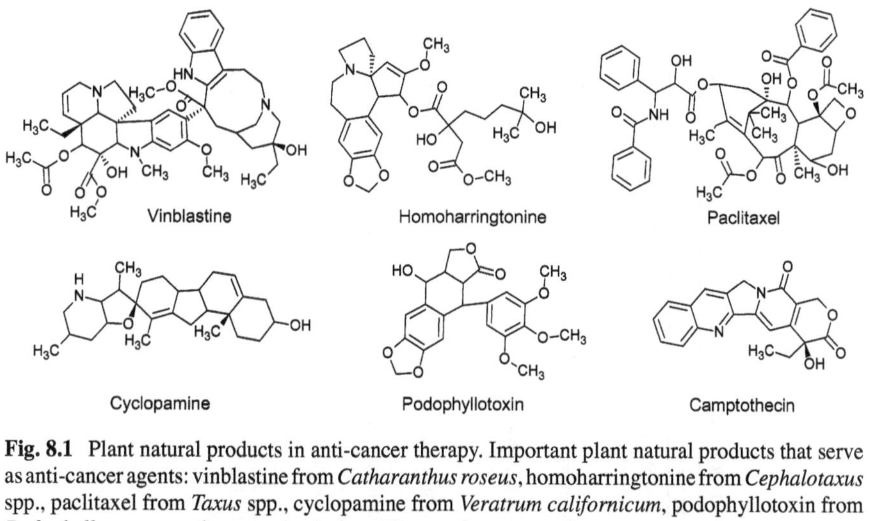

Fig. 8.1 Plant natural products in anti-cancer therapy. Important plant natural products that serve as anti-cancer agents: vinblastine from *Catharanthus roseus*, homoharringtonine from *Cephalotaxus* spp., paclitaxel from *Taxus* spp., cyclopamine from *Veratrum californicum*, podophyllotoxin from *Podophyllum* spp., and camptothecin from *Camptotheca accuminata*

Table 8.1 Natural products used in cancer therapy

Natural product	Class	Native plant	Mode of action	Cancers treated	Analogs
Vinca alkaloids (vinblastine, vincristine)	Alkaloid	*C. roseus*	Tubulin polymerization inhibitor	Leukemia, lymphoma, Hodgkin's disease, non-small-cell lung cancer	Vindesine, vinorelbine, vinflunine
Paclitaxel	Terpenoid	*Taxus* spp.	Tubulin polymerization stabilizer	Breast, lung and ovarian carcinomas; Kaposi's sarcoma	Docetaxel
Podophyllotoxin	Lignan	*Podophyllum* spp.	DNA-topoisomerase II inhibitor, tubulin polymerization inhibitor	Lung, ovarian and testicular cancer, leukemia, lymphoma, sarcoma, melanoma	Etoposide, etophos, teniposide
Camptothecin	Alkaloid	*Camptotheca accuminata*	DNA-topoisomerase I inhibitor	Ovarian cancer, small-cell lung cancer, colorectal cancer	Topotecan, irinotecan
Cephalotaxine	Alkaloid	*Cephalotaxus* spp.	Cell-cycle specific protein inhibition	Leukemia, solid tumors	Homoharringtonine
Cyclopamine	Alkaloid	*Veratrum californicum*	Hedgehog pathway inhibitor	Medulloblastoma, basal-cell carcinoma, pancreatic adenocarcinoma	IPI-926

The list includes drugs approved by the FDA and investigational drugs currently in clinical trials. Semi-synthetic chemical derivatives (i.e., analogs) are listed with their parent NP

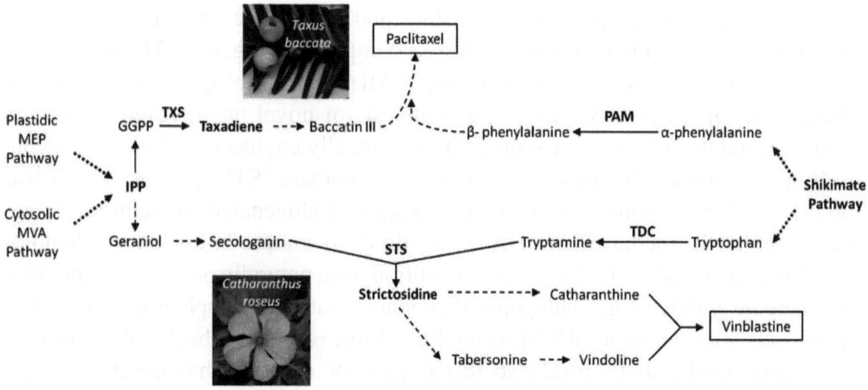

Fig. 8.2 Biosynthetic pathways of paclitaxel and vinblastine. Biosynthetic pathways of paclitaxel (terpenoid) in *Taxus* ssp. and the Vinca alkaloids (e.g., vinblastine) in *C. roseus*. *Dashed lines* indicate multiple steps and *bolded compounds* (i.e., taxadiene and strictosidine) indicate committed steps. Key enzymes are shown above reaction steps: TS (taxadiene synthase), PAM (phenylalanine aminomutase), TDS (tryptophan decarboxylase), and STS (strictosidine synthase)

metabolites in small concentrations, in the hopes that at least one will be useful against an external stressor, such as herbivorous insects (Firn and Jones 2003). In this perspective, a single plant can be compared to a naturally occurring drug discovery program, in which many compounds with slightly different functionalities are produced from simple starting structures. Plants accomplish this incredible diversity at minimal biosynthetic cost by employing promiscuous enzymes, using branched and matrix biosynthetic pathways, and generally producing minute quantities of these SMs (Fischbach and Clardy 2007).

While plants already express an incredible amount of diversity naturally, increasing the pool of NPs available for drug discovery may result in new, more effective drugs. New approaches to library formation that harness the diversity of pathways already existing in plants may lead to novel anti-cancer compounds. One promising approach is plant combinatorial biosynthesis.

8.3 Combinatorial Biosynthesis

Combinatorial biosynthesis is a drug discovery platform that seeks to harness naturally occurring biochemical diversity in living systems to generate novel compounds that have a high likelihood to possess bioactivity. Combinatorial biosynthesis has already generated many promising drug leads when applied in microbes, as exemplified by the discovery of novel indolocarbazole compounds (Sánchez et al. 2005). The unique capabilities of plants to generate novel compounds in the alkaloid and terpenoid classes have only begun to be recognized. An excellent review of combinatorial biosynthesis in plants has recently been published (Pollier et al. 2011). A current example of the power of plant combinatorial biosynthesis is the rational engineering of *C. roseus* hairy root cultures (HRCs) to create novel products.

The anti-cancer compounds vinblastine and vincristine are members of the monoterpene indole alkaloid (MIA) class of compounds (Fig. 8.2). Metabolic engineering has been successfully used to adjust MIA pharmacological properties biosynthetically in *C. roseus* HRCs. The synthesis of novel halogenated MIAs was accomplished in a stepwise fashion, first by rationally engineering the enzyme controlling entry to the MIA pathway, strictosidine synthase (STS) (Fig. 8.2). STS was modified and transformed into *C. roseus* to accept halogenated tryptamine analogs that were fed in media to the HRCs, resulting in unique halogenated alkaloids (Bernhardt et al. 2007). To decrease competition from naturally occurring tryptamine and increase yields of the halogenated products, native tryptophan decarboxylase (TDC) was silenced using RNAi technology. Most recently, a bacterial halogenase was transformed into *C. roseus* so that tryptamine could be halogenated in vivo instead of being fed through external media (Runguphan et al. 2010). These novel halogenated analogs of established pharmaceuticals can be evaluated for favorable changes in pharmacological properties. This approach demonstrates the power of plant metabolic engineering for the generation of libraries of novel NP drug leads that inhabit a "privileged space" of bioactivity. However, supply of pure compound remains a challenge that must be addressed individually for any compound of commercial interest.

8.4 Supply Routes: Clinical Trials to Commercial Production

Different amounts of pure product are needed at different stages of development. After a promising compound has been identified, milligram quantities are needed for structural determination, multiple grams are needed for clinical development, and multiple kilograms are needed for initial clinical trials (Cragg et al. 1995; Koehn and Carter 2005). If a novel compound is found to be effective in clinical trials and approved for use, supply will need to quickly ramp up to many kilograms for commercialization. As shown in the story of paclitaxel (see below), supply issues can be a major hurdle for NPs. Supply issues are especially untenable in the treatment of terminal diseases such as cancer, where patients cannot wait months or years for a supply route to be established. Fortunately, the field of NP synthesis has matured since the paclitaxel supply crisis of the 1990s and there now exist many options for supply depending on the properties of the NP and plant source.

8.4.1 Natural Harvest

Natural harvest is the method by which many plant NPs have initially been collected, and for intermediate to long-term production it is an option when (1) the plant producing the compound of interest is fast growing and easily cultivable, as in the case of *C. roseus*, or (2) when the product of interest is produced in relatively large quantities in planta, as in the case of podophyllotoxin in *Podophyllum hexandrum* rhizomes (Farkya et al. 2004). For many naturally occurring compounds neither of these conditions is met, and environmental concerns can also arise when a plant is endangered or vital to a local ecosystem. Additionally, in the case of novel NPs found through strategies such as combinatorial biosynthesis, the native plant will generally not produce the compound at all.

8.4.2 Chemical Synthesis

Chemical synthesis is an option when the compound of interest is relatively simple and easy to synthesize, as in the case of flavopiridol, an anti-cancer NP in clinical trials that is currently supplied via total synthesis (Naik et al. 1988). Complete chemical syntheses have been published for many NPs of interest, such as paclitaxel (Holton et al. 1994; Nicolau et al. 1994), but they often do not represent a viable, scaleable production method due to low yields, complexity and number of steps, as well as the use of environmentally unfriendly chemicals and solvents. Chemical semi-synthesis can be used to create NP derivatives with more favorable pharmacological properties, as in the case of the camptothecin derivatives irinotecan and topotecan, and in some cases, can be used to derive a NP from precursors that occur naturally at high levels. For example, a paclitaxel intermediate, 10-deacetylbaccatin III, was chemically modified to produce a semi-synthetic version of paclitaxel in the 1990s (Denis et al. 1988) and was used as the primary supply route for paclitaxel by Bristol-Myers Squibb (BMS) for over a decade.

8.4.3 Heterologous Synthesis in Microbes

Transfer of plant NP pathways to microbial hosts has advanced rapidly in the past decade. The driving force behind this development is the perceived advantage of microbes, such as *Escherichia coli* and *Saccharomyces cerevisiae*, as production organisms for fermentation, because they are well-studied model systems; additionally, a mature infrastructure exists for the fermentation of these microbes at large scales. One of the first success stories in this field was heterologous expression in yeast (Ro et al. 2006) and *E. coli* (Dietrich et al. 2009) of enzymes to synthesize precursors to artemesinin, an anti-malarial agent from the plant *Artemisia annua*. Under a tripartite partnership of Amyris Biotechnologies, the Institute for OneWorld Health and the pharmaceutical company Sanofi-Aventis, the fermentation for artemesinin is currently moving into the commercial phase and is expected to begin production in 2012 (Hale et al. 2007; Solomon 2011). Efforts to transfer metabolic pathways for isoflavones (Leonard et al. 2008), taxanes (Ajikumar et al. 2010), and alkaloids (Nakagawa et al. 2011) are currently underway. However, major hurdles to microbial biosynthesis still exist, including incomplete NP biosynthesis pathway knowledge and access to cloned genes, as well as the difficulty in expressing plant enzymes in microbial systems.

Despite much research, knowledge of the biosynthetic pathways of many important plant NPs remain incomplete in planta. For example, in the paclitaxel pathway there are thought to be approximately 19 steps, of which several remain completely unknown, and among the known steps, enzyme identification is incomplete (Croteau et al. 2006). The formation of the oxetane ring in particular is essential to the clinical function of paclitaxel (Kingston 1994), but enzyme candidates for this reaction have not yet been found. A special difficulty in identifying biosynthesis genes is that genes in the same biosynthetic pathway are not clustered in plants as is common in microbial systems, making identification difficult unless the entire plant genome is sequenced (Schäfer and Wink 2009). Next-generation sequencing techniques are starting to alleviate this challenge, but in the present and immediate future, lack of plant genetic knowledge and tools has been a major bottleneck in the transfer of plant pathways to microbes.

Another challenge that must be addressed on a case by case basis for microbial transfer is enzyme engineering. Adapting enzymes for optimal function in a non-native host can be difficult, particularly for plant cytochrome P450s (CYP450s). CYP450s and their associated reductase partners are heme-thiolate enzymes constituting one of the largest super families of proteins, and are found in both prokaryotes and eukaryotes. In eukaryotes, CYP450s are membrane bound, while in prokaryotes they are soluble (Werck-reichhart and Feyereisen 2000). CYP450s are responsible for many of the unusual oxidative reactions seen in the biosynthesis of plant SMs, making them essential for heterologous expression of many plant NPs. Numerous challenges exist in the expression of plant CYP450s, including: improper folding, lack of an appropriate membrane binding site in prokaryotes, and difficulties in maintaining adequate NADPH pools for reductase activity. These difficulties can be overcome on a case by case basis via protein engineering and related strategies, and multiple functioning plant CYP450s have been successfully expressed in *E. coli* (Chemler and Koffas 2008).

8.4.4 Plant Cell Culture

Heterologous production of plant NPs in microbes will continue to develop for the foreseeable future, but it is important to situate these breakthroughs in the context of current pharmaceutical supply. Microbial fermentation has been touted as a cost-effective route for large-scale production of high value plant NPs, but as mentioned above, it is not possible for certain NPs, particularly for those with undefined biosynthetic routes. PCC is an attractive option for immediate production of plant NPs when heterologous production is infeasible.

Environmentally friendly PCC has been used industrially for plant NP supply since 1984, when Mitsui Petrochemical Industries, Ltd. began using a cell suspension of *Lithospermum erythrorhizon* to produce the pigment shikonin on a 750-L bioreactor scale (Georgiev et al. 2009). Today, PCC is utilized commercially to produce a wide variety of pharmaceuticals, pigments, and food additives at scales up to 75,000-L, as in the case of Phyton Biotech, Inc.'s paclitaxel process.

Plant SMs, which form the majority of plant NPs, are often under strict and specific metabolic control; for example, some plant NPs are synthesized in response to exogenously applied elicitors such as methyl jasmonate, salicylic acid or nitric oxide, which are known to be key signaling molecules in the plant defense response (Zhao et al. 2005). Some plant NPs require specialized cell types to synthesize the desired NP; for example, the biosynthesis of vindoline, one intermediate in the vinblastine pathway (Fig. 8.2) appears to require three different mature cell types and multiple intracellular compartments (Kutchan 2005). Overcoming limitations imposed by metabolic control is a major area of research, with efforts placed towards activating inherent secondary metabolism via elicitation, as well as metabolic engineering and expression of transcription factors to rationally control plant metabolism.

8.4.4.1 Plant Cell Suspension Culture

Plant cell suspension culture (PCSC) is a subset of PCC in which cells from an intact plant or plant embryo are "de-differentiated" using phytohormones and grown in liquid media. Plant cells are totipotent, in that given correct signals they can differentiate and develop into any plant cell type or organ; this also gives them the ability to be maintained in a de-differentiated state. Given their constant growth, PCSCs must be subcultured into fresh media every 1–4 weeks depending on their specific growth rate. Importantly, cryopreservation techniques have been developed for a number of medicinally relevant plant species to preserve vitality without repeated subculture (Mustafa et al. 2011). However, recovery of PCSCs from cryopreserved cultures is a time-consuming process and does not always result in a predictable performance (Harding 2004). As a result of incomplete separation after cell division, plant cells in suspension grow as aggregates ranging in size from a few cells (100 μm diameter) to thousands of cells (2 mm diameter) (Kolewe et al. 2008). In addition, plant cells are more sensitive to shear than microbial cells, owing to their large, water-filled vacuoles. A number of novel impeller configurations and bioreactor types have been developed to meet the unique needs of PCSCs—most notably the recent invention of disposable, gas-permeable-bag wave-type bioreactors—although many plant cell types have been grown successfully in commonly available stirred tank bioreactors at culture volumes up to 70 m^3 (Eibl and Eibl 2007).

A major area of research in PCSC is the variability in metabolite accumulation of de-differentiated plant cells that have been subcultured repeatedly over a period of years. It is known that the callus culture method by which de-differentiated PCSCs are formed generally creates a heterogeneous population of cell types in suspension (Senger et al. 2006). The de-differentiated state itself sometimes leads to genomic instability over repeated subculture (Baebler et al. 2005). Additionally, the aggregated nature of plant cells in suspension may cause epigenetic changes, affecting product accumulation through unknown mechanisms (Patil et al. 2011).

The relationship between aggregate size and NP accumulation has been studied in a number of different plant systems, but results are varied, in part due to inconsistent experimental methodologies. In a recent study using a more accurate whole-culture measure of metabolite synthesis, paclitaxel accumulation in de-differentiated PCSC increased as aggregate sized decreased, suggesting aggregation as a process parameter that should be optimized (Kolewe et al. 2011).

A recent development in PCSC is the suspension culture of cambial meristematic cells (CMCs). In planta, cells in the cambial meristem layer of tissue can grow indefinitely and become any cell type. Because of this feature they have been likened to plant stem cells and defined as "un-differentiated" in contrast to the "de-differentiated" cells discussed previously. Suspension cultures of *Taxus* CMCs were made by creating callus culture from isolated *Taxus* cambial meristem tissue. The CMC cultures grew faster, formed smaller aggregates and displayed greater genetic stability than de-differentiated cultures (Lee et al. 2010).

8.4.4.2 Hairy Root Culture

CMCs are an attractive new option for PCCs; however, some plant NPs are produced preferentially in more differentiated cell types and may be produced in small quantities, or not at all, in truly un-differentiated cells (Roberts and Kolewe 2010). HRC is another subset of PCC in which root tissue from a plant of interest is genetically transformed by the soil bacterium *Agrobacterium rhizogenes*, leading to branched "hairy roots" that can grow indefinitely without exogenously supplied phytohormones needed for PCSCs. Because HRCs are differentiated tissue, they frequently have the ability to produce SMs that de-differentiated PCSCs cannot, and in some cases can even produce plant NPs that are not normally found in roots of intact plants. For instance, HRCs of *A. annua* accumulated artemisinin, whereas in whole plants artemisinin is produced only in the aerial sections (Kim et al. 2002). HRCs also appear to be more genetically stable than de-differentiated cultures and do not lose biosynthetic potential upon repeated subculture (Georgiev et al. 2010).

Despite their proven biosynthetic capabilities, HRCs have not yet been used commercially for production of any plant NPs, mostly due to the lack of proven, large-scale bioreactors. The branched and corporeal nature of HRCs make them difficult to culture in stirred tank bioreactors, and nutrient transport limitations result in slow growth and low productivity. Novel bioreactor designs may improve HRC, and among the most promising are mist bioreactors. This novel bioreactor type sprays exposed roots with a nutrient solution and it has been shown to increase growth and metabolite productivity at high root densities (Weathers et al. 2008).

8.4.4.3 Yield Enhancement Strategies

Apart from variability in product accumulation, another major area of research in PCC is enhancement of biosynthetic capabilities. This goal is realized through a number of strategies, including chemical elicitation of secondary metabolism, selection of elite cell lines, and metabolic engineering. While plant NPs are generally produced in extremely small quantities under normal conditions, in some cases SMs may constitute 20–60% of a plant's dry weight (Verpoorte et al. 1999). Activation of secondary metabolism using elicitors such as methyl jasmonate has been shown to significantly increase production of plant NPs, almost 50-fold in the case of paclitaxel in *Taxus baccata* PCSC (Yukimune et al. 1996). Elicitation pathways are highly conserved in plants, and elicitors such as methyl jasmonate have been shown to activate secondary metabolism in a wide variety of plants through similar pathways (van der Fits and Memelink 2000; Gundlach et al. 1992).

A useful technique in cell culture to improve yields is to select individual cells that have high biosynthetic productivity and culture them as "elite" cell lines. This technique is generally difficult to accomplish with plant cell lines, as cultures are composed of genetically heterogeneous aggregates that may respond differently to selection pressures. Furthermore, due to the unstable nature of de-differentiated cells, elite cell lines may lose their increased capacity over time. A successful example of using selection pressure to create stable, elite cell lines is establishment of *Lavendula vera* cultures producing high amounts of the phenolic plant NP rosmarinic acid. Selection was accomplished by feeding toxic amino acid analogs in the media with the expectation that only cells expressing high amounts of an enzyme common to both amino acid and rosmarinic acid biosynthesis would be able to detoxify the amino acid analogs and remain viable (Georgiev et al. 2006). This approach was effective in creating genetically stable, high-producing cell lines, but is limited to cases in which an effective screening procedure exists. Additionally, the heterogeneous and aggregated nature of PCSCs may dilute the effect of small numbers of high-producing cells because they are not separated from large numbers of low-producing cells. In cases where it is possible to disassociate individual cells from aggregates, labeling of the plant NP of interest and flow cytometric cell sorting is a more robust method for isolating and re-culturing elite cell lines (Naill and Roberts 2005a, b).

Manipulation of biosynthetic pathways at a genetic level offers significant potential for increasing yields of plant NPs in PCC. Recent technological developments, such as 454 pyrosequencing, have lowered the barriers to identifying genes in biosynthetic pathways, and deliver the promise of making "non-model" organisms amenable to metabolic analysis (Bräutigam and Gowik 2010). Once biosynthetic pathway genes are known, *Agrobacterium tumefaciens* transformation can be used to stably introduce or silence genes in plants, although some types of plants such as trees and crop plants can be recalcitrant to *Agrobacterium* transformation. Engineering of *Agrobacterium* strains for increased virulence as well as manipulation of plant culture conditions have resulted in a wider applicability of this technique (Gelvin 2003). Transcription factors are also targets of metabolic engineering efforts. For example, simultaneous up-regulation of the gene *G10H* and the *ORCA*3 transcription factor in HRCs of *C. roseus* resulted in a 6.5-fold increase in catharanthine, a precursor of vinblastine (Ni et al. 2011). In many cases, however, transcription factor engineering is not effective at increasing production of the desired NP, and new models of plant metabolism may better our understanding of how to manipulate metabolism on a systems level (Stitt et al. 2010).

8.5 The Paclitaxel Story

Paclitaxel is one of the most successful drugs in the history of chemotherapeutics, with annual reported sales in 2000 exceeding 1.5 billion USD (Expósito et al. 2009); however, the early history of paclitaxel was fraught with complications, and at many points it was almost discarded as a drug lead. The story of paclitaxel (summarized in Table 8.2) began in 1962, when a sample of bark from the Pacific yew, *Taxus brevifolia*, was collected by United States Department of Agriculture (USDA) workers. An extract tested positive for activity in the KB cytotoxicity assay. Following the positive result, Dr. Monroe Wall fractionated the sample and isolated the cytotoxic agent, paclitaxel, in 1971. Paclitaxel had favorable but unremarkable cytotoxic activity against common cell-based cytotoxicity screens; however, interest in paclitaxel was ignited in 1979, when Dr. Susan Horwitz showed that paclitaxel had a unique mechanism of action against tumor cells. Whereas previous spindle poisons (e.g., the vinca alkaloids vinblastine and vincristine) act by rapidly depolymerizing microtubules and preventing spindle formation in mitosis, paclitaxel instead stabilizes the tubulin polymers, preventing the cell from properly assembling its spindle and continuing through mitosis.

Paclitaxel was a poor drug lead due to low aqueous solubility, structural complexity precluding easy chemical synthesis, and the lack of a large and renewable supply of the compound. However, interest from the research community, due to its novel mode of action, helped to move it past major difficulties during its initial development, including several deaths in Phase I clinical trials due to allergic reactions with Cremophor, an emulsifying agent in its formulation. Although at many points in

Table 8.2 Timeline of paclitaxel discovery, efficacy, supply, and process development milestones

Year	Event
1962	*T. brevifolia* bark sample collected by USDA/NCI collaboration
1971	Paclitaxel isolated and identified by Dr. Monroe Wall
1979	Paclitaxel's unique mechanism of action identified by Dr. Susan Horwitz
1984	Paclitaxel enters clinical trials against ovarian cancer
1988	Semi-synthetic route to paclitaxel published
1990	Phyton Biotech, Inc. founded
1991	NCI enters into CRADA with Bristol-Myers Squibb (BMS)
1991	First patent for PCC production of paclitaxel is issued
1992	FDA approves paclitaxel for treatment of ovarian cancer
1992	Pacific Yew Act is signed into law by George Bush
1993	Phyton Biotech acquires Phyton GmbH and their 75,000 L cGMP PCC facility
1994	FDA approves semi-synthetic route to paclitaxel
1994	Total syntheses of paclitaxel published independently by two different groups
1995	Phyton Biotech licenses its PCC process to BMS for paclitaxel production
1995	Samyang Genex begins work on PCC process to produce paclitaxel, under the name Genexol®
1997	Paclitaxel produced by PCC is approved in Korea
2004	FDA approves PCC route to paclitaxel
2007	Phyton Biotech patents elicitation strategies to increase titers to 900 mg/L

paclitaxel's history the chances of it emerging as a successful chemotherapeutic appeared slim, clinical trials continued after the formulation was reworked, and extremely promising Phase II results were obtained against refractory ovarian cancer (Cragg et al. 1993).

Even as paclitaxel enjoyed success in the clinic, there was still a major hurdle to its development as a large-scale chemotherapeutic. While early research and initial clinical trials had relied on collection and extraction of bark from wild-growing *T. brevifolia*, the collection and extraction processes were environmentally damaging. *T. brevifolia*'s properties as a SM producer were a perfect storm of unfortunate specifications: the tree was slow growing, paclitaxel accumulated only in the bark, and yields were very low. In contrast to harvest of the vinca alkaloids from fast-growing *C. roseus*, natural harvest of paclitaxel was unsustainable—16,000 lb of bark from approximately 2,000 yew trees were required to produce 1 kg of paclitaxel (Cragg and Snader 1991). In the early 1990s the paclitaxel supply problem was finally recognized as a crisis. In 1991 the National Cancer Institute (NCI) entered into a Cooperative Research and Development Agreement (CRADA) with BMS to heavily fund research into alternate supply routes. Meanwhile, the destruction of trees became a high profile environmental issue and in 1992 federal legislation (The Pacific Yew Act) was passed to manage the survival of the *T. brevifolia*. Concurrently in 1992, paclitaxel was approved by the FDA to treat ovarian cancer and demand for the drug sharply accelerated. By 1993 NCI was supporting 35 grants on paclitaxel research, with funding of $4.6 million, in addition to BMS's independent research (Cragg et al. 1993).

The most immediate relief for the supply issues was a semi-synthetic method that was first developed in 1986 and later re-worked to be more efficient (Denis et al. 1988). 10-deacetylbaccatin III was extracted from the needles of various *Taxus* species and converted to paclitaxel via chemical methods. Needle harvest was more environmentally friendly than bark harvest, leaving the tree viable and intact, and in 1994 the semi-synthetic method was approved by the FDA to supply paclitaxel, now trademarked Taxol®. However, the semi-synthesis method was expensive and environmentally unfriendly due to a number of harsh chemical solvents used. Research into production alternatives continued to be a hot topic during the 1990s, and in 1994 a total synthesis of paclitaxel was first reported (Holton et al. 1994; Nicolau et al. 1994). Paclitaxel is a complex molecule containing over 11 chiral centers and a unique oxetane ring chemistry, and total organic syntheses have still not resulted in any cost-effective or industrially viable processes.

PCC was first funded by the NCI as a supply route for plant-derived anti-cancer agents as early as 1977, but the contracts were terminated prematurely in 1980 due to lack of interest at the time (Cragg et al. 1993). A number of PCC projects were funded by the NCI in response to the paclitaxel crisis, and the first patent for PCSC of *T. brevifolia* for paclitaxel production was issued in 1991, with reported yields of 1–3 mg/L (Christen et al. 1991). Phyton Biotech, Inc. was formed in 1990 near Cornell University (Ithaca, NY) in response to a renewed interest in PCC technologies to supply paclitaxel. The company grew quickly and in 1993 acquired Phyton GmbH and a 75,000 L cGMP PCC facility in Heidelberg, DE. Phyton Biotech licensed their PCC process to BMS in 1995, and continued to improve the paclitaxel process, recently filing a patent for strategies to increase broth titers to 900 mg/L (Bringi et al. 2007). FDA approved the PCC route for paclitaxel supply in 2004 and the current worldwide supply of paclitaxel is provided mainly by Phyton Biotech, which is now a subsidiary of DFB Pharmaceuticals, Inc. Samyang Genex, a Korean company, also produces paclitaxel for the international market using a PCC process (marketed as Genexol®), which they began implementing in 1995 and approved for distribution in Korea in 1997. Recent research in *Taxus* PCC focuses on improving paclitaxel yield and minimizing production variability by better understanding paclitaxel biosynthesis and regulation on a genetic level, as well as the influence of key process variables such as aggregation (Wilson and Roberts 2011).

8.6 Cyclopamine: A Case Study in the Future of Plant-Derived Anti-cancer NPs

The Hedgehog pathway (Hh) has recently become an important target in anti-cancer drug discovery. Disruption of the Hh pathway is thought to be a major regulator of tumor initiation and growth in cancers such as basal-cell carcinoma and medulloblastoma, as well as a possible factor in metastasis and recurrence of solid tumors (Rubin and de Sauvage 2006).

Cyclopamine, a plant NP from *Veratrum californicum* (corn lily), was found to be a potent inhibitor of the Hh pathway by binding directly to the protein Smoothened (Smo) (Chen et al. 2002), and has become an important pharmacological tool in developmental biology and cancer research as well as the precursor to promising anti-tumor pharmaceuticals. Cyclopamine was discovered in the 1950s, when sheep ranchers in Idaho found that ewes that had grazed on *V. californicum* bore lambs with severe birth defects, such as a single eyeball in the middle of the forehead; the responsible compound was subsequently isolated from *V. californicum* and named cyclopamine as an homage to its signature birth defect (Heretsch et al. 2010). Cyclopamine remained a little-known plant NP until 1998, when its activity against the Hh pathway was discovered (Cooper et al. 1998). Cyclopamine began to be used in the research community to test hypotheses on the role of Hh signaling in developmental biology. Concurrently, the potential importance of the Hh pathway in cancer therapy began to be realized (Booth 1999). Cyclopamine itself was considered as a drug lead, but its clinical applications were limited due to poor solubility, weak activity in vivo and acid lability, precluding oral delivery (Tremblay et al. 2008).

In 2009, the MA-based company Infinity Pharmaceuticals, Inc. published their discovery of semi-synthetic derivatives of cyclopamine that improved upon its pharmacological properties (Tremblay et al. 2009). One of these analogs, IPI-926, had favorable results in animal models; IPI-926 inhibited tumor growth in mice with medulloblastoma, delayed post-chemotherapy cancer regrowth in mice with ovarian cancer and small-cell lung cancer, and prolonged survival of mice with pancreatic cancer when used in conjunction with the chemotherapeutic gemcitabine (Read and Palombella 2012). IPI-926 entered Phase I clinical trials in 2010, and is currently in Phase I clinical trials for head and neck cancer, BCC and solid tumors, and Phase Ib/II clinical trials for pancreatic cancer and chondrosarcoma (Read and Palombella 2012).

As in the case of paclitaxel, if IPI-926 is shown to be effective, large quantities of the drug will be rapidly needed. IPI-926 utilizes cyclopamine as a chemical precursor, and current supply of cyclopamine is dependent on wild-growing *V. californicum* harvested from the Western U.S. Dried roots of *V. californicum* contain approximately 2.34 g of cyclopamine per kg of plant material, and current published extraction protocols yield 1.3 g cyclopamine per kg of plant material; however, these protocols utilize significant amounts of benzene for extraction, which will become an environmental issue at large production scales (Oatis et al. 2008). A total synthesis route for cyclopamine has also been published, but involves many steps and results in extremely low yields, making it unsuitable for commercialization (Giannis et al. 2009).

In 2011, Infinity Pharmaceuticals gathered multiple tons of wild *V. californicum* from mountainous regions of the Western U.S., where the weedy plant had once been targeted for permanent removal due to its harmful effects on grazing livestock (Havnes 2011). Infinity Pharmaceuticals is currently working on controlled cultivation strategies for the plant, and is simultaneously looking into the use of PCC as a renewable supply option. As cyclopamine began to be recognized as a potential chemotherapeutic, research was directed into creating *V. californicum* PCC protocols. In 2006, suspension and HRCs of *V. californicum* were established and optimized, with the end-goal of metabolically engineering *V. californicum* to produce large quantities of cyclopamine in PCC (Ma et al. 2006). The patented plant cell lines and optimal media formulations were shown to produce small quantities of cyclopamine, especially when cultures were exposed to light (Ritala et al. 2007).

IPI-926 is not the only Hh targeting drug in development pipelines, as there are also synthetic molecules in clinical trials, such as Genentech's GDC-0449 and Novartis's NVP LDE225, both Smo inhibitors discovered in high-throughput screening programs (Mas and Ruiz i Altaba 2010). GDC-0449 was successful in Phase II clinical trials against advanced basal-cell carcinoma and is being reviewed for approval by the FDA at the time of this review, but GDC-0449 failed in its most recent trials treating colorectal and ovarian cancers in combination with gemcitabine (Read and Palombella 2012).

8.7 Conclusions

Plant NPs are still very relevant to drug discovery in the modern age, as is seen in the case of cyclopamine and its derivatives. However, despite the vast biosynthetic capabilities of plants and their NPs, their usefulness as pharmaceuticals is only as good as our ability to foster and identify biodiversity as it pertains to clinical relevance.

If a bark sample from *T. brevifolia* had never been collected, paclitaxel might never have been discovered, and if *V. californicum* had been eradicated from the mountains of the Western U.S. as was desired 50 years ago, IPI-926 would never have been developed as a potent anti-cancer agent. The usefulness of technologies such as PCC in quickly providing stable, environmentally friendly sources of important anti-cancer drugs should not be overlooked by the pharmaceutical industry. In addition to creating access to naturally occurring plant NPs, PCC is also becoming a force in library synthesis and drug lead discovery through the use of combinatorial biosynthesis, the products of which will need to be produced in PCC by necessity.

While chemical syntheses and microbial expression of plant NPs may lead to cheaper manufacturing processes, PCC both creates and maintains diversity in drug development and manufacturing. Cancer research is constantly uncovering new pathways and targets of interest, and, as in the case of cyclopamine, the diverse pool of plant NPs can provide initial structures as drug leads far before syntheses and high-throughput screening has engineered a suitable drug lead. PCC can be used to readily scale-up supply to meet present clinical needs. The benefits of PCC also extend to large-scale commercialization, as in the case of paclitaxel, and infrastructure and technological expertise now exist for scale-up of PCC processes. Further developments in plant biology, as well as next-generation sequencing technologies are bringing about major breakthroughs in our understanding of plant NP biosynthesis and metabolic control, which will translate to increased productivity at the large scale. PCC has matured greatly as a field since the 1990s, and will continue to provide access to life-saving pharmaceuticals in the twenty-first century.

Acknowledgements The authors would like to acknowledge the support of grants from the National Science Foundation (CBET 0730779) and the National Institutes of Health (GM070852).

References

Ajikumar PK, Xiao W-H, Tyo KEJ et al (2010) Isoprenoid pathway optimization for taxol precursor overproduction in Escherichia coli. Science (New York, NY) 330:70–74. doi:10.1126/science.1191652

Baebler Š, Hren M, Camloh M et al (2005) Establishment of cell suspension cultures of yew (Taxus × media Rehd.) and assessment of their genomic stability. In Vitro Cell Dev Biol Plant 41:338–343. doi:10.1079/IVP2005642

Bernhardt P, McCoy E, O'Connor SE (2007) Rapid identification of enzyme variants for reengineered alkaloid biosynthesis in periwinkle. Chem Biol 14:888–897. doi:10.1016/j.chembiol.2007.07.008

Booth DR (1999) The hedgehog signalling pathway and its role in basal cell carcinoma. Cancer Metastasis Rev 18:261–284

Bräutigam A, Gowik U (2010) What can next generation sequencing do for you? Next generation sequencing as a valuable tool in plant research. Plant Biol (Stuttg) 12:831–841. doi:10.1111/j.1438-8677.2010.00373.x

Bringi V, Kadkade PG, Prince CL, Roach BL (2007) Enhanced production of taxol and taxanes by cell cultures of Taxus species. Patent# 7264951

Chemler JA, Koffas MAG (2008) Metabolic engineering for plant natural product biosynthesis in microbes. Curr Opin Biotechnol 19:597–605. doi:10.1016/j.copbio.2008.10.011

Chen JK, Taipale J, Cooper MK, Beachy PA (2002) Inhibition of hedgehog signaling by direct binding of cyclopamine to Smoothened. Genes Dev 16:2743–2748. doi:10.1101/gad.1025302
Cheng A-X, Y-gen L, Y-bo M et al (2007) Plant terpenoids: biosynthesis and ecological functions. J Integr Plant Biol 49:179–186. doi:10.1111/j.1672-9072.2006.00395.x
Christen A, Gibson DM, Bland J (1991) Production of taxol or taxol-like compounds in cell culture. Patent# 5019504
Cooper MK, Porter JF, Young KE, Beachy PA (1998) Teratogen-mediated inhibition of target tissue response to Shh signaling. Science 280:1603–1607. doi:10.1126/science.280.5369.1603
Cragg G, Snader K (1991) Taxol: the supply issue. Cancer Cells 3:233–235
Cragg GM, Schepartz SA, Suffness M, Grever MR (1993) The taxol supply crisis. New NCI policies for handling the large-scale production of novel natural product anticancer and anti-HIV agents. J Nat Prod 56:1657–1668
Cragg GM, Boyd MR, Grever MR et al (1995) Pharmaceutical prospecting and the potential for pharmaceutical crops. Natural product drug discovery and development at the United States National Cancer Institute. Ann Mo Bot Gard 82:47–53
Croteau R, Ketchum REB, Long RM et al (2006) Taxol biosynthesis and molecular genetics. Phytochem Rev 5:75–97. doi:10.1007/s11101-005-3748-2
Denis J-N, Greene AE, Guenard D et al (1988) A highly efficient, practical approach to natural taxol. J Am Chem Soc 110:5917–5919
Dietrich JA, Yoshikuni Y, Fisher KJ et al (2009) A novel semi-biosynthetic route for artemisinin production using engineered substrate-promiscuous P450(BM3). ACS Chem Biol 4:261–267. doi:10.1021/cb900006h
Eibl R, Eibl D (2007) Design of bioreactors suitable for plant cell and tissue cultures. Phytochem Rev 7:593–598. doi:10.1007/s11101-007-9083-z
Evans BE, Rittle KE, Bock M et al (1988) Methods for drug discovery: development of potent, selective, orally effective cholecystokinin antagonists. J Med Chem 31:2235–2246
Expósito O, Bonfill M, Moyano E et al (2009) Biotechnological production of taxol and related taxoids: current state and prospects. Anticancer Agents Med Chem 9:109–121
Farkya S, Bisaria VS, Srivastava AK (2004) Biotechnological aspects of the production of the anticancer drug podophyllotoxin. Appl Microbiol Biotechnol 65:504–519. doi:10.1007/s00253-004-1680-9
Firn RD, Jones CG (2003) Natural products? A simple model to explain chemical diversity. Nat Prod Rep 20:382. doi:10.1039/b208815k
Fischbach MA, Clardy J (2007) One pathway, many products. Nat Chem Biol 3:353–355
Gelvin SB (2003) Improving plant genetic engineering by manipulating the host. Trends Biotechnol 21:95–98
Georgiev M, Pavlov A, Ilieva M (2006) Selection of high rosmarinic acid producing Lavandula vera MM cell lines. Process Biochem 41:2068–2071. doi:10.1016/j.procbio.2006.05.007
Georgiev MI, Weber J, Maciuk A (2009) Bioprocessing of plant cell cultures for mass production of targeted compounds. Appl Microbiol Biotechnol 83:809–823. doi:10.1007/s00253-009-2049-x
Georgiev MI, Ludwig-Muller J, Bley T (2010) Hairy root culture: copying nature in new bioprocesses. In: Arora R (ed) Medicinal plant biotechnology. CABI, Cambridge, MA, pp 156–175
Giannis A, Heretsch P, Sarli V, Stössel A (2009) Synthesis of cyclopamine using a biomimetic and diastereoselective approach. Angew Chem Int Ed Engl 48:7911–7914. doi:10.1002/anie.200902520
Gundlach H, Muller MJ, Kutchan TM, Zenk MH (1992) Jasmonic acid is a signal transducer in elicitor-induced plant cell cultures C. Plant Biol 89:2389–2393
Hale V, Keasling JD, Renninger N, Diagana TT (2007) Microbially derived artemisinin: a biotechnology solution to the global problem of access to affordable antimalarial drugs. Am J Trop Med Hyg 77:198–202
Harding K (2004) Genetic integrity of cryopreserved plant cells: a review. Cryoletters 25:3–22
Hartmann T (2007) From waste products to ecochemicals: fifty years research of plant secondary metabolism. Phytochemistry 68:2831–2846. doi:10.1016/j.phytochem.2007.09.017

Havnes M (2011) Plant may yield weapon in the war on cancer. Salt Lake Tribune
Heretsch P, Tzagkaroulaki L, Giannis A (2010) Cyclopamine and hedgehog signaling: chemistry, biology, medical perspectives. Angew Chem Int Ed Engl 49:3418–3427. doi:10.1002/anie.200906967
Holton RA, Somoza C, H-baik K et al (1994) First total synthesis of taxol. 1. Functionalization of the B ring. J Am Chem Soc 116:1597–1598
Kim Y, Wyslouzil BE, Weathers PJ (2002) Secondary metabolism of hairy root cultures in bioreactors. In Vitro Cell Dev Biol Plant 38:1–10. doi:10.1079/IVP2001243
Kingston DGI (1994) Taxol: the chemistry and structure-activity relationships of a novel anticancer agent. Trends Biotechnol 12:222–227
Koehn FE, Carter GT (2005) The evolving role of natural products in drug discovery. Nat Rev Drug Discov 4:206–220. doi:10.1038/nrd1657
Kolewe ME, Gaurav V, Roberts SC (2008) Pharmaceutically active natural product synthesis and supply via plant cell culture technology. Mol Pharm 5:243–256. doi:10.1021/mp7001494
Kolewe ME, Henson MA, Roberts SC (2011) Analysis of aggregate size as a process variable affecting paclitaxel accumulation in Taxus suspension cultures. Biotechnol Prog 27:1365–1372. doi:10.1002/btpr.655
Kutchan TM (2005) A role for intra- and intercellular translocation in natural product biosynthesis. Curr Opin Plant Biol 8:292–300. doi:10.1016/j.pbi.2005.03.009
Lee E-K, Jin Y-W, Park JH et al (2010) Cultured cambial meristematic cells as a source of plant natural products. Nat Biotechnol 28:1213–1217. doi:10.1038/nbt.1693
Leonard E, Yan Y, Fowler ZL et al (2008) Strain improvement of recombinant Escherichia coli for efficient production of plant flavonoids. Mol Pharm 5:257–265. doi:10.1021/mp7001472
Lipkus AH, Yuan Q, Lucas KA et al (2008) Structural diversity of organic chemistry. A scaffold analysis of the CAS registry. J Org Chem 73:4443–4451. doi:10.1021/jo8001276
Ma R, Ritala A, Oksman-Caldentey K-M, Rischer H (2006) Development of in vitro techniques for the important medicinal plant Veratrum californicum. Planta Med 72:1142–1148. doi:10.1055/s-2006-946697
Mas C, Ruiz i Altaba A (2010) Small molecule modulation of HH-GLI signaling: current leads, trials and tribulations. Biochem Pharmacol 80:712–723. doi:10.1016/j.bcp. 2010.04.016
Mustafa NR, de Winter W, van Iren F, Verpoorte R (2011) Initiation, growth and cryopreservation of plant cell suspension cultures. Nat Protoc 6:715–742. doi:10.1038/nprot.2010.144
Naik R, Kattige SL, Bhat SV et al (1988) An antiinflammatory cum immunomodulatory piperidinylbenzopyranone from dysoxylum binectariferum: isolation, structure and total synthesis. Tetrahedron 44:2081–2086
Naill MC, Roberts SC (2005a) Culture of isolated single cells from Taxus suspensions for the propagation of superior cell populations. Biotechnol Lett 27:1725–1730. doi:10.1007/s10529-005-2738-1
Naill MC, Roberts SC (2005b) Flow cytometric identification of paclitaxel-accumulating subpopulations. Biotechnol Prog 21:978–983. doi:10.1021/bp0495441
Nakagawa A, Minami H, Kim J-S et al (2011) A bacterial platform for fermentative production of plant alkaloids. Nat Commun 2:326. doi:10.1038/ncomms1327
Newman DJ, Cragg GM (2007) Natural products as sources of new drugs over the last 25 years. J Nat Prod 70:461–477. doi:10.1021/np068054v
Ni X, Wen S, Wang W, et al. (2011) Enhancement of camptothecin production in Camptotheca acuminata hairy roots by overexpressing ORCA3 gene. Journal of Applied Pharmaceutical Science 01:85–88
Nicolau KC, Yang Z, Liu JJ et al (1994) Total synthesis of taxol. Nature 367:630–634
Oatis JE, Brunsfeld P, Rushing JW et al (2008) Isolation, purification, and full NMR assignments of cyclopamine from Veratrum californicum. Chem Cent J 2:12. doi:10.1186/1752-153X-2-12
Ortholand J-Y, Ganesan A (2004) Natural products and combinatorial chemistry: back to the future. Curr Opin Chem Biol 8:271–280. doi:10.1016/j.cbpa.2004.04.011
Patil RA, Kolewe ME, Normanly J et al (2011) Taxane biosynthetic pathway gene expression in Taxus suspension cultures with different bulk paclitaxel accumulation patterns & a molecular

approach to understand variability in paclitaxel accumulation. Biotechnol J 1–10. doi:10.1002/biot.201100183
Pollier J, Moses T, Goossens A (2011) Combinatorial biosynthesis in plants: a (p)review on its potential and future exploitation. Nat Prod Rep 1897–1916. doi:10.1039/c1np00049g
Read MA, Palombella VJ (2012) The hedgehog signaling pathway in cancer pathogenesis and therapy. In: Frank DA (ed) Signaling pathways in cancer pathogenesis and therapy. Springer, New York, NY, pp 55–79
Ritala A, Rischer H, Oksman-Caldentey K-M (2007) Plant cell lines established from the medicinal plant Veratrum californicum. Patent# US 2009/0305338 A1
Ro D-K, Paradise EM, Ouellet M et al (2006) Production of the antimalarial drug precursor artemisinic acid in engineered yeast. Nature 440:940–943. doi:10.1038/nature04640
Roberts S, Kolewe M (2010) Plant natural products from cultured multipotent cells. Nat Biotechnol 28:1175–1176. doi:10.1038/nbt1110-1175
Rubin LL, de Sauvage FJ (2006) Targeting the hedgehog pathway in cancer. Nat Rev Drug Discov 5:1026–1033. doi:10.1038/nrd2086
Runguphan W, Qu X, O'Connor SE (2010) Integrating carbon-halogen bond formation into medicinal plant metabolism. Nature 468:461–464. doi:10.1038/nature09524
Sánchez C, Zhu L, Braña AF et al (2005) Combinatorial biosynthesis of antitumor indolocarbazole compounds. Proc Natl Acad Sci U S A 102:461–466. doi:10.1073/pnas.0407809102
Schäfer H, Wink M (2009) Medicinally important secondary metabolites in recombinant microorganisms or plants: progress in alkaloid biosynthesis. Biotechnol J 4:1684–1703. doi:10.1002/biot.200900229
Schnur DM, Beno BR, Tebben AJ, Cavallaro C (2011) Methods for Combinatorial and Parallel Library Design. Chemoinformatics and Computational Chemical Biology. doi:10.1007/978-1-60761-839-3
Senger RS, Phisalaphong M, Karim MN, Linden JC (2006) Development of a culture sub-population induction model: signaling pathways synergy and taxanes production by Taxus canadensis. Biotechnol Prog 22:1671–1682. doi:10.1021/bp0602552
Solomon LD (2011) Synthetic biology: Science, Business, and Policy. 128–129
Stitt M, Sulpice R, Keurentjes J (2010) Metabolic networks: how to identify key components in the regulation of metabolism and growth. Plant Physiol 152:428–444. doi:10.1104/pp. 109.150821
Tremblay MR, Nevalainen M, Nair SJ et al (2008) Semisynthetic cyclopamine analogues as potent and orally bioavailable hedgehog pathway antagonists. J Med Chem 51:6646–6649. doi:10.1021/jm8008508
Tremblay MR, Lescarbeau A, Grogan MJ et al (2009) Discovery of a potent and orally active hedgehog pathway antagonist (IPI-926). J Med Chem 52:4400–4418. doi:10.1021/jm900305z
van der Fits L, Memelink J (2000) ORCA3, a jasmonate-responsive transcriptional regulator of plant primary and secondary metabolism. Science 289:295–297. doi:10.1126/science.289.5477.295
Verpoorte R, Heijden RVD, Hoopen HJG, Memelink J (1999) Metabolic engineering of plant secondary metabolite pathways for the production of fine chemicals. Biotechnol Lett 21:467–479
Weathers P, Liu C, Towler M, Wyslouzil B (2008) Mist reactors: principles, comparison of various systems, and case studies. Biotechnol Bioeng 3:29–37
Werck-reichhart D, Feyereisen R (2000) Protein family review cytochromes P450: a success story. Genome 1:1–9
Wilson RM, Danishefsky SJ (2006) Small molecule natural products in the discovery of therapeutic agents: the synthesis connection. J Org Chem 71:8329–8351. doi:10.1021/jo0610053
Wilson SA, Roberts SC (2011) Recent advances towards development and commercialization of plant cell culture processes for the synthesis of biomolecules. Plant Biotechnol J 1–20. doi:10.1111/j.1467-7652.2011.00664.x
Yukimune Y, Tabata H, Higashi Y, Hara Y (1996) Methyl jasmonate-induced overproduction of paclitaxel and baccatin III in Taxus cell suspension cultures. Nat Biotechnol 14:1129–1132
Zhao J, Davis LC, Verpoorte R (2005) Elicitor signal transduction leading to production of plant secondary metabolites. Biotechnol Adv 23:283–333. doi:10.1016/j.biotechadv.2005.01.003

Chapter 9
Natural Products as Tools for Discovering New Cancer Targets

John A. Beutler

Abstract In addition to being a major source of novel cancer drugs, natural products have provided important insights into cancer mechanisms and cellular targets. There are many natural products which have been reported to have empirical activity in inhibiting cancer cell proliferation, carcinogenesis, angiogenesis, and metastasis. By developing an understanding of the mechanisms by which these empirically active compounds exert their effects, novel targets for cancer therapy may be identified. This review describes prominent classes of anticancer natural products for which mechanisms are not currently known, discusses experimental approaches to elucidating the targets, and offers some recent examples of targets that have been elucidated.

9.1 Introduction

9.1.1 Cancer

Since cancer encompasses a multitude of diseases by any measure (histologic, genetic, organ of origin), rational treatment demands a correspondingly large number of therapeutic approaches, each tuned to the specific nature of the cancer.

J.A. Beutler (✉)
Molecular Targets Laboratory, Center for Cancer Research,
National Cancer Institute, Frederick, MD 21702, USA
e-mail: beutlerj@mail.nih.gov

F.E. Koehn (ed.), *Natural Products and Cancer Drug Discovery*, Cancer Drug Discovery and Development, DOI 10.1007/978-1-4614-4654-5_9,

The current approach to this problem entails such undertakings as (1) The Cancer Genome Atlas (TCGA) project (Vockley and Ozenberger 2012), which aims to genetically analyze hundreds of samples from different tumor types with a goal of identifying potential therapeutic and diagnostic targets; (2) high throughput biochemical screening of validated molecular cancer targets against very large libraries (>1,000,000) of synthetic compounds; and (3) personalized medicine, which attempts to match a patient's tumor with therapy customized to its biology.

While these are worthy approaches, an enormous amount of work is required to translate identified genes, screening hits, and specific clinical information into useful therapies. A complementary approach would be to investigate compounds which have demonstrated *empirical* biological activity in assays relevant to cancer processes, to determine how their effects are mediated. Examples of empirical activity include inhibition of cell growth, angiogenesis, or metastasis, as well as induction of cell death. These cellular endpoints are cell fate processes which integrate a vast system of subcellular constituents, including molecular targets and signaling pathways. Because the cell represents the fundamental unit of biological organization, it is arguable that these cell fate endpoints are more relevant to targeting a disease of disregulated cells than targeting molecules in isolation.

Natural products are an attractive starting point for such an endeavor, since they have been shaped by evolutionary processes either to antagonize biological processes of competitor organisms or to communicate between organisms (Beutler 2009; Li and Vederas 2009). The constraints placed on their evolution favor the selection of molecules that can diffuse in aqueous environments of the cellular milieu, and cross cell membranes. In addition, natural products have been a major source for anticancer drug development in the past (Butler 2008; Newman and Cragg 2007). They have contributed to the elucidation of cellular targets for many types of processes (Dixon et al. 2007) including cancer (Newman and Cragg 2006).

9.1.2 *Defining Terms*

At the outset, it may be useful to distinguish four terms that are widely used but poorly defined. These are molecular target, mechanism of action, mode of action, and pathway. The most specific term is *molecular target*, which can be defined as a biological macromolecule to which a drug binds and through which it exerts its desired effects. This includes proteins, especially enzymes, but can also include carbohydrates and nucleic acids. Thus, topoisomerase I, a protein with enzymatic activity, is said to be the molecular target of camptothecin (Hsiang et al. 1985).

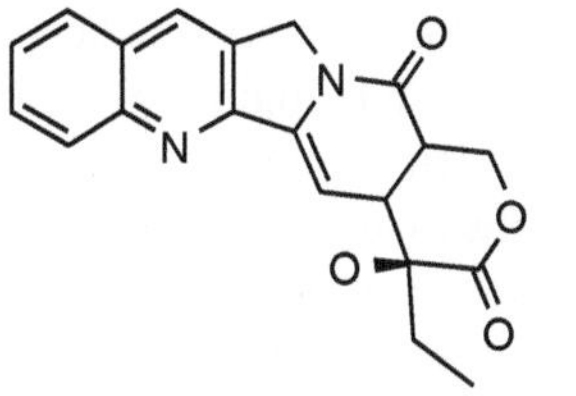

Camptothecin

More precisely, camptothecin covalently binds DNA as part of a ternary "cleavable complex" consisting of topoisomerase I, DNA, and camptothecin (Kerrigan and Pilch 2001; Redinbo et al. 1998). Using the same example, the *mechanism of action* of camptothecin can be described as inhibition of topoisomerase I enzymatic activity (Hsiang et al. 1985). The *mode of action* is induction of DNA single strand breaks and inhibition of DNA synthesis (Hsiang et al. 1985). The affected *pathway* is DNA repair. Thus, camptothecin affects a molecular target within the DNA repair pathway, inducing phenomena which have a substantial biological effect, namely inhibition of tumor growth.

9.1.3 Validation of Targets

Many compounds can be shown by biophysical methods such as calorimetry, surface plasmon resonance, or fluorescence anisotropy to bind to purified proteins or other macromolecules. Such methods have been used in reductionist screening models to identify hits for drug targets. However, starting with reductionist models carries the downstream burden of showing that the binding event causes a measurable biochemical event in the target (e.g., enzyme inhibition), then showing that the biochemical event also occurs in a cellular context, before testing for a biological response in an animal model. If large numbers of hits are generated in the primary screen, this can place large demands on resources for the secondary assays necessary to complete these steps.

In a targeted screening program, the proposed molecular target must be validated to justify the expenditure of resources required to find compounds which affect the target. What this means is that modulation of the target must be shown to have the desired biological effect in a higher-order model, preferably an animal model. Knockout mouse models are one avenue to demonstrating the functional importance of inhibition of a target, although if a simple germ line knockout is lethal, conditional knockouts may be required to properly understand the importance of a cancer target, especially if the target is also involved in early development (Vidalin et al. 2009). Knockout models are also prone to adaptation of other pathways to make up for germline deletions, which complicates interpretation of the results. In cells, siRNA can be a useful tool for evaluating the importance of a particular gene product as long as the function of the gene is understood. Thus, inhibition of cancer cell growth caused by a specific siRNA may be considered valuable, especially if treated cells lose the ability to form tumors in xenografts (Vidalin et al. 2009). The best validation of a molecular target is to discover a proof-of-principle compound that is shown to directly interact with the target, and has the desired effect in humans or animals, even if the compound is not suitable for development as a marketed drug (Frye 2010).

With elucidation of the target of a compound identified in an empirical screen, the same validation criteria apply, although in this case the investigator does not choose the target, it is determined by experimentation, and validation is performed after the fact.

9.1.4 Is There a Demarcation Between Targeted and Empirical Screens?

A cell-free binding assay screening model and a simple cancer cell growth assay can be considered to be at opposite poles of a continuum between targeted and empirical screening; there are currently many screening models which mix aspects of both poles. For example, the Vogelstein group constructed a screen using an isogenic pair of cell lines, one harboring a mutant p21 gene, and the other with wild type p21. The endpoint of the assay was preferential inhibition of cell growth in the mutant cell line (Torrance et al. 2001). Our group has conducted a similar screen using a particular *c-kit* mutation (Henrich et al. 2007). While identical in endpoint to an empirical screen, a particular molecular target is interrogated. Hits from these screens must be carefully evaluated to determine the mechanism by which they modulate the target, if they do, and to remove hits which are a result of experimental artifacts such as compound autofluorescence or generic protein synthesis inhibition. Thus, there are very few screens which are purely empirical or purely targeted.

9.2 Intriguing Classes of Natural Products Active in Empirical Cancer Assays with Poorly Defined Targets

The following examples are not exhaustive, but are intended to give a sampling of empirically active anticancer natural products in which the molecular target responsible for the activity has not been convincingly determined. In some cases targets have been proposed, while in others there is no mechanistic information. All appear to be worthy of further study.

9.2.1 Aplyronine A

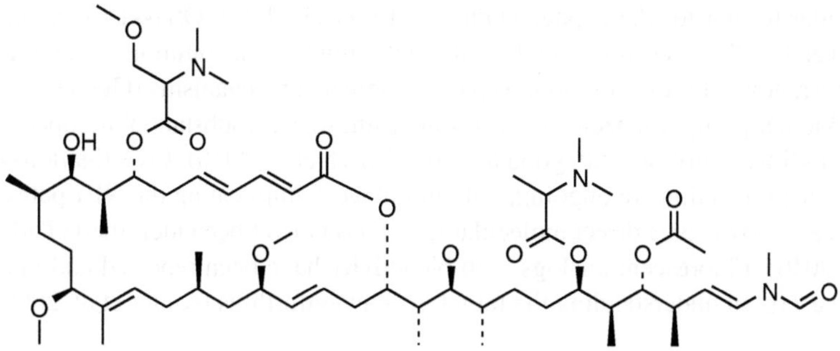

Aplyronine A

Aplyronine A is a macrolide isolated from the sea hare *Aplysia kurodai*, a Japanese marine invertebrate (Yamada et al. 1993). Its potent cancer cell growth inhibition stimulated research into its mechanism, and depolymerization of actin was proposed (Saito et al. 1996). A number of analogs were prepared which also depolymerized actin (Kigoshi et al. 1996; Suenaga et al. 1997), however the cytotoxicity and actin depolymerization activity were not highly correlated. In fact, in vitro actin polymerization activity for aplyronine A itself has been found to be in the micromolar range, while cell growth inhibition has been found to be in the low nanomolar range. Thus, there would seem to be another mechanism responsible for the observed cell growth inhibition. Supporting this, many other known actin inhibitors have similar profiles in the NCI 60-cell screen (Wrona et al. 2009) but aplyronine does not cluster with them. Further investigation to attempt to separate actin depolymerization activity from cell growth inhibition in this series would appear to be warranted.

9.2.2 Schweinfurthins

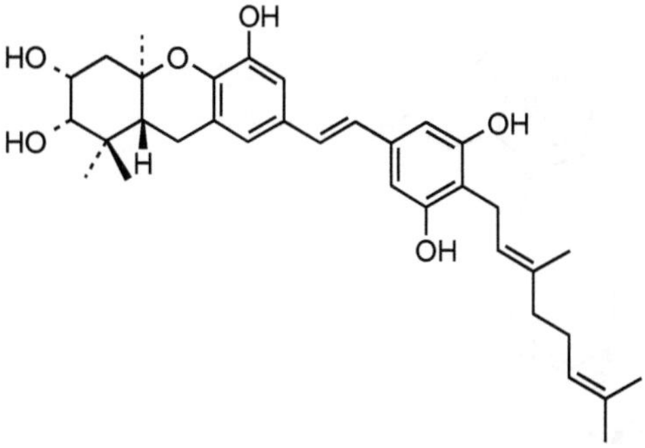

Schweinfurthin A

The schweinfurthins are a group of prenylstilbene plant products which show potent and selective cell growth inhibition against CNS, renal, and leukemia cell lines in the NCI 60-cell screen (Beutler et al. 1998; Yoder et al. 2006). The profile of activity is similar to that for the cephalostatins (Pettit et al. 1988), OSW-1 (Mimaki et al. 1997) and stellettins (McCormick et al. 1996), implying a common mode of action, however, none of these compounds have established mechanisms (Guo et al. 1999). The Wiemer group has been successful in synthesizing a substantial number of analogs of schweinfurthins with good activity (Ulrich et al. 2010). Investigations of the mechanism of action are ongoing, with initial work implicating the NF1 pathway in CNS cells, however, a direct molecular target has not yet been identified (Turbyville et al. 2010). Fluorescent analogs with bioactivity have been reported and may also prove useful in understanding the target of schweinfurthins (Kuder et al. 2009).

9.2.3 Ellipticines

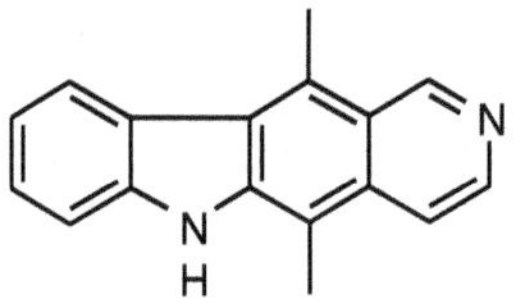

Ellipticine

Ellipticines were identified as CNS cancer-selective agents in the NCI 60-cell screen, and cellular uptake mechanisms were found to be critical for their activity (Acton et al. 1994; Shi et al. 1998; Vistica et al. 1994, 1996). More recently, ellipiticine itself has been shown to increase the nuclear localization of p53 in colon cancer cells, with a resultant increase in transactivation of the p21 promoter (Xu et al. 2008). Further studies to explore the generality of these results in CNS cancer cells appear to be warranted.

9.2.4 Diketopiperazines

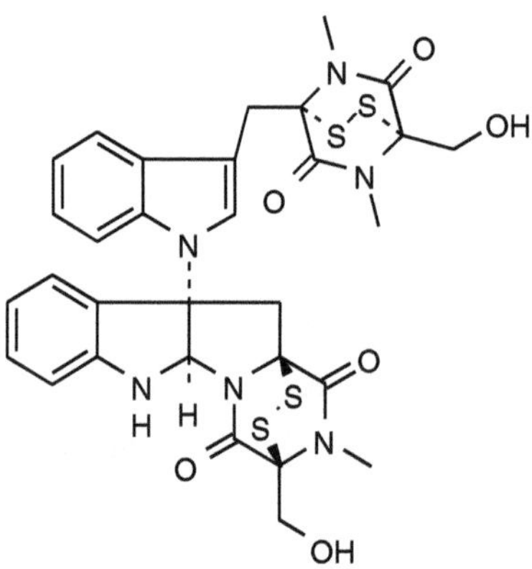

Chaetomin

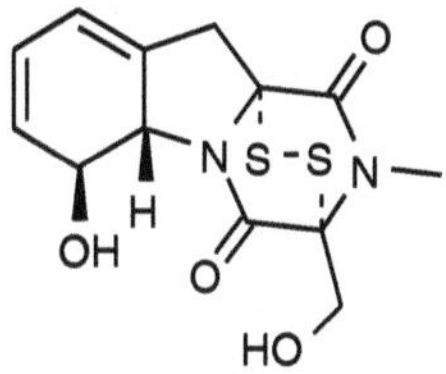

Gliotoxin

Diketopiperazines are fungal and lichen metabolites formed by the cyclization of two amino acids. Most interesting are those diketopiperazines wherein an additional disulfide bridge is added, known as epidithiodiketopiperazines. Chaetomin and gliotoxin were the earliest reported members of this class (Bell et al. 1958; McInnes et al. 1976; Waksman and Bugie 1944; Xu et al. 2008), and many of its members have been reported to potently inhibit cell growth (Ernst-Russell et al. 1999; Onodera et al. 2004; Williams et al. 1998). Many mechanisms of action for these compounds have been proposed, including topoisomerase I inhibition (Funabashi et al. 1994), epidermal growth factor receptor antagonism (Hegde et al. 1997), inhibition of mitochondrial ATP synthesis (Moerman et al. 2003), and stimulation of calcium influx (Hurne et al. 2002). More recently, the HIF-1 pathway has been implicated, with chaetomin blocking the interaction between HIF-1a and p300 via a zinc ejection mechanism (Cook et al. 2009; Kung et al. 2004). Whether these diverse effects are relevant to inhibition of cancer cell growth remains to be sorted out.

9.2.5 *Quassinoids*

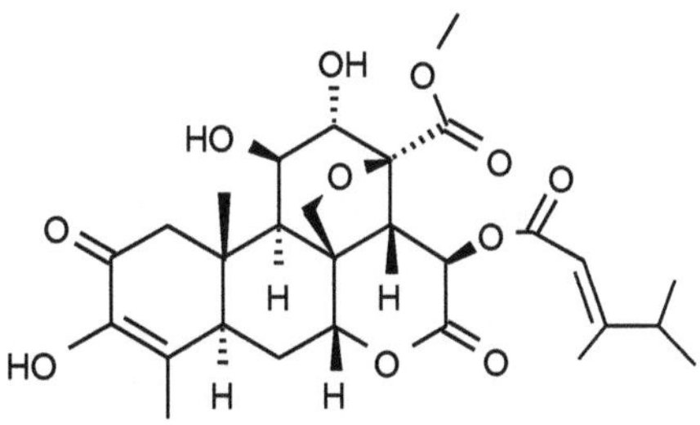

Bruceantin

Quassinoids are nontriterpene plant bitter principles which were discovered to have empirical activity in leukemias and other cancers (Kupchan et al. 1973). The quassinoid bruceantin advanced as far as phase II clinical trials (Wiseman et al. 1982). The mechanism of action has long been considered to be inhibition of protein synthesis (Fukamiya et al. 2005), however, our group identified several quassinoids as inhibitors of the transcription factor AP-1, and found poor correlation between AP-1 activity and cell growth inhibition (Beutler et al. 2009), while others have detected several other potential mechanisms (Cuendet et al. 2004; Morre et al. 1998; Rosati et al. 2004). Thus, there may be an opportunity to develop agents in this class which do not depend on protein synthesis inhibition for their activity.

9.2.6 *Cucurmin*

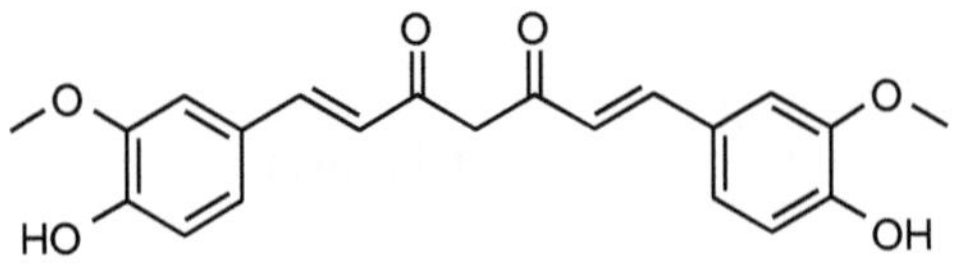

Curcumin

Curcumin is a phenolic pigment isolated from the spice turmeric, which is widely used in Asian cuisine. Several studies have shown an empirical cancer chemopreventative effect in animal models of carcinogenesis, as well as angiogenesis (Maheshwari et al. 2006). Since it is ingested in large amounts in food, it is noteworthy that some of these effects have been detected in colorectal tumor models (Perkins et al. 2002). The molecular basis for these activities is unclear; many pathways and targets have been found to be affected by curcumin, including AP-1, NFκB, TNF-α, and IL-1β (Maheshwari et al. 2006). Nonetheless, the strong empirical data supports continued efforts to uncover a molecular explanation.

9.3 Two Cautionary Tales

Not all empirically active compounds investigated necessarily lead to a clear-cut target, and not all clear-cut targets of empirically discovered compounds are valid targets for cancer therapy. Two examples illustrate these points.

9.3.1 *Resveratrol*

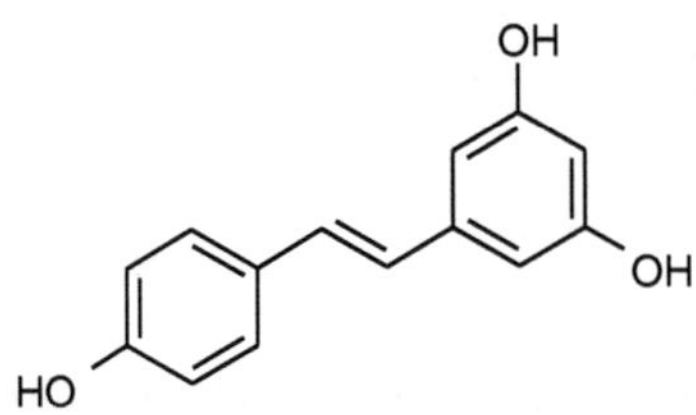

Resveratrol

Resveratrol is a simple stilbene found in a wide variety of plants, including such foods as grapes, peanuts, and mulberries. It was identified as having cancer chemopreventative activity, interfering with tumor initiation, progression, as well as inflammation (Jang et al. 1997). Since the seminal paper, dietary supplement makers have reaped great benefits from these findings. The scientific literature has also been active, with over 1,000 papers concerning resveratrol and cancer entering the literature. Literally dozens of mechanistic hypotheses have been posed, but none has been widely accepted (Goswami and Das 2009). Given its small molecular size and apparent affinity for many potential targets, it is likely that in vitro assays will not provide a meaningful answer to mechanistic questions about the compound. However, abundant in vivo data supports its use and importance in chemoprevention (Goswami and Das 2009).

9.3.2 Cardiac Glycosides

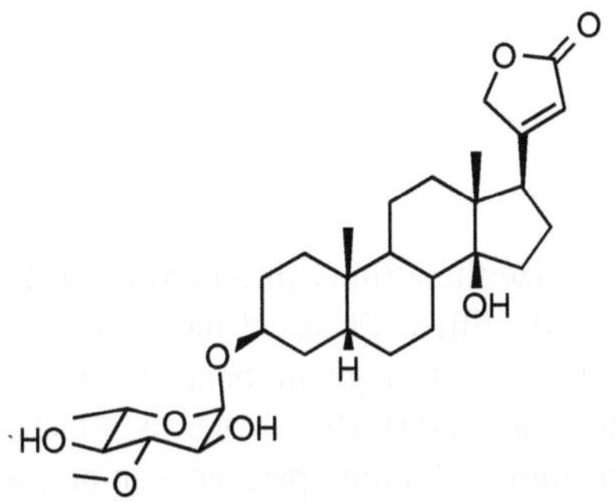

Neriifolin, a cardiac glycoside

Originally isolated from the plant *Digitalis purpurea* as cardiovascular agents, cardiac glycosides from a wide variety of plant sources have also been found to inhibit growth of cancer cell lines (Decosterd et al. 1994; Jolad et al. 1981). Mice and rats are much less sensitive to the toxicity of these compounds than humans (Gupta et al. 1986); when cardiac glycosides are tested in xenografts of human cell lines in mice, there is a substantial therapeutic window. However, actual therapy of cancer requires treatment of human tumors in human hosts, where no such window exists. Thus, reliance on xenografts can lead to poor estimates of the potential of cardiac glycosides in cancer therapy. As far as can be ascertained, the potency of various cardiac glycosides to inhibit cell growth entirely parallels their ability to inhibit Na/K ATPase, and no other convincing mechanism has been proposed to account for their cell growth inhibition. Nonetheless, enthusiasm for the compound class as anticancer agents still reappears regularly (Mijatovic et al. 2007).

9.4 Experimental Approaches to Target Identification

No single method can be relied on to give an accurate indication of the molecular target or pathway affected by a small molecule. In most cases a judicious combination of techniques is required to make the case for a target.

9.4.1 *Microscopy and Fluorescent Tags*

The localization of compounds within cells can provide clues to site of action and mechanism. By adding a fluorescent moiety to the compound to be studied, its movement through the cell can be monitored by microscopy. A generic strategy has been proposed for fluorescent labeling of bioactive natural products by La Clair in which modular fluorophore chemistries are proposed with several points of attachment to the natural product (Alexander et al. 2006), since care must be taken in using fluorescently labeled small molecules to see that the bioactivity is not impaired by the tag. Red-shifted fluorescent tags are preferred to shorter wavelengths to avoid interference from cellular constituents.

9.4.2 *Affinity Methods*

Various methods have been developed to utilize the affinity of a compound for its target to isolate the target so that it can be identified. Classical methods utilizing columns have largely been superseded by the use of magnetic beads with suitable surface chemical modifications (Sakamoto et al. 2009). By covalently attaching a small molecule to the beads, affinity purification of binding partners can be carried out from cell lysates. Alternatively, beads coated with streptavidin can be used if the small molecule is suitably conjugated to a biotin moiety. Both of these approaches demand a thorough understanding of the structure-activity relationships (SAR) of the molecule so that linkers can be attached to a noncritical position of the molecule. Analysis of the bound and eluted partners can be carried out by gel electrophoresis, although proteomic approaches are more powerful analytical techniques (Sleno and Emili 2008).

A more challenging technique is the use of photoaffinity tags. This requires precise knowledge of SAR since the moieties to be attached may lead to undesirable steric or electronic consequences. The probe is incubated with the cell or cell lysate, and then irradiated, forming a covalent bond with whatever biomolecule is in close proximity. The technique is most often used to map precise binding sites of molecules on known targets (Kuroda et al. 2006; Ojima et al. 1995).

9.4.3 *Computational Methods*

9.4.3.1 COMPARE

The NCI 60-cell screening assay is a completely empirical system, with multiple human cell lines in nine organ panels. It is a 2-day cell growth assay with a sulforhodamine B protein stain endpoint. The dose response curves for all cell lines are analyzed for selectivity and presented in a "mean-graph" format (Paull et al. 1989).

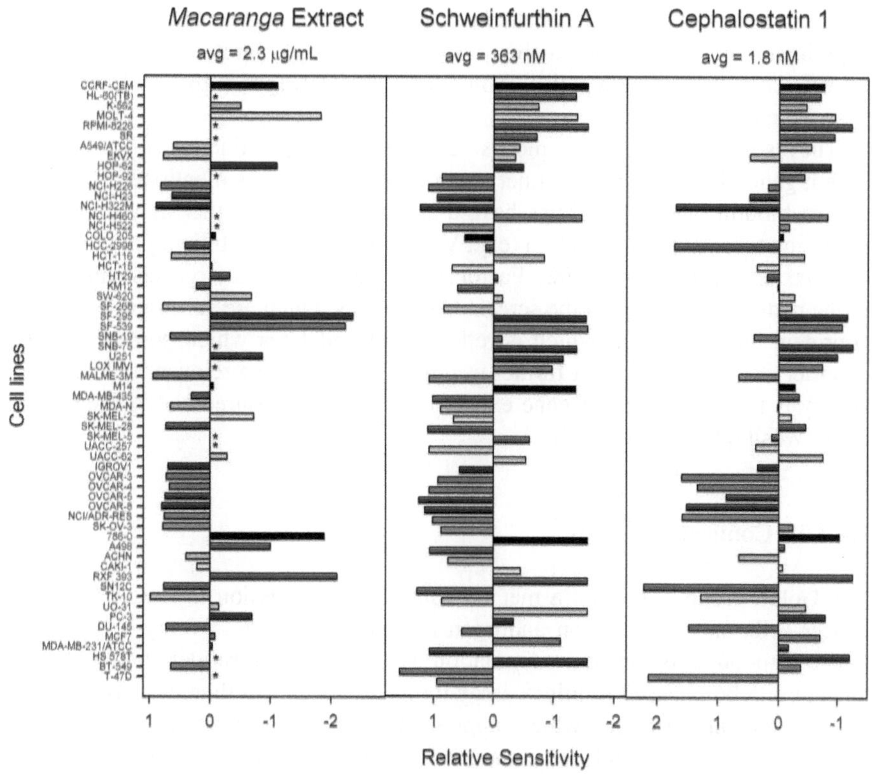

Fig. 9.1 A comparison of the mean bar graphs at the GI_{50} level for NCI 60 cell tests for (**a**) an organic solvent extract of *Macaranga schweinfurthii* Muell.Arg., [N17023], for which the average GI_{50} value was 2.3 μg/mL; (**b**) the purified compound schweinfurthin A from that extract, for which the mean GI_{50} value was 363 nM; (**c**) the unrelated natural product cephalostatin 1 for which the mean GI_{50} value was 1.8 nM. Asterisks indicate that cell line data was not available for the test. Cell line data is not shown where only one sample was tested against that cell line. Replicates were $n=1$ for the extract, $n=3$ for schweinfurthin A, and $n=9$ for cephalostatin 1. Pearson correlation coefficients at this level were 0.73 between (**a**) and (**b**), and 0.59 between (**b**) and (**c**)

While the initial intent was to discover compounds selective for all the cell lines in a particular organ panel, it was soon realized that the patterns of cell line selectivity were reflective of the mode of action of compounds (Shoemaker 2006). For instance, all compounds which were known to interact with tubulin possessed a similar pattern at the total growth inhibition (TGI) level of analysis (Paull et al. 1992). Similarly, protein synthesis inhibitors generated a consistent profile in a recent study (Chan et al. 2004). More sophisticated clustering analyses soon made global comparisons possible (Keskin et al. 2000; Monks et al. 1997; Weinstein et al. 1992, 1997). In the case of natural products extracts, it has been shown that purified active compounds from the extract almost always posses a profile similar to that of the extract (Fig. 9.1a vs. b), with rare exceptions (Beutler et al. 2000).

Thus it is possible to map all of the data for crude extracts and pure compounds together. Where pure compounds with known mode of action occur, extracts within that region may be suspected of sharing the same mode, and that hypothesis can be experimentally tested. Where there is no "signpost," one is limited to grouping similar extracts, purifying the active compounds, and stating that the mode of action appears to be novel.

The resolution of the data is modest; that is to say, fine distinctions of mechanism do not generally result in a distinct pattern. For example, as mentioned above, all known tubulin interacting drugs have a similar pattern, whether they operate by inhibiting tubulin polymerization (e.g., Vinca alkaloids) or stabilizing microtubules (e.g., taxanes) (Paull et al. 1992). Factors other than molecular target also play a role in the patterns generated in the screen. Compounds which are substrates for p-glycoprotein have a pattern which is reflective of cell lines which posses the drug exporter (Wu et al. 1992). A further use of NCI 60-cell data relevant to target discovery is the correlation of gene expression patterns measured across the 60 cell lines (Wosikowski et al. 1997).

9.4.3.2 Connectivity Map

The Golub group published a method using gene expression in LnCAP prostate cancer cells and focusing on androgen receptor activation. Twenty-seven genes which responded to androgen stimulation or deprivation were identified as a gene signature. A bead-based Luminex assay for all 27 genes was then used as a screening tool to identify samples which suppressed the androgen signature. In addition to several known drugs, multiple compounds in the celastrol and gedunin classes of triterpenoid natural products were identified (Hieronymus et al. 2006). Their activity was traced to inhibition of the HSP90 pathway by reference to Connectivity Map data, a tool for systematically matching gene expression profiles to those of known drugs (Lamb et al. 2006; Lamb 2007). Others have identified similar activity for gedunin analogs (Brandt et al. 2008). The Connectivity Map database initially contained gene expression profiles from 4 human cell lines with 164 bioactive small molecules, and currently contains more than 7,000 expression profiles representing 1,309 compounds (Lamb 2010).

9.4.4 Proteomic Methods

In the same fashion as with gene expression data, proteomic analysis can be used to generate hypotheses about the molecular target of natural products. Proteomic analysis of affinity experiments (vide infra) can be useful as a high throughput readout of affinity-isolated samples from experiments. In work on the Annonaceous acetogenin squamocin, this led to the identification of alternative targets to mitochondrial complex I for its pro-apoptotic activity (Derbre et al. 2008). Alternatively, analysis of differential protein expression between treated and control cells has been done. Choice of drug concentration and timing in gene expression experiments are critical, and apply to protein expression as well. Thus, a Brazilian group exposed

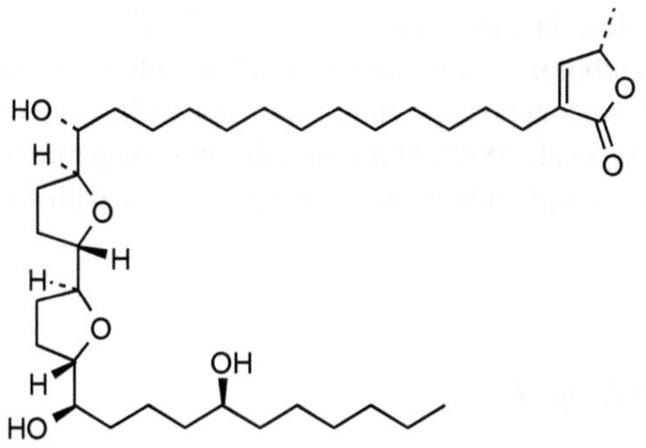

Squamocin A

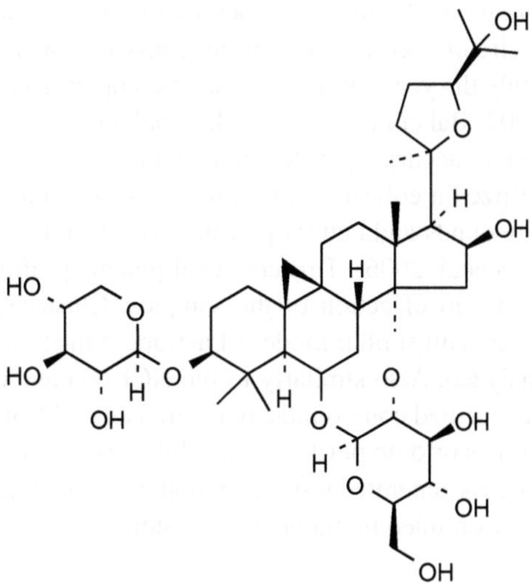

Astragaloside IV

glioblastoma cells to perillyl alcohol for six different treatment times, though at a single dose (1.8 mM). Early events included stimulation of phosphorylation of glycogen synthesis kinase and inhibition of ERK phosphorylation after 10 min (Fischer et al. 2010). Similarly, the human liver cancer cell line HepG2 was treated with the triterpene glycoside astragaloside IV for 3 days without causing cytotoxicity, but with inhibition of colony forming activity. A number of proteins showed expression changes, with the Vav3.1 oncogene down-regulated, supporting potential cancer chemopreventative use of the compound (Qi et al. 2010).

9.4.5 Metabolomic Methods

The measurement of large numbers of cellular metabolites is a more recent approach to mechanistic investigation than genomics or proteomics. While it holds promise for understanding the mechanism and mode of action of anticancer compounds, most of the published literature in cancer concerns analysis of toxicity rather than efficacy. For example, a metabolomic study of rats treated with aconitine alkaloids identified changes in 27 endogenous metabolites, particularly reductions in glutamine and creatinine (Sun et al. 2009). Metabolomics may be particularly useful in understanding anticancer compounds which affect glycolytic pathways (Vander Heiden et al. 2009).

9.4.6 *Yeast Genetic Methods*

The yeast *three* hybrid technique involves ligation of the target molecule to another molecule with a known affinity (e.g., methotrexate) for a protein (i.e., dihydrofolate reductase). A complementation circuit is completed for gene products which interact with the test molecule, and only the yeast strain containing complemented genes are able to grow (Abida et al. 2002; Baker et al. 2003). Mammalian versions of the screen have been developed to more accurately reflect human biology.

The Boone laboratory has utilized a collection of viable yeast haploid deletion mutants to screen both pure compounds and natural product extracts for hypersensitivity to the test sample (Parsons et al. 2006). This chemical-genetic profiling was undertaken to explore the mechanism of action of the samples. Clustering techniques were used to group samples with similar modes of action, in much the same way that NCI 60-cell data is analyzed. Also similarly to the NCI-60 data, extracts revealed the same patterns as the purified compounds that were isolated from them, enabling classification of the extracts prior to purification. While yeast genes are not identical to human genes, the process generated testable hypotheses about modes of action that could be productively examined in mammalian systems.

9.5 Recent Examples of Cancer Target Identification for Natural Products

9.5.1 *Cucurbitacins*

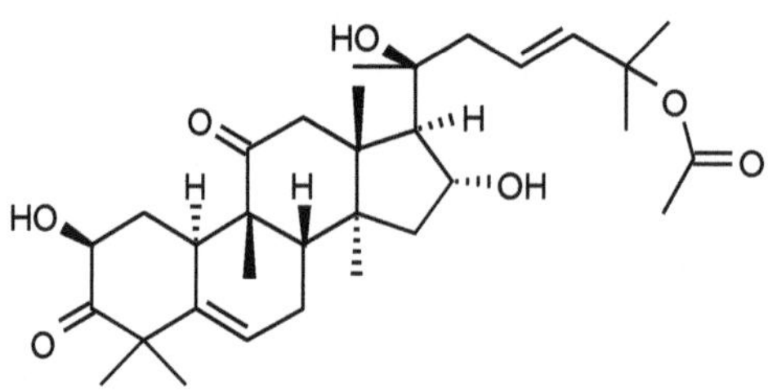

Cucurbitacin B

Cucurbitacins are a large group of highly oxygenated triterpenes bitter principles isolated primarily from the plant family Cucurbitaceae (cucumbers). Some cucurbitacins were shown many years ago to inhibit cancer cell growth and to have antitumor effects in mouse models (Gitter et al. 1961), and are often reported as cell growth inhibitors. Cucurbitacin glycosides generally have much weaker cell growth inhibition. Cucurbitacins B, D, E, and I have similar patterns of cell growth inhibition in the NCI 60-cell assay, suggesting that they may possess a common mechanism (Fuller et al. 1994). Many mechanisms have been proposed for this activity of cucurbitacins, however, the most convincing proposal involves inhibition of the JAK/STAT3 signaling pathway (Blaskovich et al. 2003; Sun et al. 2005). It should be noted that the effect is indirect, and that the precise molecular target remains unidentified (Yue and Turkson 2009).

9.5.2 V-ATPase Inhibitors

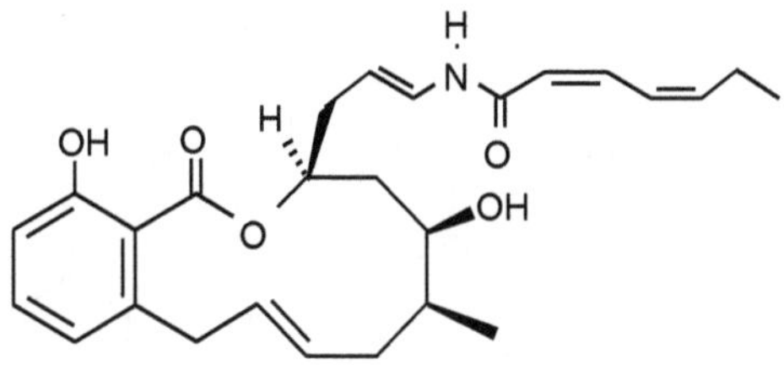

Salicylihalamide A

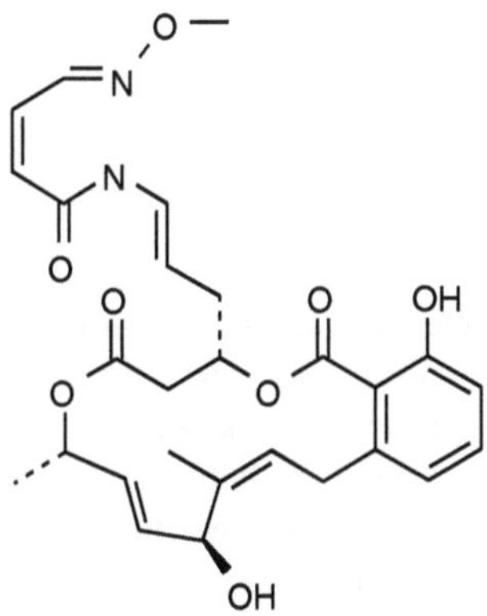

Lobatamide A

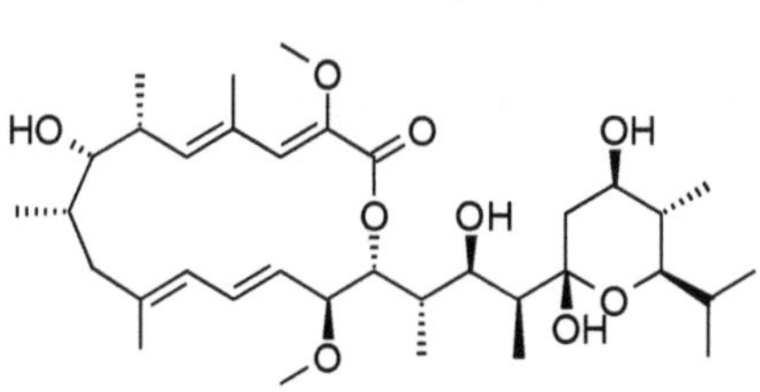

Bafilomycin A_1

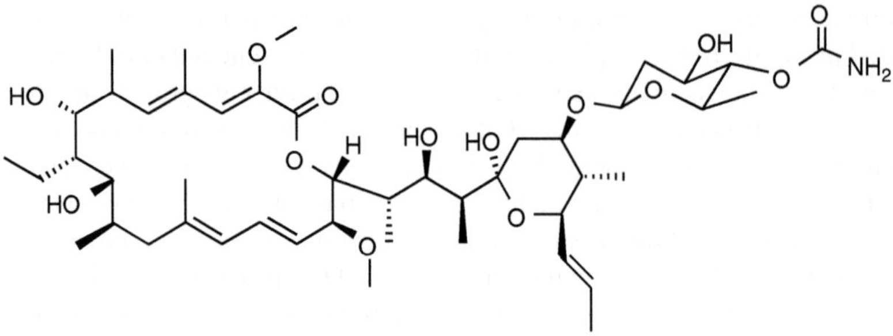

Concanamycin A

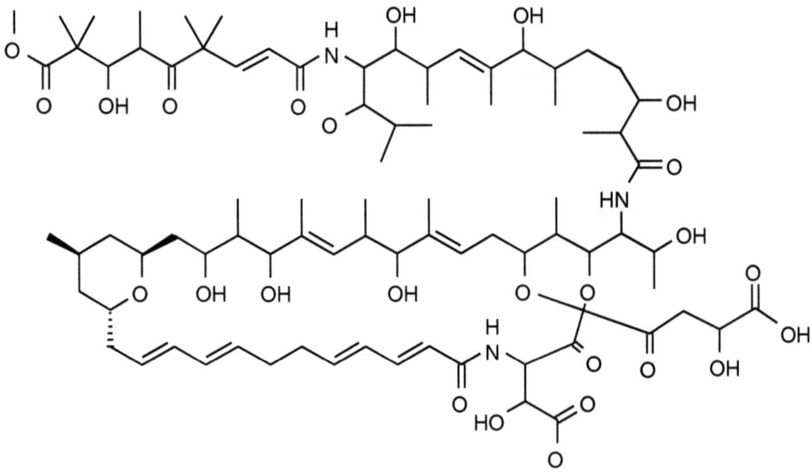

Chondropsin A

A number of novel macrocyclic compounds have been identified on the basis of potent cell growth inhibition and later found to inhibit vacuolar ATPase. In our laboratory, the NCI 60-cell patterns for the salicylihalamides (Erickson et al. 1997) and lobatamides (McKee et al. 1998) were found to correlate to bafilomycins and concanamycins, agents already known to inhibit V-ATPase (Bowman and Bowman 2002). Experimental verification of this mechanism showed that mammalian V-ATPase was inhibited, but the homologous fungal enzyme was not (Boyd et al. 2001). A further structural class of inhibitors, the chondropsins, was later found to inhibit both fungal and mammalian V-ATPases (Bowman et al. 2003; Cantrell et al. 2000). Whether V-ATPase is a valid target for cancer therapy remains an unsettled question (Torigoe et al. 2002).

9.5.3 Cyclopamine

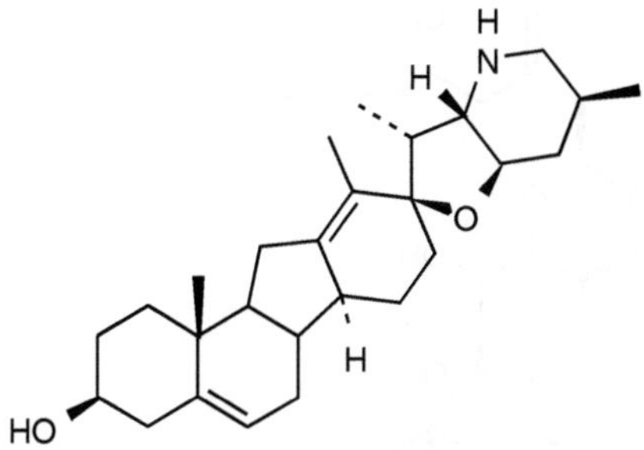

Cyclopamine

Cyclopamine is a steroidal alkaloid discovered on the basis of its teratogenicity to sheep and other animals (Keeler and Binns 1968). Research on developmental processes identified the Sonic hedgehog pathway as a potential nexus for cyclopamine action (Incardona et al. 1998), which led cancer researchers to investigate the mechanism further. Beachy's group reported that cyclopamine could reverse oncogenic mutations in both the Smoothened and Patched genes (Taipale et al. 2000). Direct binding of cyclopamine to the Smoothened protein was demonstrated using photoaffinity and fluorescent tagged cyclopamine derivatives (Chen et al. 2002). It is not clear whether or not the "on-target" effects of cyclopamine on stem cells will prohibit application in cancer therapeutics (Hovhannisyan et al. 2009).

9.5.4 Cortistatin

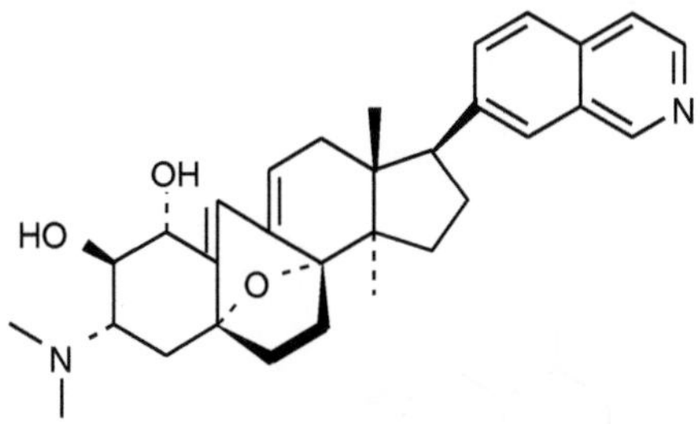

Cortistatin A

The marine sponge-derived cortistatins are steroidal alkaloids of unique structure with potent and selective inhibition of growth of endothelial cells. They may be of use in controlling angiogenesis (Aoki et al. 2006). By screening cortistatin A against a panel of 402 (!) protein kinases in a binding assay format, Nicolau's group found that the compound avidly bound and potently inhibited ROCK, CDK8 and CDK11 (Cee et al. 2009). Other groups are exploiting these findings to develop anti-angiogenesis drug candidates (Czakó et al. 2009).

9.5.5 *Spongistatin*

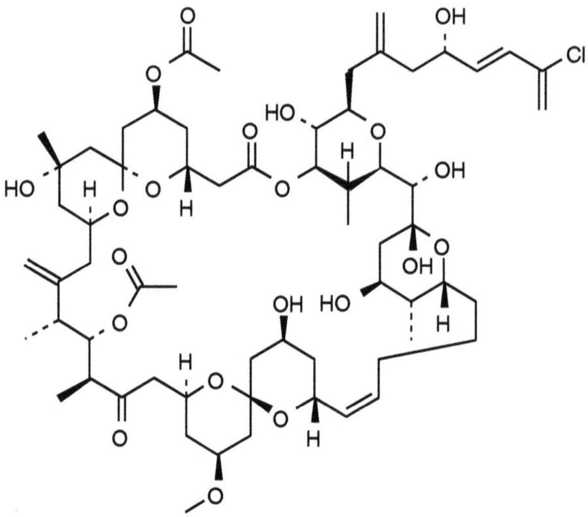

Spongistatin 1

The Pettit group isolated spongistatins from an Eastern Indian Ocean sponge as potent cell growth inhibitors (Pettit et al. 1993a, b). Spongistatin 1 was found to bind to tubulin at the vinca site and inhibit microtubule assembly (Bai et al. 1993). More recently, the Vollmar group has demonstrated that picomolar concentrations of spongistatin 1 are capable of inducing degradation of X-linked inhibitor of apoptosis (XIAP) in leukemia cells, thereby leading to apoptotic cell death (Schyschka et al. 2008). Since mitotic figures were observed at picomolar concentrations in the earlier work (Bai et al. 1993), it seems likely that the two phenomena are linked in some fashion, with apoptosis being a downstream result of mitotic inhibition.

9.5.6 *Withaferin*

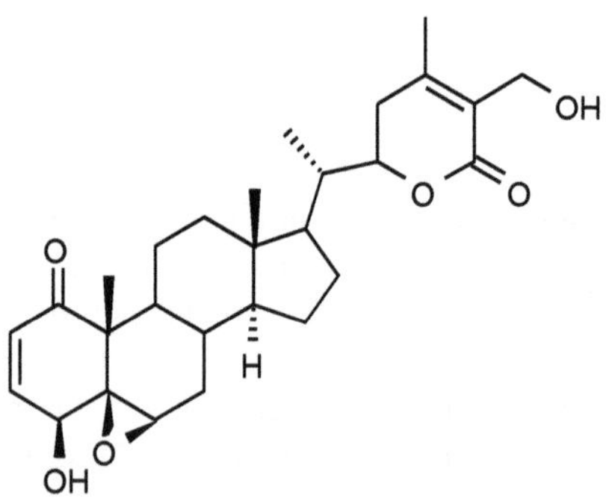

Withaferin A

Withaferin A is a steroidal lactone isolated from ashwaganda, the Ayurvedic medicinal plant *Withania somnifera* (Lavie et al. 1965). Ashwaganda has many indications, and it and its constituents have been found to have antitumor effects (Chakraborti et al. 1974). Early work implicated inhibition of RNA synthesis as a mode of action for withaferin A (Chowdhury and Neogy 1975). More recently, it was found that withaferin A inhibits the chymotrypsin-like activity of the 20S proteasome (Yang et al. 2007). Other studies found covalent binding to annexin II to be important (Falsey et al. 2006), as well as HSP90 (Yu et al. 2010).

9.5.7 Avrainvillamide

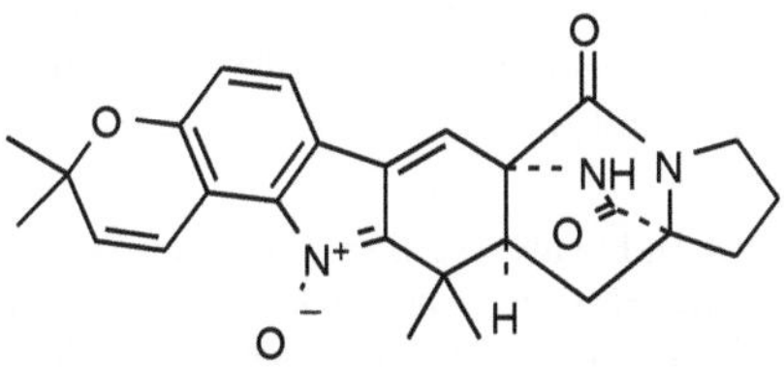

Avrainvillamide

The fungal product avrainvillamide was isolated from a species of *Aspergillus* on the basis of its antibacterial properties (Sugie et al. 2001). Affinity reagents generated in the course of total synthetic studies were used to show that the nuclear chaperone nucleophosmin binds avrainvillamide, and further cellular studies confirmed that the binding was biologically relevant (Wulff et al. 2007).

9.5.8 Pladienolides

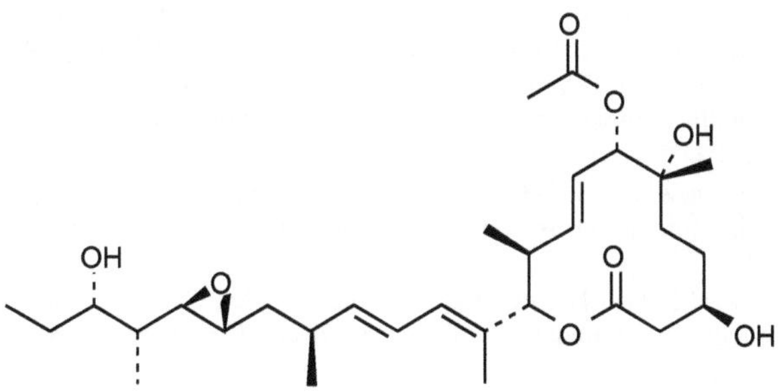

Pladienolide B

Pladienolides are novel 12-membered macrolides isolated from cultures of *Streptomyces platensis* on the basis of inhibition of the HIF-1 pathway under hypoxia (Mizui et al. 2004). The splicing factor SF3b was identified using affinity methods as the proximate target of pladienolides (Kotake et al. 2007). Given that pladienolides have shown activity in vivo, it will be interesting to see if splicing factors can be validated as useful cancer targets.

9.6 Conclusions

Natural products identified through empirical cancer assays are a viable source for the identification of novel cancer therapeutic targets, even if the natural product itself is not a viable development candidate. New techniques and approaches in the elucidation of targets, mechanisms and pathways affected by natural products have been developed that make the task more tractable. This route is a worthwhile alternative to the currently dominant paradigm of high throughput genomics and high throughput screening, and can provide a complementary pipeline to meet the need for novel cancer therapeutics in an efficient manner.

Acknowledgements I thank Drs. Thomas Turbyville and Barry O'Keefe for their thoughtful comments. Funded by the Intramural Research Program of the NIH, Center for Cancer Research, National Cancer Institute. The content of this publication does not necessarily reflect the views or policies of the Department of Health and Human Services, nor does mention of trade names, commercial products, or organizations imply endorsement by the U.S. Government.

References

Abida WM, Carter BT, Althoff EA et al (2002) Receptor-dependence of the transcription read-out in a small-molecule three-hybrid system. Chembiochem 3:887–895

Acton EM, Narayanan VL, Risbood PA et al (1994) Anticancer specificity of some ellipticinium salts against human brain tumors in vitro. J Med Chem 37:2185–2189

Alexander MD, Burkart MD, Leonard MS et al (2006) A central strategy for converting natural products into fluorescent probes. Chembiochem 7:409–416

Aoki S, Watanabe Y, Sanagawa M et al (2006) Cortistatins A, B, C, and D, anti-angiogenic steroidal alkaloids, from the marine sponge *Corticium simplex*. J Am Chem Soc 128:3148–3149

Bai R, Cichacz ZA, Herald CL et al (1993) Spongistatin 1, a highly cytotoxic, sponge-derived, marine natural product that inhibits mitosis, microtubule assembly, and the binding of vinblastine to tubulin. Mol Pharmacol 44:757–766

Baker K, Sengupta D, Salazar-Jimenez G et al (2003) An optimized dexamethasone-methotrexate yeast 3-hybrid system for high-throughput screening of small molecule-protein interactions. Anal Biochem 315:134–137

Bell MR, Johnson JR, Wildi BS et al (1958) The structure of gliotoxin. J Am Chem Soc 80:1001

Beutler JA (2009) Natural products as a foundation for drug discovery. Curr Protoc Pharmacol 46:9.11.1–9.11.21

Beutler JA, Shoemaker RH, Johnson T et al (1998) Cytotoxic geranyl stilbenes from *Macaranga schweinfurthii*. J Nat Prod 61:1509–1512

Beutler JA, McCall KL, Herbert K et al (2000) Novel cytotoxic diterpenes from *Casearia arborea*. J Nat Prod 63:657–661

Beutler JA, Kang M-I, Robert F et al (2009) Quassinoid inhibition of AP-1 function does not correlate with cytotoxicity or protein synthesis inhibition. J Nat Prod 72:503–506

Blaskovich MA, Sun J, Cantor A et al (2003) Discovery of JSI-124 (cucurbitacin I), a selective Janus kinase/signal transducer and activator of transcription 3 signaling pathway inhibitor with potent antitumor activity against human and murine cancer cells in mice. Cancer Res 63: 1270–1279

Bowman BJ, Bowman EJ (2002) Mutations in subunit c of the vacuolar ATPase confer resistance to bafilomycin and identify a conserved antibiotic binding site. J Biol Chem 277:3965–3972

Bowman EJ, Gustafson KR, Bowman BJ et al (2003) Identification of a new chondropsin class of antitumor compound that selectively inhibits V-ATPases. J Biol Chem 278:44147–44152

Boyd MR, Farina C, Belfiore P et al (2001) Discovery of a novel antitumor benzolactone enamide class that selectively inhibits mammalian vacuolar-type (H+)-ATPases. J Pharmacol Exp Ther 297:114–120

Brandt GE, Schmidt MD, Prisinzano TE et al (2008) Gedunin, a novel hsp90 inhibitor: semisynthesis of derivatives and preliminary structure-activity relationships. J Med Chem 51: 6495–6502

Butler MS (2008) Natural products to drugs: natural product-derived compounds in clinical trials. Nat Prod Rep 25:475–516

Cantrell CL, Gustafson KR, Cecere MR et al (2000) Chondropsins A and B, novel tumor cell growth-inhibitory macrolide lactams from the marine sponge *Chondropsis* sp. J Am Chem Soc 122:8825–8829

Cee VJ, Chen DY, Lee MR et al (2009) Cortistatin A is a high-affinity ligand of protein kinases ROCK, CDK8, and CDK11. Angew Chem Int Ed Engl 48:8952–8957

Chakraborti SK, De BK, Bandyopadhyay T (1974) Variations in the antitumour constituents of *Withania somnifera* Dunal. Experientia 30:852–853

Chan J, Khan SN, Harvey I et al (2004) Eukaryotic protein synthesis inhibitors identified by comparison of cytotoxicity profiles. RNA 10:528–543

Chen JK, Taipale J, Cooper MK et al (2002) Inhibition of hedgehog signaling by direct binding of cyclopamine to smoothened. Genes Dev 16:2743–2748

Chowdhury K, Neogy RK (1975) Mode of action of Withaferin A and Withanolide D. Biochem Pharmacol 24:919–920

Cook KM, Hilton ST, Mecinovic J et al (2009) Epidithiodiketopiperazines block the interaction between hypoxia-inducible factor-1alpha (HIF-1alpha) and p300 by a zinc ejection mechanism. J Biol Chem 284:26831–26838

Cuendet M, Christov K, Lantvit DD et al (2004) Multiple myeloma regression mediated by bruceantin. Clin Cancer Res 10:1170–1179

Czakó B, Kürti L, Mammoto A et al (2009) Discovery of potent and practical antiangiogenic agents inspired by cortistatin A. J Am Chem Soc 131:9014–9019

Decosterd LA, Gustafson KR, Cardellina JH II et al (1994) The differential cytotoxicity of cardenolides from *Thevetia ahouia*. Phytother Res 8:74–77

Derbre S, Gil S, Taverna M et al (2008) Highly cytotoxic and neurotoxic acetogenins of the Annonaceae: new putative biological targets of squamocin detected by activity-based protein profiling. Bioorg Med Chem Lett 18:5741–5744

Dixon N, Wong LS, Geerlings TH et al (2007) Cellular targets of natural products. Nat Prod Rep 24:1288–1310

Erickson KL, Beutler JA, Cardellina JH II et al (1997) Salicylihalamides A and B, novel cytotoxic macrolides from the marine sponge *Haliclona* sp. J Org Chem 62:8188–8192

Ernst-Russell MA, Chai CLL, Hurne AM et al (1999) Structure revision and cytotoxic activity of the scabrosin esters, epidithiopiperazinediones from the lichen *Xanthoparmelia scabrosa*. Aust J Chem 52:279–283

Falsey RR, Marron MT, Gunaherath GMKB et al (2006) Actin filament aggregation induced by withferin A is mediated by annexin II. Nat Chem Biol 2:33–38

Fischer JD, Liao L, Carvalho PC et al (2010) Dynamic proteomic overview of glioblastoma cells (A172) exposed to perillyl alcohol. J Proteomics 73(5):1018–1027

Frye SV (2010) The art of the chemical probe. Nat Chem Biol 6:159–161

Fukamiya N, Lee KH, Muhammad I et al (2005) Structure-activity relationships of quassinoids for eukaryotic protein synthesis. Cancer Lett 220:37–48
Fuller RW, Cardellina JH II, Cragg GM et al (1994) Cucurbitacins; differential cytotoxicity, dereplication and first isolation from *Gonystylus keithii*. J Nat Prod 57:1442–1445
Funabashi Y, Horiguchi T, Iinuma S et al (1994) TAN-1496 A, C and E, diketopiperazine antibiotics with inhibitory activity against mammalian DNA topoisomerase I. J Antibiot 47: 1202–1218
Gitter S, Gallily R, Shohat B et al (1961) Studies on the antitumor effect of cucurbitacins. Cancer Res 21:516–521
Goswami SK, Das DK (2009) Resveratrol and chemoprevention. Cancer Lett 284:1–6
Guo C, LaCour TG, Fuchs PL (1999) On the relationship of OSW-1 to the cephalostatins. BioMed Chem Lett 9:419–424
Gupta RS, Chopra A, Stetsko DK (1986) Cellular basis for the species differences in sensitivity to cardiac glycosides (digitalis). J Cell Physiol 127:197–206
Hegde VR, Dai P, Patel M et al (1997) Novel thiodiketopiperazine fungal metabolites as epidermal growth factor receptor antagonists. Tetrahedron Lett 38:911–914
Henrich CJ, Goncharova EI, Wilson JA et al (2007) Natural products active in aberrant c-Kit signaling. Chem Biol Drug Des 69:321–330
Hieronymus H, Lamb J, Ross KN et al (2006) Gene expression signature-based chemical genomic prediction identifies a novel class of HSP90 pathway modulators. Cancer Cell 10:321–330
Hovhannisyan A, Matz M, Gebhardt R (2009) From teratogens to potential therapeutics: natural inhibitors of the Hedgehog signaling network come of age. Planta Med 75:1371–1380
Hsiang YH, Hertzberg R, Hecht S et al (1985) Camptothecin induces protein-linked DNA breaks via mammalian DNA topoisomerase I. J Biol Chem 260:14873–14878
Hurne AM, Chai CL, Moerman K et al (2002) Influx of calcium through a redox-sensitive plasma membrane channel in thymocytes causes early necrotic cell death induced by the epipolythiodioxopiperazine toxins. J Biol Chem 277:31631–31638
Incardona JP, Gaffield W, Kapur RP et al (1998) The teratogenic Veratrum alkaloid cyclopamine inhibits sonic hedgehog signal transduction. Development 125:3553–3562
Jang M, Cai L, Udeani GO et al (1997) Cancer chemopreventative activity of resveratrol, a natural product derived from grapes. Science 275:218–220
Jolad SD, Hoffman JJ, Cole JR et al (1981) 3′-O-Methylevomonoside: a new cytotoxic cardiac glycoside from *Thevetia ahouia* A.DC (Apocynaceae). J Org Chem 46:1946–1947
Keeler RF, Binns W (1968) Teratogenic compounds of *Veratrum californicum* (Durand). V. Comparison of cyclopean effects of steroidal alkaloids from the plant and structurally related compounds from other sources. Teratology 1:5–10
Kerrigan JE, Pilch DS (2001) A structural model for the ternary cleavable complex formed between human topoisomerase I, DNA, and camptothecin. Biochemistry 40:9792–9798
Keskin O, Bahar I, Jernigan RL et al (2000) Characterization of anticancer agents by their growth-inhibitory activity and relationships to mechanism of action and structure. Anticancer Drug Design 15:79–98
Kigoshi H, Suenaga K, Mutou T et al (1996) Aplyronine A, a potent antitumor substance of marine origin, aplyronines B and C and artificial analogues: total synthesis and structure-cytotoxicity relationships. J Org Chem 61:5326–5351
Kotake Y, Sagane K, Owa T et al (2007) Splicing factor SF3b as a target of the antitumor natural product pladienolide. Nat Chem Biol 3:570–575
Kuder CH, Neighbors JD, Hohl RJ et al (2009) Synthesis and biological activity of a fluorescent schweinfurthin analogue. Bioorg Med Chem 17:4718–4723
Kung AL, Zabludoff SD, France DS et al (2004) Small molecule blockade of transcriptional coactivation of the hypoxia-inducible factor pathway. Cancer Cell 6:33–43
Kupchan SM, Britton RW, Ziegler MF et al (1973) Bruceantin, a new potent antileukemic simaroubolide from *Brucea antidysenterica*. J Org Chem 38:178–179
Kuroda T, Suenaga K, Sakakura A et al (2006) Study of the interaction between actin and antitumor substance aplyronine A with a novel fluorescent photoaffinity probe. Bioconjug Chem 17:524–529

Lamb J (2007) The Connectivity Map: a new tool for biomedical research. Nat Rev Cancer 7:54–60

Lamb J (2010) Connectivity Map 02. http://www.broadinstitute.org/cmap/. Accessed 23 Feb 2010

Lamb J, Crawford ED, Peck D et al (2006) The Connectivity Map: using gene-expression signatures to connect small molecules, genes, and disease. Science 313:1929–1935

Lavie D, Glotter E, Shvo Y (1965) Constituents of *Withania somnifera* Dun. Part IV. The structure of withaferin A. J Chem Soc 50:7517–7531

Li JW, Vederas JC (2009) Drug discovery and natural products: end of an era or an endless frontier? Science 325:161–165

Maheshwari RK, Singh AK, Gaddipati J et al (2006) Multiple biological activities of curcumin: a short review. Life Sci 78:2081–2087

McCormick JL, McKee TC, Cardellina JH II et al (1996) Cytotoxic triterpenes from a marine sponge, *Stelletta* sp. J Nat Prod 59:1047–1050

McInnes AG, Taylor A, Walter JA (1976) The structure of chetomin. J Am Chem Soc 98:6741

McKee TC, Galinis DL, Pannell LK et al (1998) The lobatamides, novel cytotoxic macrolides from southwestern Pacific tunicates. J Org Chem 63:7805–7810

Mijatovic T, Van QE, Delest B et al (2007) Cardiotonic steroids on the road to anti-cancer therapy. Biochim Biophys Acta 1776:32–57

Mimaki Y, Kuroda M, Kameyama A et al (1997) Cholestane glycosides with potent cytostatic activities on various tumor cells from *Ornithogalum saundersiae* bulbs. BioMed Chem Lett 7:633–636

Mizui Y, Sakai T, Iwata M et al (2004) Pladienolides, new substances from culture of *Streptomyces platensis* Mer-11107. III. In vitro and in vivo antitumor activities. J Antibiot 57:188–196

Moerman KL, Chai CL, Waring P (2003) Evidence that the lichen-derived scabrosin esters target mitochondrial ATP synthase in P388D1 cells. Toxicol Appl Pharmacol 190:232–240

Monks A, Scudiero DA, Johnson GS et al (1997) The NCI anti-cancer drug screen: a smart screen to identify effectors of novel targets. Anticancer Drug Des 12:533–541

Morre DJ, Grieco PA, Morre DM (1998) Mode of action of the anticancer quassinoids—inhibition of the plasma membrane NADH oxidase. Life Sci 63:595–604

Newman DJ, Cragg GM (2006) Natural products from invertebrates and microbes as modulators of antitumor targets. Curr Drug Targets 7:279–304

Newman DJ, Cragg GM (2007) Natural products as sources of new drugs over the last 25 years. J Nat Prod 70:461–477

Ojima I, Duclose O, Dormán G et al (1995) A new paclitaxel photoaffinity analog with a 3-(4-benzoylphenyl)propanoyl probe for characterization of drug-binding sites on tubulin and p-glycoprotein. J Med Chem 38:3891–3894

Onodera H, Hasegawa A, Tsumagari N et al (2004) MPC1001 and its analogues: new antitumor agents from the fungus *Cladorrhinum* species. Org Lett 6:4101–4104

Parsons AB, Lopez A, Givoni IE et al (2006) Exploring the mode-of-action of bioactive compounds by chemical-genetic profiling in yeast. Cell 126:611–625

Paull KD, Shoemaker RH, Hodes L et al (1989) Display and analysis of patterns of differential activity of drugs against human tumor cell lines: development of mean graph and COMPARE algorithm. J Natl Cancer Inst 81:1088–1092

Paull KD, Lin CM, Malspeis L et al (1992) Identification of novel antimitotic agents acting at the tubulin level by computer-assisted evaluation of differential cytotoxicity data. Cancer Res 52:3892–3900

Perkins S, Verschoyle RD, Hill K et al (2002) Chemopreventive efficacy and pharmacokinetics of curcumin in the min/+ mouse, a model of familial adenomatous polyposis. Cancer Epidemiol Biomarkers Prev 11:535–540

Pettit GR, Inoue M, Kamano Y et al (1988) Isolation and structure of the powerful cell growth inhibitor cephalostatin 1. J Am Chem Soc 110:2006–2007

Pettit GR, Cichacz ZA, Gao F et al (1993a) Isolation and structure of the remarkable human cancer cell growth inhibitors spongistatins and 3 from an eastern Indian Ocean *Spongia* sp. J Chem Soc Chem Commun 14:1166–1168

Pettit GR, Cichacz ZA, Gao F et al (1993b) Isolation and structure of spongistatin 1. J Org Chem 58:1302–1304

Qi H, Wei L, Han Y et al (2010) Proteomic characterization of the cellular response to chemopreventive triterpenoid astragaloside IV in human hepatocellular carcinoma cell line HepG2. Int J Oncol 36:725–735

Redinbo MR, Stewart L, Kuhn P et al (1998) Crystal structures of human topoisomerase I in covalent and noncovalent complexes with DNA. Science 279:1504–1513

Rosati A, Quaranta E, Ammirante M et al (2004) Quassinoids can induce mitochondrial membrane depolarisation and caspase 3 activation in human cells. Cell Death Differ 11:S216–S218

Saito S, Watabe S, Ozaki H et al (1996) Novel actin depolymerizing macrolide aplyronine A. J Biochem 120:552–555

Sakamoto S, Kabe Y, Hatakeyama M et al (2009) Development and application of high-performance affinity beads: toward chemical biology and drug discovery. Chem Rec 9:66–85

Schyschka L, Rudy A, Jeremias I et al (2008) Spongistatin 1: a new chemosensitizing marine compound that degrades XIAP. Leukemia 22:1737–1745

Shi LM, Myers TG, Fan Y et al (1998) Mining the National Cancer Institute anticancer drug discovery database: cluster analysis of ellipticine analogs with p53-Inverse and central nervous system-selective patterns of activity. Mol Pharmacol 53:241–251

Shoemaker RH (2006) The NCI60 human tumour cell line anticancer drug screen. Nat Rev Cancer 6:813–823

Sleno L, Emili A (2008) Proteomic methods for drug target discovery. Curr Opin Chem Biol 12:46–54

Suenaga K, Kamei N, Okugawa Y et al (1997) Cytotoxicity and actin depolymerizing activity of aplyronine A, a potent antitumor macrolide of marine origin, and the natural and artificial analogs. BioMed Chem Lett 7:269–274

Sugie Y, Hirai H, Inagaki T et al (2001) A new antibiotic CJ-17,665 from Aspergillus ochraceus. J Antibiot 54:911–916

Sun J, Blaskovich MA, Jove R et al (2005) Cucurbitacin Q: a selective STAT3 activation inhibitor with potent antitumor activity. Oncogene 24:3236–3245

Sun B, Wu S, Li L et al (2009) A metabolomic analysis of the toxicity of Aconitum sp. alkaloids in rats using gas chromatography/mass spectrometry. Rapid Commun Mass Spectrom 23: 1221–1228

Taipale J, Chen JK, Cooper MK et al (2000) Effects of oncogenic mutations in smoothened and patched can be reversed by cyclopamine. Nature 406:1005–1009

Torigoe T, Izumi H, Ise T et al (2002) Vacuolar H(+)-ATPase: functional mechanisms and potential as a target for cancer chemotherapy. Anticancer Drugs 13:237–243

Torrance CJ, Agrawal V, Vogelstein B et al (2001) Use of isogenic human cancer cells for high-throughput screening and drug discovery. Nat Biotechnol 19:940–945

Turbyville TJ, Gürsel DB, Tuskan RG et al (2010) Schweinfurthin A selectively inhibits proliferation and Rho signaling in glioma and neurofibromatosis type 1 tumor cells in an NF1-GRD dependent manner. Mol Cancer Ther 9(5):1234–1243

Ulrich NC, Kodet JG, Mente NR et al (2010) Structural analogues of schweinfurthin F: probing the steric, electronic and hydrophobic properties of the D-ring substructure. Bioorg Med Chem 18:1676–1683

Vander Heiden MG, Cantley LC, Thompson CB (2009) Understanding the Warburg effect: the metabolic requirements of cell proliferation. Science 324:1029–1033

Vidalin O, Muslmani M, Estienne C et al (2009) In vivo target validation using gene invalidation, RNA interference and protein functional knockout models: it is the time to combine. Curr Opin Pharmacol 9:669–676

Vistica DT, Kenney S, Hursey ML et al (1994) Cellular uptake as a determinant of cytotoxicity of quaternized ellipticines to human brain tumor cells. Biochem Biophys Res Commun 200: 1762–1768

Vistica DT, Kenney S, Hursey ML et al (1996) Role of membrane potential in the accumulation of quaternized ellipticines by human tumor cell lines. J Pharmacol Exp Ther 279:1018–1025

Vockley J, Ozenberger B (2012) The Cancer Genome Atlas. http://cancergenome.nih.gov/. Accessed 29 Aug 2012

Waksman SA, Bugie E (1944) Chaetomin, a new antibiotic substance produced by chaetomium cochliodes: I. Formation and properties. J Bacteriol 48:527–530

Weinstein JN, Kohn KW, Grever MR et al (1992) Neural computing in cancer drug development: predicting mechanism of action. Science 258:447–451

Weinstein JN, Myers TG, O'Connor PM et al (1997) An information-intensive approach to the molecular pharmacology of cancer. Science 275:343–349

Williams DE, Bombuwala K, Lobkovsky E et al (1998) Ambewelamides A and B, antineoplastic epidithiapiperazinediones isolated from the lichen *Usnea* sp. Tetrahedron Lett 39:9579–9582

Wiseman CL, Yap HY, Bedikian AY et al (1982) Phase II trial of bruceantin in metastatic breast carcinoma. Am J Clin Oncol 5:389–391

Wosikowski K, Schuurhuis D, Johnson K et al (1997) Identification of epidermal growth factor receptor and c-erbB2 pathway inhibitors by correlation with gene expression patterns. J Natl Cancer Inst 89:1505–1515

Wrona IE, Lowe JT, Turbyville TJ et al (2009) Synthesis of a 35-member stereoisomer library of bistramide A: evaluation of effects on actin state, cell cycle and tumor cell growth. J Org Chem 74:1897–1916

Wu L, Smythe AM, Stinson SF et al (1992) Multidrug-resistant phenotype of disease-oriented panels of human tumor cell lines used for anticancer drug screening. Cancer Res 52: 3029–3034

Wulff JE, Siegrist R, Myers AG (2007) The natural product avrainvillamide binds to the oncoprotein nucleophosmin. J Am Chem Soc 129:14444–14451

Xu GW, Mawji IA, Macrae CJ et al (2008) A high-content chemical screen identifies ellipticine as a modulator of p53 nuclear localization. Apoptosis 13:413–422

Yamada K, Ojika M, Ishigaki T et al (1993) Aplyronine A, a potent antitumor substance, and the congeners aplyronines B and C isolated from the sea hare Aplysia kurodai. J Am Chem Soc 115:11020–11021

Yang H, Shi G, Dou QP (2007) The tumor proteasome is a primary target for the natural anticancer compound Withaferin A isolated from "Indian winter cherry". Mol Pharmacol 71:426–437

Yoder B, Norris A, Miller JS et al (2006) Cytotoxic prenylated stilbenes and flavonoids from *Macaranga alnifolia* from the Madagascar rain forest. J Nat Prod 70:342–346

Yu Y, Hamza A, Zhang T et al (2010) Withaferin A targets heat shock protein 90 in pancreatic cancer cells. Biochem Pharmacol 79:542–551

Yue P, Turkson J (2009) Targeting STAT3 in cancer: how successful are we? Expert Opin Investig Drugs 18:45–56

Index

F.E. Koehn (ed.), *Natural Products and Cancer Drug Discovery*, Cancer Drug Discovery and Development, DOI 10.1007/978-1-4614-4654-5,

MIX
Papier aus verantwortungsvollen Quellen
Paper from responsible sources
FSC® C105338

If you have any concerns about our products,
you can contact us on
ProductSafety@springernature.com

In case Publisher is established outside the EU,
the EU authorized representative is:
Springer Nature Customer Service Center GmbH
Europaplatz 3, 69115 Heidelberg, Germany

Printed by Libri Plureos GmbH
in Hamburg, Germany